Aquatic Insects
in the Vicinity of the Black Hills,
South Dakota and Wyoming

Grant D. De Jong

Aquatic Insects
in the Vicinity of the Black Hills,
South Dakota and Wyoming

Grant D. De Jong

2018

First Printing: 2018

ISBN 978-1-387-87230-5

Grant D. De Jong
6835 Rickwood Drive
Pensacola, Florida 32526

Distributed by Lulu Press, Inc.
627 Davis Drive, Suite 300
Morrisville, North Carolina 27560

Cover photo: Annie Creek just downstream of confluence with Lost Camp Gulch and Ross Valley after a rainstorm. The exposed mudstone rockface is where I have collected fossilized "worm tubes," *Palaeophycus* isp.

Abstract

The aquatic insect fauna known or expected to exist in the Black Hills and their immediate vicinity are listed, based on primary and secondary literature sources, as well as biomonitoring samples collected between 1985 and 2011. There are a total of 93 family-, 355 genus-, and 447 species-level aquatic insect taxa known or anticipated in the Black Hills listed in this draft catalog; this number is *definitely* expected to increase as further studies are done in the Black Hills and this checklist is refined over time. Additional information for each taxon includes literature references for diagnosis of each taxon, general distributional information, specific mention of its presence in or around the Black Hills, and selected ecological traits. The dystiscid beetle *Dytiscus marginicollis* (LeConte), the elmid beetle *Lara avara* LeConte, and the hydroscaphid beetle *Hydroscapha natans* LeConte are reported as new state records for South Dakota. No taxonomic changes are made.

Table of Contents

ii

List of Figures

List of Tables

List of Acronyms and Abbreviations

AWWQRP	Arid West Water Quality Research Program
B&B	bed and breakfast
BU	burrower (habit/behavior group)
CB	climber (habit/behavior group)
CL	clinger (habit/behavior group)
CONABIO	Comisión Nacional para el Conocimiento y Usa de la Biodiversidad
CPOM	coarse particulate organic matter
DI	diver (habit/behavior group)
DSA	Dragonfly Society of the Americas
EMAP	Environmental Monitoring and Assessment Program
EPA	U.S. Environmental Protection Agency
EPT	Ephemeroptera, Plecoptera, Trichoptera
FC	filter-collector (functional feeding group)
FFG	functional feeding group
FPOM	fine particulate organic matter
FWS	U.S. Fish and Wildlife Service
GC	gather-collector (functional feeding group)
GEI	GEI Consultants, Inc.
GPA	Game Production Area
HBI	Hilsenhoff Biotic Index
IDEQ	Idaho Department of Environmental Quality
ITIS	Integrated Taxonomic Information System
mm	millimeters
μm	micrometers
NAWQA	National Water-Quality Assessment Program
NOAA	U.S. National Oceanic and Atmospheric Administration
NPS	U.S. National Park Service
NPWRC	Northern Plains Wildlife Research Center

PA	parasite (functional feeding group)
PI	piercer (functional feeding group)
PL	plankton (habit/behavior group)
PMSW	preliminary mean stream width
PR	predator (functional feeding group)
QA	quality assurance
SDDENR	South Dakota Department of Environment and Natural Resources
SC	scraper (functional feeding group)
SH	shredder (functional feeding group)
SK	skater (habit/behavior group)
SOP	standard operating procedure
SP	sprawler (habit/behavior group)
STE	standard taxonomic effort
SW	swimmer (habit/behavior group)
TSN	Taxonomic Serial Number
USDA	U.S. Department of Agriculture
USGS	U.S. Geological Survey
USNM	National Museum of Natural History, Smithsonian Institution
WSII	Wyoming Stream Integrity Index

Preface

As an ecologist and invertebrate taxonomist working in a fairly well-defined, restricted geographic area, I often found it useful to build a list of the organisms that are known to exist there or would be reasonably expected to occur there. Sometimes it was fairly easy; other times it was tedious to track down the gray literature that might hold pertinent records. The activity is not unique to me; I know there are many others who share it, whether they work professionally with the taxonomic group in question or are amateur naturalists. For taxonomists, an existing taxa list to work from is invaluable; for ecologists, a taxa list with associated ecological data is golden.

This book represents my compilation of the aquatic insects found in the Black Hills of South Dakota and Wyoming, along with some ecological information. For the most part, it was done on my own time, since my work as an invertebrate ecologist and taxonomist kept me busy at the microscope or writing reports and not writing books. GEI Consultants, Inc. (GEI), the engineering consulting firm where I was employed for 22 years, had several gold mining clients in the northern Black Hills for which the ecology division primarily provided biomonitoring services. GEI's relationship with these clients began prior to my hiring at GEI and persisted after I left.

The existence of these clients should not imply that they had anything to do with this work except provide me an impetus to assemble the data. They did not have any say whatsoever into what went in. That said, I do regret starting this project later than I should have. Had I started earlier, I may have been able to work with those clients and obtain permission to use some of the biomonitoring data GEI collected on their behalf to supplement the ecological data I found in the published literature. But, I delayed. So the book is being published without those data – maybe somedy I will be able to negotiate the use of those data.

I trust that the introduction on the use of aquatic insect ecological information in biomonitoring is useful as a resource. While some of those views were adopted from work and honed at GEI, some may be also be contrary to GEI's official opinion. GEI did not have a lot of say on what went into this book, either.

I would like to thank those at GEI who gave me the opportunity to work on the biomonitoring projects. These included James Chadwick (now retired), Steve Canton, Don Conklin, Lee Bergstedt, and Craig Wolf. That opportunity not only gave me the experience of assembling and using this information, but annual field trips to those projects over the course of 22 years gave me a wealth of first-hand ecological information. That experience is likewise useful now as I teach biological science courses to undergraduate students at Pensacola Christian College, a private, liberal arts college in the Florida Panhandle. Others who contributed to this work include my family – wife Nikki, and daughters Abby and Katie – who knowingly or not persisted while I worked on this.

I know that this work cannot be the final, definitive word on what aquatic insect taxa live in the Black Hills, since I point out lots of knowledge gaps that need to be filled (like the total lack of records of the alderfly family Sialidae in South Dakota despite the fact that all eight surrounding states boast multiple species). Maybe I should request co-authorship on papers based on those knowledge gaps! Anyway, this book is dedicated to those scientists who will carry on faunal and ecological studies in the Black Hills and make real use of this book. I hope their peer review keeps me up to date on any mistakes or omissions I made, for which I take full responsibility, and that they let me know of their advances in the field for a possible future edition.

Finally, I trust that this work exemplifies my personal motto:

<< Δια ευσχήμου και ταξεως προσ ή δοξα του κτιστου ό θεος μου >>.

Introduction

The Black Hills represent a relatively small island of mountainous terrain in the vast prairie grasslands of North America. These isolated mountains in far northeastern Wyoming and southwestern South Dakota (Figure 1) are surrounded by the Great Plains. The Black Hills obtained their name from a literal translation of the Lakota Sioux name for the hills, Pahá Sápa, apparently so called because the dense forests covering the hills look black from a distance. Archaeological and historical evidence indicates that, in addition to the Sioux, the Arapaho, Cheyenne, and Kiowa tribes also made use of the Black Hills

Figure 1:   Relative location of the Black Hills in the United States.

Geologically, the Black Hills represent an uplift due to volcanic activity, called the Laramide Orogeny, which is thought to have occurred during the Tertiary. Geologic strata laid down earlier were pushed up and buckled during the uplift, exposing those strata like a target. The bullseye of the target is a granite core, emplaced by magma in the Trans-Hudson orogeny, when sedimentary rocks were folded and twisted and metamorphosed. Various layers of sedimentary rocks were laid down over the granite core and have since eroded. During the "Wisconsin glaciation", ice and perennial snow extended as far south as the Black Hills.

Byers (1961), working particularly with the crane-fly genus *Dolichopeza*, suggested that this glaciation geoclimatic event may have helped shape the biological communities in the Black Hills. The presence of three species of *Dolichopeza* in the Black Hills caused him to question the existing biogeography paradigm and the effectiveness of the Great Plains as a barrier to dispersal. He concluded that the genus inhabited a boreal forest at the southern front of the glaciers. The retreat of the glaciers left the forest intact only at the higher, cooler, wetter elevations in the Black Hills, later isolated as the lower elevations became dry forests and eventually grasslands. Furthermore, he attributed the lack of speciation within *Dolichopeza*, despite geographic isolation from other populations of the species, to the inherent variability within the genus and its species, since they are distributed from Alaska and northwestern Canada to the Florida peninsula, from sea level to over 1,800 m elevation. Similar stories could probably be told of many other taxa in the Black Hills, since there appear to be very few endemic species.

The Black Hills spread across all or parts of Butte, Custer, Fall River, Lawrence, Meade, Pennington, and Shannon counties in South Dakota and Crook and Weston counties in Wyoming. The Black Hills National Forest covers 21,043 km^2 (8,125 mi^2) of land within the Black Hills, and the region is home to Devils Tower National Monument, Jewel Cave National Monument, Mount Rushmore National Memorial, Wind Cave National Park, Harney Peak (at 2,207 m [7,242 ft] above sea level, the highest point in North America east of the Rocky Mountains), Custer State Park (the largest state park in South Dakota), Bear Butte State Park, and the Crazy Horse Memorial (the largest sculpture in the world, still under construction). The major city in the region is Rapid City (metropolitan population ~125,000), and other towns scattered throughout the Black Hills and immediate environs include Spearfish, Lead, Sturgis, Custer, Keystone, Hot Springs, Deadwood, Hill City, and Newcastle, supporting a total of nearly 250,000 people.

After 1876, the economy of the Black Hills centered on mining and timber harvesting. Currently, tourism (including gambling), mining, ranching, timber, and some manufacturing support most of the economy. Gold was first discovered near Custer, but the large placer gold deposits of Deadwood Gulch were discovered in 1875. Eponymic Black Hills Gold must be designed in a traditional tri-gold, grape leaf motif and be manufactured in the Black Hills. Currently (2018), only the Wharf Mine is still mining gold in the Black Hills, since the huge Homestake Mine closed in 2003. The Homestake's deep shafts are now being converted into the Sanford Underground Research Facility, with neutrino physics laboratories located up to 1,585 m (5,200 ft) underground and more planned to 2,440 m (8,000 ft) underground, away from any potential solar interference. Ellsworth Air Force Base is located just outside Rapid City and supports about 8,000 military members, family members, and civilian employees.

The Black Hills are divided into three level IV ecoregions (Figure 2; Omernik 1987, Bryce et al. 1996): 17a – the Black Hills Foothills, 17b – the Black Hills Plateau, and 17c, the Black Hills Core Highlands. Ecoregion 43g – the Semiarid Pierre Shale Plains borders the Black Hills extensively on the north and east, and with a thin strip on the west; Ecoregion 43w – Powder River Basin extends westward from Ecoregion 43g. Ecoregion 43e – Sagebrush Steppe borders the Black Hills on the south, with a small area near the northwest corner of the Black Hills (Bryce et al. 1996, Chapman et al. 2004). These ecoregions are defined as areas where internal ecosystems are generally similar, based on data from multiple U.S. agencies that analyzed patterns of biotic and abiotic phenomena (such as geology, soils, vegetation, climate, land use, wildlife, and hydrology; Omernik 1987).

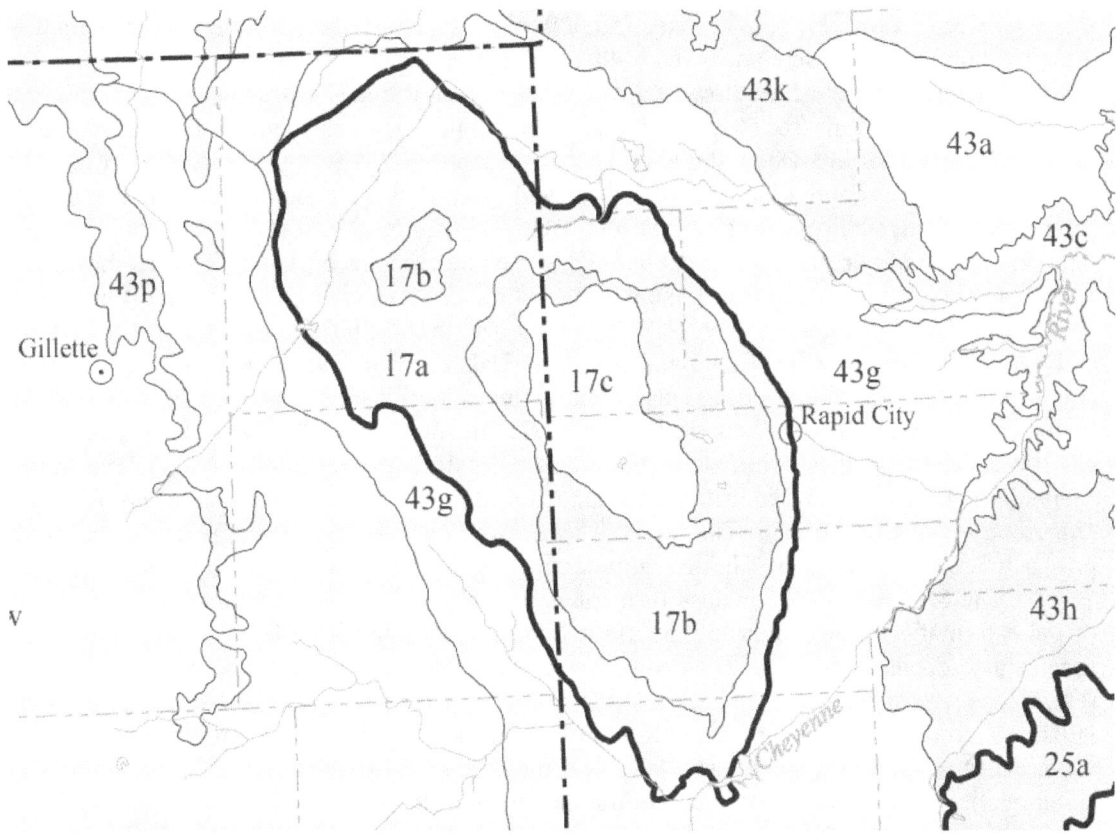

Figure 2: Level IV ecoregions in and around the Black Hills, from ftp://newftp. epa.gov/EPADataCommons/ORD/Ecoregions/reg8/epa_reg8_f.pdf .

Major streams draining the northern Black Hills include tributaries of Beaver Creek, Redwater Creek, Spearfish Creek, and Whitewood Creek, flowing north to confluence with the Belle Fourche River. Bear Butte Creek, Elk Creek, and Boxelder Creek drain to the northeast, north of Rapid City. Castle Creek originates in the heart of the Black Hills, joining Rapid Creek, which flows eastward through Rapid City. Spring Creek, Battle Creek, French Creek and Beaver Creek flow southeast of Rapid City. The Cheyenne River flows along the southern boundary of the Black Hills. On the western front, much smaller streams, such as Cold Springs Creek, Stockade Creek, and Beaver Creek, flow in Wyoming.

Feeding these major streams are numerous tributaries, many of which are ephemeral or intermittent in nature. For definition, ephemeral streams are at all times above the water table and flow only in response to precipitation or snowmelt; intermittent streams flow primarily in response to precipitation that had raised the water table above the level of the stream bed. Perennial streams flow year-round, although flows may be extremely low (and even subsurface in places) during dry times of the year. Most headwaters are ephemeral or intermittent, unless they are spring-fed. It is not uncommon for even these tributaries, and even some of the major streams, to become intermittent or ephemeral drainages as they exit the Black Hills and flow onto the plains. This phenomenon is due to the local aquifer geology, described below, and is especially noticeable as one drives south of Rapid City on State Highway 79 and crosses numerous dry washes.

Conversely, these washes can flow spectacularly in response to large thunderstorm activity. On June 9-10, 1972, flooding caused by a large mesoscale convective system was seen in the central Black Hills extending to the eastern flanks (Driscoll et al. 2012). In Spring Creek, flows were 617 m^3/sec (21,800 cfs), with approximately a 400 year recurrence interval (i.e., a "400-year flood event"), and in

Boxelder Creek, flows hit 1,430 m³/sec (50,500 cfs), exceeding the 500-year recurrence interval. At least 238 human deaths were attributed to that storm. Yet, even those flows pale in comparison to two nearly unbelievable prehistoric flows that were detected on Rapid Creek using stratigraphic analysis: a flood of at least 1,812 m³/sec (64,000 cfs) occurred about 1,000 years ago, and a nearly unimaginable flood of at least 3,625 m³/sec (128,000 cfs) occurred about 440 years ago (Driscoll et al. 2012).

There are approximately 70 lakes and major named ponds within the Black Hills (http://www.goingoutside.com/lakeranges/67_Black_Hills.html), most of them being small impoundments. Major lakes and impoundments include Iron Creek Lake, Roubaix Lake, Deerfield Lake, Pactola Reservoir, Sheridan Lake, Horsethief Lake, Stockade Lake, and Angostura Reservoir.

The U.S. Geological Survey (USGS), along with the South Dakota Department of Environment and Natural Resources (SDDENR) and the West Dakota Water Development District (WDWDD), conducted numerous hydrologic and water quality investigations in the Black Hills from 1990 to 2002 (among the resulting publications are Greene 1993, Greene et al. 1999, Carter et al. 2002, Driscoll et al. 2002). These studies focused on the stacked, underlying aquifers (Inyan Kara, Minnekahta, Mennelusa, Madison, and Deadwood, each named for the formations in which they are found), and the numerous springs in the area. The aquifers tend to be confined by less permeable rock layers and primarily receive recharge by filtration at outcrops and stream losses.

The Madison and Minnelusa formations are very permeable, so when streams cross outcrops of these formations, they often lose all flow; hence, this setting is called the "loss zone." Artesian springs also occur in this region, since recharge can often exceed the capacity of the aquifer in the region. The largest headwater and artesian springs occur in the northern and southern areas of the Black Hills (e.g., in the Spearfish Creek and Rapid Creek basins, Cox Lake, and Crow Creek). Groundwater water quality is generally good, especially near the core of the Black Hills; a few limitations on good water quality exist due to elevated concentrations of chloride, sulfate, sodium, manganese, iron, radon, uranium, and arsenic. Surface water quality generally meet their designated beneficial uses, occasionally failing for low dissolved oxygen concentrations, high temperatures, or excessive suspended sediment, particularly in low flow conditions and along the periphery of the Black Hills.

I have found no records of fossil aquatic insects from the Black Hills. I have, however, found *Palaeophycus* isp. trace fossils in an outcrop along Annie Creek in the northern Black Hills. Although they were likely formed by polychaetes or other aquatic worms, it is possible that aquatic insects may have been involved (Pemberton and Frey 1982). Another fossil site of interest is the excellent mammoth dig site in Hot Springs. This site marks a karst sinkhole that was filled with warm artesian spring water and became a death trap over 350 to 700 years for several species of animals due to the steep, slippery sides made of Spearfish shale. As of 2008, a total of 85 animal species had been identified from the deposits, including two Diptera specimens (fragments of an adult and a puparium) and one species of carabid beetle – it cannot be determined from the remains if these were aquatic species or not. Other aquatic invertebrates included eight species of mollusks, including *Pisidium casertanum* (Poli), *P. compressum* Prime, *P. obtusale* (Lamarck) *P. walkeri* Sterki, *Fossaria parva* (Lea), *F.* cf. *dalli* (Baker), *Physa* sp., and *Gyraulus parvus* (Say). Some of these species are also common in present-day ponds and streams in the Black Hills.

Caves in the Black Hills include Wind Cave, purported to be the fourth longest natural cave in the world, with over 200 km of explored passages. Although caves often include extensive groundwater habitats, the caves in the Black Hills are generally dry, with water being restricted to a few seeps, places where surface water percolates into the cave (many are protected from this by impervious overlying formations), and low-lying areas where the cave intersects with the water table. Therefore, although the biota of caves in the Black Hills has been fairly well studied since 1959 (Walthall 1962), the aquatic fauna is very sparse due to dryness and the general arthropod fauna is sparse due to low energy inputs (Moore et al. 1996; Culver 1999).

The composition of the aquatic macroinvertebrate fauna of the Black Hills is, if not poorly known, not compiled into a single resource until now. Very few broad checklists of invertebrates have been created for the region, including only the Ephemeroptera and Plecoptera and scattered reports from other orders. With the exception of biomonitoring projects by state agencies and private consulting companies on behalf of the mines, there have been very few reports on the aquatic invertebrate communities (Jurgens 1968; Drewes 1984). Use of some of these older publications in determining the aquatic insect communites can be difficult, since numerous taxonomic changes have been made in the interim, and many studies used a coarser level of taxonomy (genus or family-level identification). In Shearer's (2006) attempt to define a macroinvertebrate multimetric index for bioassessment of streams in the Black Hills based on 88 samples from 64 sites, he listed a total of 138 taxa, mostly identified to the genus level, from 27 orders and 74 families.

This present work is intended to provide species-level checklists, when such information is attainable, for all aquatic insect groups in the Black Hills. In the future, I intend to produce regional identification keys to supplement the checklists. Keys could probably be produced for some groups now, but I am hoping that this work spurs research into the Black Hills aquatic insect fauna and we scientists fill in the gaps for those groups that are less well known in the Black Hills.

The species accounts have citations to descriptions of various life stages, when available, and a broad characterization of the species' distribution. Reports of each species from the Black Hills are summarized, usually including stream names or specific localities, but sometimes only indicating county-level records. Any specific observations I could add without referring to confidential biomonitoring data, including reported estimates of relative abundance or density from biomonitoring studies or data from my own personal collections, are also included for each species. I have attempted to be careful to record only samples and reports of presence; given the limited areal extent of sampling within a site and the potential for further data reduction during laboratory subsampling, I do not conclude that absence from a sample equals absence from a site or a stream.

Additional species trait information is tabulated for each taxon at or below the family level. These ecological characterizations include tolerance values (for calculation of indices of the Hilsenhoff Biotic Index form), functional feeding groups, habit/behavior group, temperature preferences, lotic, lentic, current and substrate preferences, and voltinism. Better descriptions of the various species traits, as well as data on how I acquired the information are detailed below.

Data Sources

South Dakota Department of Game Fish and Parks produced a brochure of "Common Water Insects" that provides very limited identification of insects, mostly to the order or family levels. Insects that can be identified using this brochure include dragonflies, mayflies, stoneflies, damselflies, caddisflies, mosquitoes, horseflies/deerflies, water boatmen, water scavenger beetles, predaceous diving beetles, water striders, midges, giant water bugs, whirligig beetles, and backswimmers. Also included with each group is an account of their food and a "fun fact." Likewise, the University of Wyoming Biodiversity Institute produced a small, spiral-bound field guide (Tronstad and Barber 2014) to "Wyoming's Stream Macroinvertebrates", which provides identifications generally to the family level. At the order level, there are notes for identification and adult emergence charts by major basin. For each family-level taxon, the guide provides common names, some key features, habitat, functional feeding group, pollution tolerance (sensitive, moderate, tolerant), voltinism, habit, maximum larval size, and a few natural history notes, as well as a map showing distribution by major basin. While certainly a good resource for aspiring young entomologists and perhaps the dedicated angler, these publications do not provide the taxonomic resources necessary for serious, in-depth scientific study.

King (1993) provided identification keys to the orders of common benthic invertebrates, including insects, in the State of Wyoming. A brief description of one to four representatives within each order was given, although specific identifications of the taxa and keys to lower taxonomic levels were not provided. At the back of the King (1993) manual is a list of functional feeding groups and tolerance values for 444 family-, genus-, and species-level benthic invertebrate taxa that may occur in Wyoming. Since the Black Hills are only a small portion of Wyoming, this work could only provide confirmation that a particular taxon that I have identified from the Black Hills does in fact occur in the region.

James (2013), during a reconnaissance survey for the invasive New Zealand mud snail, *Potamopyrgus antipodarum* Gray, in Bear Butte, Elk, and Rapid creeks, sampled the benthic invertebrate communities of these streams. The good news is that those streams still remained negative for New Zealand mud snail; the fair news is that the invertebrate communites were identified only to the family level, so they were not very useful for the purposes of this study.

Shearer (2006) took the first steps to define a macroinvertebrate multimetric index for bioassessment of streams in the Black Hills. Seven metrics that reliably separated impacted streams from reference condition streams were gleaned from a large list of candidate metrics. Scoring of metrics and derivation of a final multimetric index following Barbour et al. (1999) were not part of the study and has not yet been completed althought the State of South Dakota began the process in early 2016. In the appendix to Shearer's (2006) report, a list of 138 aquatic invertebrate taxa from the Black Hills, nearly all identified to the genus level, from 27 orders and 74 families, can be found.

Utah State University hosts the Western Center for Monitoring and Assessment of Freshwater Ecosystems (USU WCMAFE), which, jointly with the National Aquatic Monitoring Center, maintain a database of biological and environmental data from streams all over the United States. The public can query this database using tools available on the USU WCMAFE website, https://qcnr.usu.edu/wmc/. I have found numerous records for Black Hills aquatic insects using this database.

Several aquatic ecology and engineering consultants have worked in the Black Hills, conducting biomonitoring surveys for various clients. Work has focused on the northern Black Hills, particularly on streams potentially affected by the extensive gold mining operations around Lead and Deadwood.

Sporadic records of some taxa were found in published checklists, taxonomic revisions, etc. Particularly useful were state-level checklists (e.g., Guenther and McCafferty 2008 for mayflies). In many cases, though, records were found for areas outside the Black Hills, but few, if any, records were from within the Black Hills proper, primarily due to lack of collecting effort. The Black Hills has apparently not been considered to be an aquatic bug collector's paradise! Hopefully, this work will help change that perception, as it appears that a diverse mix of taxa from northern, southern, eastern, and western biogeographic affinities converges in the Black Hills.

I had hoped to include data from the extensive biomonitoring efforts that GEI Consultants, Inc. had done from about 1985 onward in the northern Black Hills for numerous mining clients, because I thought it was likely to include records that were not reflected in the literature and it included valuable ecological data and population-level data like average and maximum densities. I have not received permission to publish those data; perhaps a future edition may be able to include stream names and example densities as I had intended. The only distributional record in biomonitoring samples collected by GEI that wasn't reported in the literature was the elmid beetle *Lara avara* LeConte, based on a single larva.

Those clients should not be faulted for that – mining companies (and for that matter, most private companies) tend to be wary of increased government oversight, and publishing data such as the density of organisms in a stream based on their own proprietary biomonitoring data is perceived as one more way in which government agencies can look over their shoulder. For example, based on teleconferencing sessions held prior to the manuscript submission for De Jong and Canton (2012) and De Jong et al. (2013) to release very select biomonitoring data, the managers at mines and other water dischargers (and their lawyers, especially) actually envision the following dictate: "So, De Jong's book says that Species X occurs in stream biomonitoring samples at an average of 32 individuals/m^2, and your results this year at that one site are way below average at 26 individuals/m^2. You must be killing them, so we are going to set up an expensive 10-year monitoring program in which you must demonstrate that you are adequately protecting Species X." And, based on my experience, some mining companies in the United States and abroad should be worried (due to their own harmful practices that push the limits of ecological damage or due to overzealous and uninformed agency personnel), while others should not worry (because they are not causing significant problems and most agency personnel recognize and appreciate the inherent variability in aquatic invertebrate community data).

Instead, and hopefully with sufficient vagueness so as not to get myself in trouble, I have included a column called "Seen" in the traits table at the end of each order treatment. In that column, a plus sign (+) indicates that I have seen the taxon specifically in stream biomonitoring samples, a minus sign (-) indicates that I have not seen it specifically in stream biomonitoring samples, and a slash (/) indicates that it would not have been possible to identify to that taxon based on life stage and/or existing literature. If a taxon was not seen in stream biomonitoring samples, it should not necessarily be considered rare; it may simply prefer habitats other than those sampled in the stream biomonitoring programs. My personal collections are discussed in the text and not reflected in the "Seen" column in the traits tables. Furthermore, a minus sign should not be construed that I have not ever seen the taxon – I've seen nearly all of these at some point or other. I guess it was fortuitous that I occasionally collected bugs for my own personal collection, whether at porch lights at the hotels I stayed at or at the biomonitoring sites or on my own "expeditions".

All told, the records in this book hail from about 120 individual localities. Sixteen sites are in Wyoming, the rest are in South Dakota. Stream sites represent about half of the locations, lakes and reservoirs another dozen or so, while the rest of the records are from towns or campgrounds or other landmarks.

Field Sampling and Laboratory Procedures

There are many reviews of sampling devices, their targeted groups, benefits and problems, and additional considerations (see list in Merritt et al. 2008). Sampling can be qualitative (e.g., with a kick net, where the results are a general assessment of the taxa present with some relative abundance information), quantitative (e.g., with a Surber sampler, where a numeric estimate of population parameters can be obtained), or semi-quantitative (e.g., where there are limited density estimates because a kick net was used over a specified area or time frame). Quantitative sampling is often replicated to be able to obtain a statistical confidence on the estimates.

Different equipment is used depending on whether the target organisms are terrestrial adults (which may require malaise traps, emergence traps, etc.) or aquatic larvae (which may require a kick net, Surber or Hess sampler, Petit Ponar, etc.) There is also the possibility that some aquatic insects inhabit the hyporheic zone and require such equipment as a Bou-Rouche pump or other technique to access them.

The decision as to which sampling plan and the sampler to use should depend on the objectives of the study. A study that is used to determine the effects of a potential point source impact in a limited area should use replicated, quantitative methods, because they provide measurable results that can be statistically compared to a reference site. A state-wide or region-wide survey for establishing a multimetric Index of Biotic Integrity (IBI) for regulatory use classification issues may use semi-quantitative methods on hundreds of streams. Comparison of subsequent test sites to the resulting IBI for bioassessment purposes would necessitate taking samples in accordance with the methods that had been used to establish the IBI. Of course, there are also opportunistic sampling regimes, whereby insects are collected more-or-less haphazardly, without a specified sampling plan, even though technical equipment, such as kick nets, Hess samplers, Malaise traps, black lights, etc. may be used.

Most aquatic invertebrate sampling that has been done in the Black Hills has primarily been opportunistic or for biomonitoring or bioassessment purposes. Consulting firms have been operating in the Black Hills for many years. The states of South Dakota and Wyoming have a vested interest in both biomonitoring and bioassessments as they attempt to enforce the state's portions of the federal Clean Water Act. One of the requirements of the Clean Water Act is that each state develops standards for their waters to ensure that beneficial uses, such as swimming and fishing, are protected.

A water quality standard defines the water quality goals set for a water body, or a portion of the water body. The water quality standards regulations establish the use or uses to be made of a water body, set criteria necessary to protect those uses, and establish policies to maintain and protect water quality. Both South Dakota and Wyoming have developed surface water quality standards for all waters of the states, as required by the Clean Water Act.

The Administrative Rules of South Dakota (ARSD 74:51:01, :02, and :03) contain South Dakota's surface water quality standards. Chapter 74:51:01contains both the numeric and narrative criteria to protect the uses of the state's water bodies. Chapters 74:51:02 and 74:51:03 designate the beneficial uses assigned to each specific water body in the state. Wyoming's Surface Water Quality Standards are included in Chapter 1 of the Water Quality Rules and Regulations (https://rules.wyo.gov/).

The South Dakota Department of Environment and Natural Resources (SDDENR) has prescribed sampling regimes for most stream biomonitoring projects in the South Dakota portion of the Black Hills, using the U.S. Environmental Protection Agency's (EPA) Environmental Monitoring and Assessment Program (EMAP) protocols (Peck et al., unpublished draft) as a guideline. This protocol was updated in South Dakota's Standard Operating Procedures (South Dakota Department of Environment and Natural Resources 2005a, b) and is available online at http://denr.sd.gov/dfta/wp/Vol2SOP.pdf.

Briefly, the South Dakota EMAP protocol prescribes that an area of 0.1 m^2 (in the original, pre-2005 version, the area was 1 ft^2, or 0.09 m^2) is sampled using a kick net with 500 μm mesh size at each of 11 transects in the stream. To determine transect spacing, a preliminary mean stream width (PMSW) is calculated by measuring water width (or streambed width if low flows confine widths to a small portion of the streambed) at ten locations. The average width of these measurements is the PMSW. Eleven transects are established, three PMSWs apart, or at least 10 m apart, for a total distance of at least 100 m. One benthic invertebrate sample is collected at each transect at 25, 50, or 75 percent of the wetted width from the left bank, determined by random assignment of the first transect, and rotated at subsequent transects. In small streams (< 1.0 m wide), the sample is taken from the thalweg. In erosional habitat, loose rocks and large substrates are scrubbed by hand and finer substrates are kicked vigorously for 30 seconds to dislodge organisms into the net. In depositional habitats, the same techniques are used, except that the net is dragged through the standing water within the 0.10 m^2 area to capture suspended organisms. The collected organisms are then combined into a single, "reach-wide" composite sample for each site. Field methods for data used in Shearer (2006) followed a modified EMAP protocol, wherein eight transects were used instead of the prescribed eleven transects.

Other methods have also been used for biomonitoring, including replicated, quantitative methods. For example, a modified Hess sampler, which encloses an area of 0.086 m^2 and has a net mesh size of 500 μm (Canton and Chadwick 1984), or the Surber sampler, which encloses an area of 0.093 m^2 and also has a net mesh size of 500 μm, may be used, depending on water depths. Replicated sampling with the Hess or Surber samplers involves choosing riffle habitats that are of similar depth, substrate composition, and water velocity – these three variables must be similar in order to approximate true replication. Riffle habitat refers to the portion of a stream in which moderate velocity and substrate roughness produce turbulent conditions which break the surface tension of the water.

In addition to the replicated, quantitative sampling, qualitative samples were sometimes also collected at each site with a kick net sampler to provide additional information on community composition for comparison with historical qualitative data. These are sometimes referred to as "sweep" samples to distinguish them from some of the more semi-quantitative samples also collected with kick nets. Although the riffles and runs are generally the most diverse and productive habitat type in mountain streams such as those in the Black Hills, they do not represent the overall habitat and diversity present in the streams. Pools and glides are other habitat types that often have interesting assemblages. An example of kick net use in these habitats was described by Horning and Pollard (1978). Timed kick samples (a semi-quantitative technique) can be collected at each site by vigorously kicking the substrate upstream of the sampler for 30 seconds or other time frame and allowing the dislodged aquatic invertebrates to drift into the sampler.

Laboratory processing of samples may be as simple as standard curation of opportunistically collected organisms to regimented sorting, identification, and enumeration of sampled material by a production taxonomy laboratory. The production taxonomy laboratory will generally identify an organism to a "standard taxonomic level" that is based on available literature and the age and condition of the organism, and is subject to time constraints. Time constraints are usually less of a concern for the university or museum taxonomy laboratory, and specimens may be identified further than the laboratory's standard taxonomic level. Plus, the type of sample influences the laboratory processing time: in my experience, samples collected using the EMAP transect technique generally requires 1.2 to 1.8 times the amount of time to process in the laboratory than Hess samples, while qualitative kick net samples usually require less time.

In consultant laboratories, organisms are sorted from the debris from Hess/Surber and kick net samples according to a rigid Standard Operating Procedure (SOP). If the sample contains excessive numbers of organisms, considering time constraints, the sample could be subsampled (Vinson and Hawkins 1996; Carter and Resh 2001). Chironomids and oligochaetes are often mounted on glass microscope slides and cleared prior to identification and counting. The sorted invertebrates would then be identified to the lowest practical taxonomic level (Carter and Resh 2001) using available taxonomic keys and counted by taxon. Quality assurance (QA) should be conducted on as many steps of the sorting and identification process as possible to ensure data quality (Stribling et al. 2003). Taxonomic certification by the Society for Freshwater Sciences (formerly, the North American Benthological Society) at the genus level for aquatic inverterat identifications has been suggested as indicative that the technician and laboratory are capable of providing accurate identifications.

Samples that provided the data for Shearer (2006) were processed at Valley City State University in Valley City, North Dakota. A fixed count of 300 individuals was sorted and all macroinvertebrates were identified to the lowest practical taxonomic level (usually genus/species), except Chironomidae, which were identified to subfamily.

Taxonomic Serial Numbers (TSNs)

Taxonomic Serial Numbers (TSNs) are provided for nearly all species-, genus-, and family-level taxa that are reported herein, based on the Integrated Taxonomic Information System (ITIS) (http://www.itis.gov/index.html). The ITIS is the result of a partnership of federal agencies formed to satisfy their mutual needs for organized, standardized, scientifically credible taxonomic information, identified by the White House Subcommittee on Biodiversity and Ecosystem Dynamics as a research priority that is fundamental to ecosystem management and biodiversity conservation. With over 787,000 taxa in the database (as of May 2018), ITIS has been, and continues to be, a significant contribution to the scientific infrastructure that is fundamental to the description, conservation, and management of the nation's biodiversity.

Federal agencies using and contributing to the ITIS and TSNs include the U.S. Department of Agriculture (USDA), EPA, U.S. Fish and Wildlife Service (FWS), USGS, U.S. National Oceanic and Atmospheric Administration (NOAA), U.S. National Park Service (NPS), the National Museum of Natural History (USNM) at the Smithsonian Institution, Agriculture and Agri-Food Canada, and Mexico's Comisión Nacional para el Conocimiento y Uso de la Biodiversidad (CONABIO). Consistent use of ITIS and TSNs will facilitate researchers in both the private and public sectors as they share biological information by providing a common framework for taxonomic data.

ITIS is primarily updated by volunteers who are experts in their respective taxonomic groups; as such, new names, synonymies, and other taxonomic works render the database slightly out of date at any given time, unless the researcher making the taxonomic action also simultaneously updates the ITIS database. Therefore, some recently described or recently synonymized taxa do not have TSNs assigned to them.

Tolerance Values

The first benthic macroinvertebrate indices to use tolerance values (TV) were Chutter (1972) in South Africa, followed by Hilsenhoff (1978, 1987, 1988) in Wisconsin. Hilsenhoff evaluated the measured physical and chemical properties of stream water to determine the relative degree of organic and nutrient pollution in each stream and assigned tolerance values to the resident fauna accordingly. Although the original Hilsenhoff Biotic Index (HBI) referred to the method and tolerance values published by Hilsenhoff (1978, 1987, 1988) and was derived for identification of organic or nutrient pollution, any index that has the same mathematical form and uses tolerance values in a similar manner is often referred to as the HBI, even if they are not designed specifically to identify organic pollution. Such generic HBI's have been widely adopted within North America and throughout the world as a part of the analysis of benthic macroinvertebrate population data.

The HBI is calculated as

$$HBI = \frac{\sum_{i=1}^{n} x_i TV_{n_i}}{x},$$

where n = the total number of taxa, x_i = the density or relative abundance of the ith taxon, TV_{ni} = the taxon-specific TV for taxon n_i, and X = the total macroinvertebrate density for which TVs are available (because TVs for species and genera are extrapolated from higher taxonomic levels, $X = \Sigma x$). Accordingly, it is an abundance-weighted average of the TVs and is expected to have higher values in "impacted" sites, since the community would be comprised of more "tolerant" organisms. Most TVs (and subsequently calculated HBI values) range from 0 (intolerant organisms) to 10 (tolerant organisms), although a few other ranges exist (e.g., $1 - 10$, $0 - 5$, etc.). The resultant HBI value gives an indication of the water quality at a site (Figure 3).

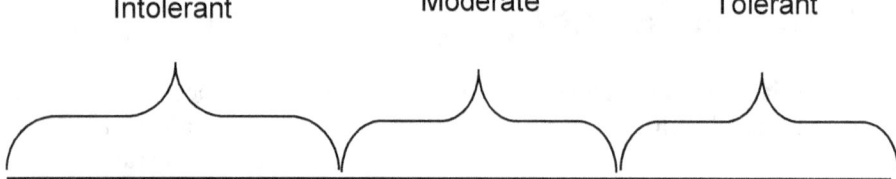

Intolerant   Moderate   Tolerant

Figure 3: Illustration of the range of tolerance values across a gradient of streams (neither photo is of a stream in the Black Hills).

It is important to recognize the semantics of "tolerance values" (TVs), which, as they are most frequently used today, do not really indicate "tolerance" but rather may suggest a "preference" for a particular level of water quality (*sensu* Grotheer et al. 1994). What is portrayed as a "tolerance" value is usually the stressor level or condition which yields the largest number of occurrences for a taxon. As such, the TV for a stressor may actually represent the least offensive conditions for that taxon, even though they may tolerate much worse; conversely, the TV for an ecological condition may actually represent a preference for that condition, with ability to tolerate conditions outside that level. TVs should be determined by the ends of the range of conditions in which the taxon is found. Nevertheless, in this document, I use TVs in their conventional sense.

Some researchers have used ways of utilizing TVs other than plugging them into the HBI formula. For example, some do not weight the index by abundance, but rather provide the average TV for the taxa present in a sample, i.e., not weighted by density. Some identify and isolate taxa with particular TVs (e.g., "sensitive" taxa with TV ≤ 3 or "tolerant" taxa with TV ≥ 7) and then calculate the proportion of the community comprised by sensitive or tolerant taxa or individuals.

Since Hilsenhoff (1978, 1987, 1988), several other researchers have provided lists of TVs. Most are based on tolerance to "general perturbations", although some are still targeted at a specific form of disturbance, e.g., pH, ionic concentration, siltation, organic/nutrient pollution, metals, etc. (Huggins and Moffitt 1988; Grotheer et al. 1994). Some TVs were derived by professional judgment, some by summarizing values provided by other researchers (as I admit to doing here), and some by novel methods. One of the more promising methods has recently been used by Yuan (2006) and Carlisle et al. (2007), wherein benthic invertebrate data are collected simultaneously with habitat and water quality data, as in state biomonitoring program data or data from the USGS National Water Quality Assessment Program. The habitat and water quality data are plugged into a principal components analysis to derive primary axes of water quality; the axes are subsequently normalized to a scale of 0 to 10. Abundances of each invertebrate taxon are then plotted along the scale to identify the optimal value for that taxon (i.e., where the abundance was highest, suggesting that the invertebrates "preferred" that level of water quality). I think that this method could be very useful in that the water quality components can be restricted to certain classes, such as nutrients or metals or field parameters; this aspect of the method was actually used in Carlisle et al. (2007) to illustrate differential tolerances to different classes of stressors. Of course, sufficient and appropriate co-collected data need to be available to conduct such calculations.

Another potential corollary to this method might be to identify not only the optimal value for each taxon, but also the breadth of values over which a particular taxon is collected. Taxa that are abundant at a certain level of water quality but can be found across the board are not very informative for deciphering a specific value for the water quality. Taxa that have a very narrow range of tolerances could be used to better delineate a level of water quality (rather than being diluted by ubiquitous taxa) or there could be a ratio of discriminating organisms to more broadly tolerant organisms within a water body. This method reflects my opinion on the topic of tolerance values, but to my knowledge, it has not been investigated.

EPA (2011) used an interesting approach to investigation of the limits of tolerance in aquatic invertebrates in West Virginia. It used state bioassessment data from 2,210 samples to plot the individual occurrences of 163 benthic invertebrate genera in relation to stream conductivity values. From these plots, EPA (2011) calculated the "XC_{95}" value, defined as "the conductivity value below which 95% of the observations of the genus occur and above which only 5% occur." This value in essence provided a reasonable bound to the conductivity levels tolerated by each genus – or would have if the stressor-response values of each taxon had been appropriately considered (Canton et al. 2011). Some genera such as the mayfly genus *Ephemerella* demonstrated the expected decreasing dose-response curve to conductivity levels, with higher levels producing fewer mayfly occurrences; the XC_{95} for *Ephemerella* was 299 µS/cm, indicating that Ephemerella apparently prefers not to live in waters with conductivity levels above that value. About 44% of the genera used by EPA (2011) exhibited the expected decreasing stressor-response to conductivity. Other genera, however, such as the empidid fly genus *Hemerodromia*, demonstrated the opposite response in which occurrence of the genus increased with increasing conductivity levels, suggesting that the upper tolerance to elevated conductivity levels had not been reached in the streams sampled for bioassessment purposes (>9,790 μS/cm), while low conductivity levels were not preferred by those genera; this occurred in ~20% of the genera used by EPA (2011). A beautiful, unimodal dose-response was demonstrated by the water penny beetle genus *Psephenus*, in which a clear optimum is visible, with reduced occurrences at both high (~9,119 μS/cm) and low (~75 μS/cm) conductivity values. The rest of the calculations that EPA (2011) conducted to derive a conductivity benchmark for the central Appalachian region (Cormier and Suter 2013, Cormier et al. 2013) are controversial (Canton et al. 2011, Roark et al. 2013), but I do think that calculations similar to their XC_{95} (and maybe an XC_{05}?), standardized along a 0 to 10 range, could have a place in investigating taxon-specific TVs and tolerance ranges for specific stressors.

TVs specific to the Black Hills aquatic invertebrates have not previously been generated, and as was previously stated, there is no *de novo* generation of TVs here, either. Blocksom and Winters (2006) recommend that tolerance values be generated using field-collected abiotic parameters to rank taxon sensitivity, but these have not been collected in most cases for Black Hills streams. Instead, the TVs I list here were assigned for each taxon from the Northwest (ID) Regional Tolerance Value database as presented in Appendix B of Barbour *et al.* (1999). These values represent data compiled by the Idaho Department of Environmental Quality (IDEQ) and are indicative of a taxon's tolerance of perturbation in a stream (Grafe 2002). The taxonomic composition of streams in the Black Hills is more similar to the benthic communities found in Idaho than to benthic communities in Ohio, North Carolina, Wisconsin, the mid-Atlantic highlands (the other regions listed in Barbour et al. 1999), or most other regions from which extensive tolerance lists are available. King (1993) provided TVs for 377 aquatic insect taxa from the State of Wyoming, but many of the taxa are not found in the Black Hills, being found in the main Rocky Mountain cordillera and the large basin and Plains areas throughout the state. Additionally, the list is far from comprehensive. Therefore, the Northwest Regional Tolerance Value database in Barbour et al. (1999) was deemed to be the most appropriate database from which to obtain taxon-specific tolerance values.

If a taxon was not listed in Appendix B of Barbour *et al.* (1999) or a TV was not presented for that taxon, I used the following extrapolation technique to derive a TV. All species within a genus were assigned the TV for that genus unless a TV already existed at the species level. For example, a species-level TV did not exist in Appendix A of Barbour et al. (1999) for the baetid mayfly *Pseudocloeon propinquum*, so the genus-level TV for *Pseudocloeon* (4) was assigned to that species. Similarly, all genera within a family were assigned the TV for that family unless a TV already existed at the genus level. Such extrapolation continued up to the ordinal level.

Finally, it should be recognized that TVs, and indices derived from their use, can be used to infer water quality, but they are not a direct measure of water quality. A water body may have a high value (supposedly indicating poorer water quality) because the value itself is diluted by ubiquitous, but high scoring, taxa; therefore, if organisms with low tolerance values are present, it is prima fasciae evidence that the water is of sufficient quality to support those organisms.

Functional Traits

Because taxonomy of immature aquatic insects can be difficult, Cummins (1973) asked whether ecological questions involving aquatic insects had to wait until the taxonomy was resolved or if alternative methods could be developed to analyze communities based on their functions within the ecosystem. In an attempt to address the second option, functional feeding group (FFG) and habit/behavior classifications were developed, and subsequently, a whole suite of functional traits have been identified. Because trait-based approaches theoretically provide a mechanistic basis between morphology or behavior and functional processes, they may help us make intuitive predictions about insect community composition. Sometimes, traits-based approaches provide good explanations for the behavior of aquatic insect communities (e.g., Poff et al. 2006); however, traits-based approaches do not always work out as predicted, possibly because of the general nature of traits or because the wrong suite of traits was investigated (e.g., Walters 2011).

Merritt et al. (2008) is a good source for FFG and habit/behavior classification, since these were assigned for nearly all North American aquatic insect genera. Poff et al. (2006) provided a table of 20 traits (59 trait states) for 311 North American lotic insect genus or family-level taxa. Traits included in Poff et al. (2006) included voltinism, developmental rates, synchronization of emergence, length of adult life span, adult's ability to exit the water, larval ability to survive desication, adult female dispersal/vagility, adult flying strength, larval occurrence in drift, maximum larval crawling rate, larval swimming ability, mode of larval attachment to the substrate, larval armoring, larval shape, mode of larval respiration, size of mature larva, larval rheophily, larval thermal preference, and habit/behavior. These two documents provide the basis for much of the trait information provided herein. Vieira et al. (2006) built a database of most of the same lotic invertebrate traits based on literature citations, and I used some data from that resource, as described below.

Detailed trait information is not available for most North American invertebrate species-level taxa. In fact, only spotty information exists for most species-level taxa. Also, although traits can be determined for individual species, most existing data are aggregated at the genus or family level because congeneric species often appear to be functionally equivalent and could be viewed as ecologically redundant (Wiggins and Mackay 1978). Nevertheless, using data provided from Poff et al. (2006), Vieira et al. (2006), and Merritt et al. (2008) – and it should be noted that I did not use all of it, ignoring some traits such as armoring, timing and synchronization of emergence, and adult life spans – I have populated the traits tables in this work for each taxon, regardless of functional redundancy among species within genera or even families.

Functional Feeding Groups and Habit/Behavior Groups

Based on mouthpart structure, specialized leg structures, construction of nets, etc., aquatic insects were classified into the following FFGs in Merritt et al. (2008):
- Shredders (SH) – detritivores that feed on coarse particulate organic matter (CPOM), herbivores that chew and mine in living vascular hydrophytes, and gougers that excavate galleries in submerged wood.
- Collectors – detritivores that decompose fine particulate organic material (FPOM); divided into two subgroups
- Collector-gatherers (GC) – feed on surface films and organic deposits on the substrate
- Collector filterers (FC) – filter FPOM that is suspended in the water column
- Scrapers (SC) – herbivores grazing on periphyton (attached algae) on mineral and organic surfaces
- Piercers (PI) – herbivores that pierce cells and tissues of living vascular hydrophytes to suck fluids
- Predators (PR) – carnivores that feed on living animal tissue; some divide these into engulfers that ingest whole animals or parts thereof and piercers that pierce the cells and tissues to suck fluids
- Parasites (PA) – internal or external parasites or parasitoids of eggs, larvae, and pupae of other aquatic invertebrates

Aquatic insect habits describe their mode of locomotion, attachment, or concealment. Their habits are usually inextricably linked to their subsequent behaviors; for example, an aquatic insect that has a hydrodynamic shape (torpedo-like) can usually swim easily, even in flowing water. These habits/behaviors include the following classifications:
- Skaters (SK) – Run or glide about on the water surface where they feed as scavengers or predators on organisms trapped in the surface film.

- Plankton (PL) – swim in the open-water zones of lentic habitats, occasionally exhibiting a diurnal vertical migration in the water column to take advantage of changing oxygen levels or temperatures.
- Divers (DI) – Swim in lentic habitats and lotic pools by rowing with their legs; usually they come to the surface for respiration or feeding, then dive to the bottom to rest or avoid predators.
- Swimmers (SW) – Swim in lentic and lotic habitats using and undulating fish-like motion; may cling to submerged objects to rest, but rarely breach the surface.
- Clingers (CL) – Have morphological structures (such as suckers or sucker-like gill arrangements) to attach to the substrate even in strong currents; some may also build fixed retreats in which to hide and stay along the substrate.
- Sprawlers (SP) – "Float" on leaves and fine sediments in the substrate of lentic habitats and lotic pools.
- Climbers (CB) – Live primarily on the vertical, submerged stems and roots of vascular plants or woody debris.
- Burrowers (BU) – Inhabit fine sediments of pools and lentic habitats, often with constructed tubes that extend above and below the substrate surface. Leaf, root, and stem miners are included in this category.

The FFG and habit/behavior classification was assigned for each taxon as the first FFG or habit/behavior mentioned for the taxon in Merritt *et al.* (2008); as in Appendix B of Barbour *et al.* (1999), this is assumed to be the taxon's primary trait state. If classifications were not listed for a taxon, the classifications were extrapolated from higher taxonomic levels, as described for the tolerance values. King (1993) provided FFGs for approximately 350 aquatic insect taxa from the State of Wyoming. Some habit/behavior classes are excluded from benthic invertebrate data sets in a few states because the trait is not necessarily "benthic"; skaters and divers are examples of excluded classes; however, if that classification adequately characterizes them, I have included it herein.

Extraction of Data from Vieira et al. (2006)

Vieira *et al.* (2006) is a database of North American lotic invertebrate life history traits, which were compiled from the literature as a joint venture between the USGS National Water-Quality Assessment (NAWQA) Program and Colorado State University. A total of 14,127 records for over 2,200 species, 1,165 genera, and 249 families were entered into the database from 967 publications, texts and reports. Data were entered into the database, according to statements in the literature; for example, Stewart and Stark (2002) state, "*Claassenia* nymphs exhibit a 2-3-yr life cycle in moderate to large streams of the Rocky Mountains . . .", and these data were translated into the database as "merovoltine," "lotic," and "5+ order streams/rivers" in the case row for *Claassenia* and Stewart and Stark (2002). Each literature source for each taxon had its own case row, so some taxa had abundant, sometimes conflicting data. Nevertheless, this outstanding database is the first attempt at a comprehensive list of North American lotic invertebrate life history traits.

Poff et al. (2006) relied on this database in compiling their list of traits for 311 lotic insect taxa, and I felt that Vieira et al. (2006) would also be a good starting point for bringing together life history data on the invertebrates of the Black Hills. However, because the data were presented as being derived from multiple literature sources, I distilled the data as follows: if a single category was assigned for a taxon, that category was used; when multiple categories were assigned to the same taxon, I calculated a weighted average of fuzzy coded variables from all pertinent records for that taxon to derive a singular category at both the species and generic levels. Examples are provided below.

Temperature preferences

Because aquatic invertebrates are poikilothermic ("cold-blooded"), temperature plays a very direct role in their distribution and life history. The upper thermal death point for most freshwater invertebrates is around 30-40°C, but some insects have been found in hot springs above 40°C. In his excellent review article on hot springs faunas, Pritchard (1991) concluded that no insects live above 50°C; however, a small assemblage of invertebrates has been discovered in a hot springs in Idaho surviving at 52°C (De Jong et al. 2005, Meier et al. 2010), and I suspect that similar assemblages could be found in hot spring environments in the Black Hills, such as those near the town of Hot Springs.

Conversely, many insects can also tolerate very cold temperatures, even being able to grow and develop at temperatures near 0°C. Some insects, such as capniid stoneflies and diamesine midges, can even emerge as adults during the winter and early spring to crawl about on ice and snow banks. Several experimenters have found a select few insects, particularly chironomid midges, capable of being frozen solid at temperatures below -20°C and being revived afterwards.

Temperature preferences for most aquatic invertebrates are not very definable, since they are usually able to tolerate a wide range of temperatures. This is likely because of the wide range of temperatures a single water body, especially in the Black Hills, can experience over the course of a year or even a day. Although some attempts have been made to determine exact temperature preferences for some aquatic invertebrates (Grafe et al. 2002), it has been more common to indicate broad categories of preferences. As such, aquatic invertebrates have generally been assigned to broad categories such as stenotherms and eurytherms, with cold, cool, warm, and hot water preferences.

Stenothermy involves having a relatively narrow range of temperatures over which a taxon is found. In contrast, eurythermy involves a relatively broad range of temperatures over which a taxon is found. I have provided abbreviations for the various iterations as described here. A few invertebrates have clear preferences for cold water; these are classified as stenothermal cold (S: cold) invertebrates or (for those which prefer extremely cold water) stenothermal hypercold (S: hypercold) invertebrates. Most of the cool, warm, and hot water invertebrates have eurythermal tendencies, so there are only eurythermal cool (E: cool), eurythermal warm (E: warm), and eurythermal hot (E: hot) classifications. Figure 4 illustrates the relative ranges and sometimes significant overlap of many of these categories; note that there is no objective scale for these designations. Figure 4 is a conceptual diagram of temperature preference classifications that I made several years ago and often use to illustrate this concept.

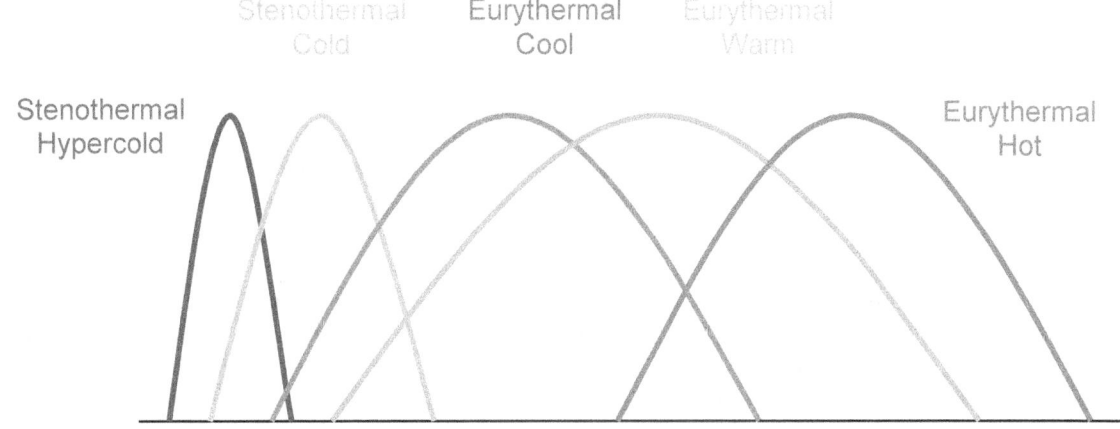

Figure 4: Relative ranges of temperature classifications

The temperature preferences I have provided are generally from the list derived by IDEQ (Grafe et al. 2002). As with tolerance values, FFGs, and habit/behaviors, I extrapolated values from higher taxonomic levels, if necessary. Vieira et al. (2006) also compiled data on temperature preferences based on statements in the literature they reviewed. Because the IDEQ data are based on observations, I chose to use the IDEQ data instead of the data provided in Vieira et al. (2006), except in very rare cases when IDEQ data are very incomplete (e.g., *Stenacron interpunctatum* and *Hydropsyche*, s.l.).

Lotic, lentic, current, and substrate preferences

Essentially any aquatic habitat, large or small, lotic or lentic, shallow or deep, can serve as habitat for some aquatic invertebrate. Many taxa, however, are restricted to only one or a few kinds of habitats. In standing water (lentic habitats), invertebrates may inhabit the surface, the edges, the open water, or the deep water substrate. They may also inhabit vegetation that grows in the water. Lentic water bodies can be very clean or stagnant, or even wetlands. Downstream of lakes and impoundments, stream communities can be affected by the presence of that impoundment; Cooper and Troelstrup (2015) looked at the impact that Pactola Resrevoir has on the aquatic insects in Rapid Creek.

Streams and rivers (lotic habitats) have margins, the thalweg, large woody debris, pocket pools, riffles, runs, and so on. Certain invertebrates may be found in any of these habitats. Most biomonitoring projects that I have been concerned with have involved lotic habitats.

Temporary streams/ponds can also provide habitat when water is flowing. Some invertebrates have life history characteristics that allow them to survive when the water dries up by entering cryptobiotic, dehydrated life stages, burrowing into the moist substrate, drifting downstream to more permanent water. In fact, sometimes the biodiversity of ephemeral streams can rival that of nearby permanent water resources. But some organisms attempt to colonize such habitats in vain, as the ephemeral resource dries up before they can complete their life cycle or reach a stage that can tolerate drying (AWWQRP 2006).

As stated above, I relied heavily on Vieira et al. (2006) for life history traits; however, the data in Vieira et al. (2006) are based on statements from the literature they reviewed, so actual numeric values do not appear to exist for most aquatic invertebrates. I compiled the data from Vieira et al. (2006) by creating a numeric index for the various lotic, lentic, current, and substrate preferences using the fuzzy coding method, described here.

Water body preference for each taxon was determined by calculating a weighted average, rounded to one decimal point, of the occurrences noted within Vieira *et al.* (2006). Fuzzy coding for lotic water bodies was as follows:1 = "1st order streams/headwaters", 3 = "2nd - 4th order streams", 5 = "5+ order streams/rivers". Fuzzy coding for lentic water bodies was as follows: 1 = "wetlands", 3 = "ponds", 5 = "lakes". Each value is a continuous variable ranging from 1 (has been most commonly reported from small water bodies) to 5 (has been most commonly reported from large water bodies). Since some organisms have been reported from both lentic and lotic habitats, values may exist for both water body types.

Current preference for each taxon was determined by calculating a weighted average, rounded to one decimal point, of the occurrences noted within Vieira *et al.* (2006). Fuzzy coding for flow was as follows: 0 = "quiet", 1 = "slow flow", 2 = "fast laminar flow", 3 = "fast turbulent flow". The resulting value is a continuous variable ranging from 0 (has been most commonly reported from habitats with no flow) to 4 (has been most commonly reported from habitats with fast turbulent flow).

Substrate preference for each taxon was determined by calculating a weighted average, rounded to one decimal point, of the occurrences noted within Vieira *et al.* (2006). Fuzzy coding for substrate was based on an approximate log$_2$ scale, modified as follows: 0 = "fines/silt" ($< 2^0$ mm), 1 = "sand" (2^0 - $2^{2.4}$ mm), 4 = "gravel" ($2^{2.5}$ - $2^{5.4}$ mm), 7 = "cobble" ($2^{5.5}$ - $2^{8.4}$ mm), 10 = "boulder" ($> 2^{8.5}$ mm). The resulting value is a continuous variable ranging from 0 (has been most commonly reported from fine, silty, substrates) to 10 (has been most commonly reported from boulder substrates). Essentially, the value reported is approximately -φ in the Wentworth Scale (Bundt and Abt 2001), with classifications smaller than "fines/silt" grouped into a single category.

Voltinism

Voltinism refers to the length of time an organism requires to complete its life cycle from egg to adult. Many insects have a one-year life cycle, with adults laying eggs in the summer, larvae hatching and growing through the autumn, larvae or pupae aestivating during the winter and spring, and adults emerging in the summer to mate and continue the life cycle. Voltinism, even within a species, can be highly influenced by elevation, latitude, water temperature, and other factors. For example, a species that requires two years to complete its life cycle in New Hampshire may be able to do so in one year in Florida. The voltinism data provided herein do not necessarily reflect the specific voltinism category for each species within the Black Hills; in almost all cases, such data simply do not exist (but we can start somewhere).

Most voltinism data presented herein are based on Table 5A in Huryn et al. (2008), supplemented by data in Finn and Poff (2005) and Poff et al. (2006). The Table in Huryn et al. (2008) uses the terms multivoltine (>3 generations/year), trivoltine (3 generations/year), bivoltine (2 generations/year), univoltine (1 generation/year), semivoltine (2 years/generation), and merovoltine (>2 years/generation) to describe each taxon. Instead of being so specific, I have limited my usage to include simply multivoltine (>1 generation/year), univoltine, and merovoltine (>1 year/generation) designations.

Potential Data Analysis

One primary purpose in the compilation of this list of taxa from the Black Hills is to inspire interest in collecting in the Black Hills. As such, it is recognized that this work is only the beginning of what is hoped to be a lot of fruitful research into basic biodiversity. This certainly is not intended to be the definitive, final word on the aquatic insects of the Black Hills; I consider it only a starting point.

However, putting biodiversity aside for the moment, a highly practical reason for summarizing the Black Hills aquatic insect fauna is for biomonitoring. In fact, tolerance values, in particular, were designed for this express purpose. In biomonitoring, samples of the stream fauna are collected, identified, and enumerated, such that ecological characteristics of the stream can be inferred from what animals are present. Those inferences are based on previous knowledge of what relative or absolute abundances, species traits, ecological niches, and physical and chemical tolerances are represented by the resident fauna.

Production laboratories generally provide species lists and estimates of abundance (number of organisms/sample) for samples collected using the EMAP protocols. In replicated, quantitative samples, the field and laboratory procedures can provide abundance estimates in terms of density of organisms/m^2, since the area sampled was rigidly defined. These taxa lists and densities, combined with the species traits identified, can be examined individually by an experienced researcher to infer ecological conditions. On the other hand, these data can also be incorporated into a multimetric index that boils what can be overwhelming amounts of ecological data into a single point.

Multimetric Indexes

There are nearly unlimited numbers of potential metrics for evaluating benthic macroinvertebrate populations (Barbour et al. 1999), and a current use of these metrics is to combine the most useful metrics into a single multimetric index. They are frequently derived for governmental agencies (such as states) to help them apply narrative water quality criteria under the auspices of the Clean Water Act. Such narrative water quality criteria are often worded as "no impairment of the biological communities", but rarely is a definition of impairment provided within the criteria. Impairment indicates that the site or stream is not in attainment of the state-assigned aquatic life use classification (indicating that the water should be capable of supporting aquatic life). Multimetric indexes can help determine whether or not a site or stream is in attainment of the aquatic life use classification.

Despite their popularity, utility, and sometimes regulatory status, multimetric indices do suffer from some shortcomings. First, reference conditions are chosen a priori by selecting physical or chemical parameters that describe the researcher's idea of reference condition. While this may be accomplished by attempting to choose sites that are relatively free from human influence, it is often done by choosing sites that have a particular physical or chemical signature. Thus, the concept of reference condition becomes biased by the researcher's definition and the selection of some reference sites follows circular reasoning. A better way may be to allow the data to organize itself and define its own reference condition. Second, the goal in derivation of a multimetric index is to calculate a threshold value (with, perhaps, a zone of uncertainty) above which a site can be considered to be in attainment. The problem is that it is calculated at a percentile of the reference condition scores, not at the whole range of reference condition scores. Thus, a certain percent of reference sites will even fall into the range of impairment.

Newman (1999) and Newman et al. (1999) built a rough multimetric index using three streams in the northern Black Hills by selecting, based on professional opinion, seven metrics from Plafkin et al. (1989), although the discriminatory ability and redundancy of those metrics was not analyzed. Bear Butte, Spearfish, and Crow creeks served as the reference streams, and Whitewood Creek, which has had a long history of mining pollution, was the test stream analyzed using this multimetric index.

The process for selection of metrics and elimination of redundant metrics for such a multimetric index was delineated by Barbour et al. (1999) and is usually accomplished on a regional or state basis, such as for the State of Idaho (Grafe 2002). Shearer (2006) followed Barbour et al. (1999), completing only the first steps in the process of building a multimetric index (and only for the Black Hills in South Dakota), in that six metrics that appeared to reliably distinguish between a priori chosen impacted and reference condition sites were distilled from a larger list of candidate metrics. Scoring for each metric, derivation of a final multimetric index, and identification of scoring thresholds to determine impairment or attainment of aquatic life uses was started in 2016 and has not yet been completed to all stakeholders' satisfaction. Recent work in the northwestern plains in South Dakota has instead used an index derived for the State of Montana (Stagliano 2006). Wyoming's multimetric index, the Wyoming Stream Integrity Index (WSII) was first developed in 2000 but was more recently redesigned including the Black Hills as a separate bioregion (Hargett 2011).

In the absence of a set of metrics defined by the State, consulting companies and others have used multiple metrics that appeared to be appropriate for the Black Hills, often using them extensively in biomonitoring projects in the Black Hills. Shearer (2006) developed a list of invertebrate community metrics for analysis using Barbour et al. (1999) as a template. Jessup et al. (2005) derived a draft multimetric index for Montana using benthic macroinvertebrates, and this index has recently been used in the State of South Dakota for bioassessments (Spagliano 2006). Feldman (2006) provided guidance for interpretation of Montana DEQ's multimetric index, including thresholds describing stream condition in the eastern Plains region (similar to the Plains north of the Black Hills in South Dakota, Montana, and Wyoming). The metrics used by some consulting companies, Shearer (2006), Jessup et al. (2005), Spagliano (2006), and Feldman (2006) are described below; sufficient information is provided in the text of this work to allow calculation of all of these same metrics, assuming appropriate data were collected for community analysis.

Richness Metrics

Taxa Richness – This metric represents the number of taxa at a site. High numbers of taxa suggest that a wide variety of environmental conditions exist and are adequate to support a diverse assemblage of different taxa (Barbour et al. 1999). Taxa richness, as "Number of Taxa", was one of the metrics used by several independent consulting companies in biomonitoring studies in the Black Hills and is expected to decrease with increased perturbation in streams.

EPT Taxa Richness – In mountain streams such as those in the Black Hills, the presence of mayfly (Ephemeroptera), stonefly (Plecoptera), and caddisfly (Trichoptera) taxa (collectively referred to as the EPT taxa) can be used as an indicator of water quality (Lydy et al. 2000). These insect groups are considered to be sensitive to a wide range of pollutants (Hynes 1970; Wiederholm 1984; Klemm et al. 1990; Barbour et al. 1999; Merritt et al. 2008). Stress to aquatic systems can be detected by comparing the EPT taxa richness between unimpacted reference sites and potentially impacted sites. Although Shearer (2006) rejected this metric as a potential metric for biomonitoring studies in the Black Hills because of high correlations with other good potential metrics (e.g., Number of Plecoptera and Number of Intolerant Taxa), number of EPT taxa has historically been used in biomonitoring studies and would be expected to be higher in unimpacted sites compared to impacted sites. The Montana DEQ used this metric in their draft index for the eastern Plains region (Jessup et al. 2005, Feldman 2006). Wyoming also included this metric for the Black Hills bioregion in the WSII (Hargett 2011).

Number of Plecoptera Taxa – Stoneflies are one of the groups that compose the EPT taxa and are considered to be generally sensitive to anthropogenic perturbations. They are often characteristic of cool, clear streams and are often negatively impacted by changes in stream temperature or increases in fine sediments (Stewart and Harper 1996). Number of Plecoptera taxa was a metric determined in Shearer (2006) to be potentially useful for biomonitoring studies in the Black Hills and would be expected to decrease with increased perturbation in streams.

Proportion of Oligochaeta and Hirudinea – Both oligochaetes and leeches can be dominant in warm, sluggish streams and ponds with high levels of nutrients and fine sediments. They are usually exceptionally tolerant of anoxic conditions. Proportion of Oligochaeta and Hirudinea was another metric determined by Shearer (2006) to be of potential use in biomonitoring of Black Hills streams and would be expected to increase with increased perturbation in streams.

Proportion of Tanypodinae – Tanypod midges represent a subfamily group within the Chrionomidae that is considered to be relatively intolerant of pollution. This is contrary to the perception of most Chironomidae, which are supposed to be relatively tolerant of perturbations. Montana DEQ used this metric in their draft multimetric index for the eastern Plains region (Jessup et al. 2005, Feldman 2006).

Proportion of Orthocladiinae within Chironomidae – The subfamily Orthocladiinae is sometimes considered to be one of the more tolerant groups of midges. Montana DEQ used this metric in their draft multimetric index for the eastern Plains region (Jessup et al. 2005, Feldman 2006).

Composition Metrics

Shannon-Weaver Diversity Index (H') – The Shannon-Weaver Diversity Index was derived from information theory in the 1940s, primarily by Claude Shannon of Bell Laboratories, and is known by various monikers, such as the Shannon Index, Shannon-Wiener Index, Shannon Entropy, etc. It provides a measure of community composition and the EPA recommends it as a measure of the effects of stress on macroinvertebrate communities (Klemm et al. 1990). It is an index that is capable of combining the very different data on density with that of taxa richness into a single value. Communities are often dominated by a few common taxa, with smaller numbers of many uncommon taxa (MacArthur 1972), but stressed sites often have low diversity communities which are strongly dominated by one or a few taxa because conditions favor those particular taxa. The original version of the index was a $\log_{10}$

based calculation, and recent versions have used the log10 calculations and then incorporated a constant to convert $\log_{10}$ to $\log_2$. The following $\log_2$ based version of the formula provided in Klemm et al. (1990) also exists to calculate the Shannon-Weaver Diversity Index:

$$H' = \frac{\left(N \log_2 N - \sum_{i=1}^{n} n_i \log_2 n_i \right)}{N},$$

where N = the total density or abundance of organisms and n_i = the density or abundance of the ith taxon. The Shannon-Weaver Diversity Index measures how evenly the density is distributed among the component taxa. This index has values ranging from 0 to $\log_2 n$, with values over 2.50 generally indicative of a healthy, balanced macroinvertebrate community (Wilhm 1970; Klemm et al. 1990).

Percent of Sensitive EPT Taxa – The percent sensitive EPT taxa metric was calculated to provide a measure of the community composition. This metric was calculated as the number of sensitive EPT taxa divided by the total number of taxa. Sensitive EPT taxa were defined as those with tolerance values of 0, 1, 2, 3, or 4 (on a scale of 0 to 10). Tolerance values are assigned as described below, regarding the Hilsenhoff Biotic Index. The percent of sensitive EPT taxa was used for biomonitoring studies in the Black Hills and would be expected to be higher in unimpacted sites, because the community would be comprised proportionately of more sensitive organisms (Wiederholm 1984; Klemm et al. 1990; Barbour et al. 1999).

Number of Univoltine taxa – The presence of long-lived organisms is assumed to indicate the permanence of suitable aquatic habitat, both in terms of presence of flow and appropriate water chemistry. Since univoltine taxa require a full year to complete their life cycle, their presence is assumed to indicate that good quality water is present for that entire year, and Shearer (2006) determined that this might be a good potential metric for use in biomonitoring studies in the Black Hills. However, recent studies (De Jong and Canton 2012; De Jong et al. 2013, 2016) have indicated that other life history traits, such as the ability of some long-lived taxa (e.g., craneflies, crayfish) to tolerate dry or low-flow conditions for a part of their life cycle, or long-term population dynamics, need to be taken into account before this assumption can be considered to be valid. Additionally, Shearer (2006) had voltinism data for only about a third of the taxa that he used in his analysis, placing a big caveat on the use of this metric in multimetric analysis. Number of univoltine taxa would be expected to be higher in streams that are less impacted, because many anthropogenic impacts can alter flow or water chemistry and extirpate local populations.

Number of Semivoltine Taxa - Similar to the Number of Univoltine Taxa metric, this metric is based on the assumption that their presence indicates the permanent presence of suitable aquatic habitat. Hargett (2011) included this metric in the WSII for the Black Hills bioregion. Merovoltine taxa, which have a life cycle at least two years long, would be an appropriate surrogate for semivoltine taxa, which have a life cycle of two years. Again, Number of Semivoltine Taxa would be expected to be higher in streams with less anthropogenic impacts.

Proportion of Semivoltine Individuals – This metric is similar to the Number of Semivoltine Taxa metric, except that it is calculated as a proportion of the total number of semivoltine individuals to the total density. Shearer (2006) determined that this metric, along with the Number of univoltine taxa, would be a good potential metric for biomonitoring studies in the Black Hills. As mentioned above, Shearer (2006) had voltinism data for only about a third of the taxa that he used in his analysis, placing a big caveat on the use of this metric in multimetric analysis. Proportion of Semivoltine Individuals would also be expected to be higher in streams with less anthropogenic impacts.

Tolerance Metrics

Hilsenhoff Biotic Index (HBI) – The HBI (Hilsenhoff 1987) and the TVs that are used in its calculation are described above. This metric is included in the WSII for the Black Hills bioregion (Hargett 2011).

According to the original Hilsenhoff (1987) document, benthic macroinvertebrate communities with scores of 0.00 – 3.50 are considered Excellent, 3.51 – 4.50 are considered Very Good, 4.51 – 5.50 are considered Good, 5.51 – 6.50 are considered Fair, 6.51 – 7.50 are considered Fairly Poor, 7.51 – 8.50 are considered Poor, and 8.51 – 10.00 are considered Very Poor. It is not known if these categories accurately reflect conditions in Black Hills streams.

Percent Tolerant Taxa – Tolerant taxa are those which have tolerance values of 7 to 10, as assigned for the HBI calculation. The percent tolerant taxa metric was calculated as the number of tolerant taxa divided by the total number of taxa and is expected to be lower in unimpacted sites, as stressed sites tend to favor tolerant taxa (Barbour et al. 1999; Grafe 2002).

Percent Intolerant Taxa – Intolerant taxa are those which have tolerance values of 0 to 4. The percent intolerant taxa metric was determined by Shearer (2006) to be a potential metric for use in the Black Hills. The metric is calculated as the number of intolerant taxa divided by the total number of taxa and is expected to be higher in unimpacted sites.

Trophic Habit Metrics

Number of Predator Taxa – Functional feeding groups (predator, collector, scraper, etc.) were assigned as described above. Predators prey upon other organisms in the stream and their persistence depends on suitable water quality and the presence of prey organisms (Merritt et al. 2008). Specialized feeders such as predators are often considered to be sensitive organisms and are usually well represented in healthy stream communities (Barbour et al. 1999). Montana DEQ used this metric in their draft multimetric index for the eastern Plains region (Jessup et al. 2005, Feldman 2006). The number of predator taxa is expected to decrease with increased perturbation.

Number of Shredder Taxa – Similar to the Number of Predator Taxa, described above, this metric is based on the taxa richness of a specialized group of organisms. Shearer (2006) considered this to be a good potential metric for biomonitoring studies in the Black Hills because it was a measure of the trophic stability of a stream and impacts to the stream's riparian area might alter the input of allochthonous material, which would affect the invertebrate structure, including shredders.

Proportion of Collectors/Filterers – This metric is based on the proportion of the community comprised of collector-filterer organisms. Members of this functional feeding group are sometimes determined to be more tolerant of perturbations and can often be found in high abundances downstream of impoundments (Oswood 1978, MacKay and Waters 1986, Ward and Stanford 1987, Richardson and MacKay 1991, Merricks et al. 2007). High proportions of collector-filterers in the community can also indicate altered biological conditions. Montana DEQ used this metric in their draft multimetric index for the eastern Plains region (Jessup et al. 2005, Feldman 2006).

Proportion of Collectors/Gatherers – In a manner very similar to the Proportion of Collector/Filterers, this metric is based on the proportion of the community comprised of collector-gatherer organisms. Members of this functional feeding group, however, are very generalized, and can often be tolerant of altered biological conditions. This metric was included in the WSII (Hargett 2011).

Example – The Wyoming Stream Integrity Index

Although entire books and reports have been produced to describe the development and use of multimetric indices (e.g., Barbour et al. 1999), I will attempt to summarize the process here using the WSII, since it is a relatively simple, straightforward multimetric index. The WSII was first developed in 2000 by TetraTech but reached its latest iteration in 2011 (Hargett 2011), with information from 1,488 benthic macroinvertebrate samples collected between 1993 and 2009 by the Wyoming Department of Environmental Quality – Water Quality Division for biomonitoring efforts across the state. The WSII was redesigned from its original version to "improve its accuracy and precision through 1) integration of additional reference sites, 2) use of a defensible iterative set of quantitative techniques for model development, 3) development of a human stressor index to evaluate the WSIIs response to human disturbance and 4) development of an empirically-derived large river component for evaluating biological condition on wadeable reservoir-regulated river segments of the Bighorn and North Platte Rivers."

Because of the extreme physiographic diversity in Wyoming, the state was divided into 11 different bioregions, of which the Black Hills was identified as a single bioregion. The bioregion concept used in the WSII demonstrates that differences in the benthic macroinvertebrate assemblages across Wyoming are tied to broad-scale abiotic and biotic factors acting upon streams since the bioregions are in part defined by the benthic macroinvertebrate communities. The aforementioned 1,488 benthic macroinvertebrate population samples were collected throughout the state in streams, with representative "stressed" and "reference" sites in each bioregion. Reference and stressed sites are usually identified by abiotic parameters, such as undistributed land, low concentrations of pollutants, lack of man-caused alterations and potential inputs, etc. In the case of the WSII, these abiotic parameters, representing stressors, were conglomerated into a disturbance index along which the sites were then located.

Usually hundreds of candidate metrics are calculated from the benthic macroinvertebrate sample data, and there is a rigorous process that culls them down to a reasonable handfull. The culling process primarily uses the ability of each metric to discriminate between reference and stressed conditions, lack of correlation with other metrics, and inclusion of metrics to represent the four families of metrics. As reported above, the WSII was culled down to four metrics for the Black Hills bioregion: Number of EPT taxa, Number of Semivoltine Taxa, Proportion of Collector-Gatherers, and the HBI.

Best Standard Values (BSVs) are calculated for each metric so that they could be scored using test data generated during biomonitoring studies (Table 1). These are usually chosen as the 95th or 5th percentile of the reference site data values, depending on whether the metric is expected to increase or decrease with increased disturbance. The scores are generally calculated by computing a ratio of the metric's value from the test site to the BSV, then multiplying times 100. The scores for each metric are averaged to obtain the final WSII score for the site.

Table 1:   **Metrics, scoring equations, and best standard values for the Black Hills region in the Wyoming Stream Integrity Index**

Metric	Scoring Equation	Best Standard Value
Number of EPT Taxa	100 x X / BSV	19
Number of Semivoltine Taxa	100 x X / BSV	6
Proportion of Collector/Gatherers	100 x (1.3 – X) / (1.3 – BSV)	24.8
Hilsenhoff Biotic Index	100 x (5.85 – X) / (5.85 – BSV)	3.5

To work through an example, we can suppose that a site has 12 EPT taxa and 4 semivoltine taxa, collector/gatherers comprise 12.4 percent of the sample, and the HBI value is 4.0. Plugging these numbers into the above formulas yields scores of 63 for Number of EPT Taxa, 67 for Number of Semivoltine Taxa, 47 for Proportion of Collector/Gatherers, and 79 for the HBI. The final score would be 64. This number is compared to thresholds that have been identified at the 5th percentile of the reference site scores to determine if the site is "attaining the aquatic life use" or is "impaired." For the Black Hills bioregion, the attainment threshold is 46.1, with scores between 30.7 and 46.1 being indeterminate, and scores less than 30.7 being non-supporting/impaired. This example site would be considered to be attaining the aquatic life use and not impaired.

Wyoming streams from the Black Hills that were used in development or testing of the WSII included the following: Test streams: several sites on Beaver Creek (WSII = 72.5, 63.7, 55.3, 62.7), Beaverdam Creek (WSII = 27.9), two sites on Blacktail Creek (WSII = 35.1 and 65.0), Fawn Creek (WSII = 35.2), two sites on Inyan Kara Creek (WSII = 57.8, 14.1), Redwater Creek (WSII = 24.3), and three sites on Stockade Beaver Creek (WSII = 61.4, 51.8, 7.4); Reference Streams: Cold Springs Creek (WSII = 87.4), two sites on Little Creek (WSII = 77.0, 65.7), Redwater Creek (WSII = 31.4), five sites on Sand Creek (WSII = 64.5, 71.3, 41.7, 34.3, 48.7), Spotted Tail Creek (WSII = 54.3), and six sites on Stockade Beaver Creek (WSII = 35.7, 46.1, 46.9, 45.3, 43.2, 28.3); and Degraded streams: North Fork Sundance Creek (WSII = 12.8), North Redwater Creek (WSII = 33.1), Stockade Beaver Creek (WSII = 35.2).

RIVPACS type models

RIVPACS (River InVertebrate Prediction and Classification System) models make predictions of the community composition at a site based on the capture probability of invertebrates and natural environmental features (Clarke et al. 1996, 2003). They attempt to model the fauna expected in the absence of anthropogenic stressors and result in an index that represents the ratio of observed taxa to expected taxa (O/E Index). While these types of models have been used fairly well in islands such as Great Britain and Australia, their use in the United States is relatively new (Hawkins 2006).

The method by which a RIVPACS-type model are developed are fairly straightforward. Reference sites are characterized by environmental variables such as elevation, ecoregion, slope, aspect, etc. The proportion of reference sites in which each taxon is collected within like reference sites is considered to be that taxon's capture probability. When a test site is encountered, the environmental variables are taken into account and, based on capture probabilities, the expected community is predicted. The actual (observed) community is compared and the ratio of observed taxa to expected taxa is calculated. Variations can include weighting the taxa by their capture probabilities, adding the capture probabilities of each taxon, or other methods. A significant departure from near one (O/E = 1) should indicate that there is something going on at the site.

Hargett et al. (2007) developed a RIVPACS-type predictive model for bioassessment of wadeable streams in Wyoming. Their effort included data from several streams in the Black Hills. Most of the taxonomy was left at the genus level, so species-level records are not available. The authors state that "The model was accurate in both space and time and precise enough (S.D. of *O/E*-values for calibration data = 0.17) to detect modest alteration in biota associated with anthropogenic stressors. Our model was comparable in performance to other RIVPACS models developed in the United States and can produce effective assessments of biological condition over a broad, ecologically diverse region." Ecologists interested in bioassessment should access this paper and study it to see if it might be a useful technique for their projects.

Aquatic Collembola

Introduction

There is some controversy as to whether or not Collembola are actually insects, and current thought is that they are a sister group to the insects within the Hexapoda; however, these discussions usually focus on philosophical issues of how classification and phylogenetic approaches intertwine. I'm including them anyway.

In all Collembola, a cylindrical, midventral appendage called the collophore appears on the first abdominal segment, usually with a pair of eversible vesicles at the tip. Collembola are commonly known as springtails, based on the presence of a furcula, a jumping organ with paired distal branches, ventrally on the third and fourth abdominal segments, in many species. They have no compound eyes, having only ≤ 8 simple ocelli on each side of the head. The legs have a tibiotarsus (fused tibia and tarsus) with a single claw at the tip. The largest of these organisms is only about 8 mm long, and the group is mostly terrestrial (Christiansen 1990).

Functionally, aquatic collembolans are generally collector-gatherers and skaters, feeding on detritus floating on the surface of the water. In the literature, only a few aquatic collembolan families, genera, or species have been characterized concerning their habits, water body preferences, etc. (Merritt et al. 2008).

Overview of the Aquatic Collembola in the Black Hills

There are no known checklists of Collembola for South Dakota, Wyoming, or the Black Hills. Shearer (2006) listed only the ordinal level Collembola. The following are aquatic collembolan families that I have seen in the Black Hills: Entomobryidae, Hypogastruridae, Isotomidae, Poduridae, and Sminthuridae. Merritt et al. (2008) provides a key to families, with emphasis on the families with aquatic representatives.

Moore et al. (2005, 2006) reported *Arrhopalites caecus* (Tullberg), a terrestrial species that can inhabit damp conditions, from a pool in Wind Cave and from Jewel Cave.

Aquatic Collembola Traits Table

Table 2: Collembola Traits Table

Species	TSN	Seen	Tolerance Value	FFG	Habit
Collembola	99237	+	10	GC	--
Entomobryidae	99643	+	10	GC	--
Hypogastruridae	99917	+	10	GC	--
Isotomidae	99245	+	10	OM	SK
Poduridae	99239	+	10	GC	SK
Podura	99240	+	10	GC	SK
Podura aquatica	99241	+	10	GC	SK
Sminthuridae	100258	+	10	GC	SK

Species	Temperature Range	Lotic	Lentic	Current	Substrate	Voltinism
Collembola	--	--	--	--	--	--
Entomobryidae	--	--	--	--	--	--
Hypogastruridae	--	--	--	--	--	--
Isotomidae	--	--	--	--	--	--
Poduridae	--	--	--	--	--	--
Podura	--	3.0	--	--	--	--
Podura aquatica	--	3.0	--	--	--	--
Sminthuridae	--	--	--	--	--	--

Ephemeroptera

Introduction

Mayflies represent the quintessential group of organisms that live for a long time (sometimes years) as immatures, then become adults only to mate and die (sometimes within days or even hours). This is reflected in the name of the order Ephemeroptera, from the Greek roots "ephemera" meaning "for a short time" and "pteron" meaning "wing". In reality, many mayfly adults live more than just a few days, although their adult life truly is ephemeral in comparison to the length of the larval stage.

The larval stage, in fact, is what most people have historically been interested in, although anglers mimic both adult and immature stages for their flies. Mayflies are often an abundant, if not dominant, prey item for fish. Plus, their presence is often considered to be an indication of good water quality. Many stream mayflies are indeed intolerant of most stressors, including chemical pollution, temperature changes, dewatering, and other alterations of their habitat.

Most mayfly species in the Black Hills are stream-dwellers and prefer a decent current and good, hard substrates. A few are lentic and like silty or sandy-bottomed water bodies.

Overview of the Ephemeroptera in the Black Hills

Morihara and McCafferty (1979) included Black Hills records for a few species of Baetidae, while Allen and Edmunds (1965) reported *Ephemerella inermis* (= *E. dorothea*) from the Black Hills. The mayfly fauna specifically of the Black Hills was first reviewed in McCafferty (1990), who listed 19 species in seven families. McCafferty and Kondratieff (1999) added another 7 species, while Kondratieff (2000b) provided mayfly species records at the county level in South Dakota and Wyoming through the USGS Northern Prairie Wildlife Research Center (NPWRC), including the Black Hills counties. More recently, Guenther and McCafferty (2008) reviewed 77 species of mayflies from South Dakota, including 37 species with records from the Black Hills.

Species Accounts

Ameletidae

De Jong and Canton (2012) reported the mayfly genus *Ameletus* from numerous streams in the northern Black Hills; however, with only larvae in hand, they were unable to identify the species. Likewise, Jurgens (1968) reported *Ameletus* at the genus level from Castle Creek. Until larvae are associated with adult material, it is impossible to know how many and which species of *Ameletus* are represented in the Black Hills.

Ameletus Eaton

Collectively, *Ameletus* are known to anglers as the brown duns, without species-level distinction.

Baetidae

Baetidae, the small minnow mayflies, is the most speciose family of mayflies in the Black Hills, being represented by 18 species. Most of the baetid species in the Black Hills are widespread across the U.S. and Canada; however, the Black Hills do represent the boundaries for some species. *Baetis bicaudatus*, *B. magnus*, and *Anafroptilum album* have only been reported from one collection each in the Black Hills in the published literature (McCafferty and Kondratieff 1999, Guenther and McCafferty 2008). Jurgens (1968) and Drewes (1984) reported *Baetis* at the genus level from several streams in the northern Black Hills, but, given the chaotic nature of baetid taxonomy at the time, it is unknown which current species or genus those *Baetis* now belong to.

Acentrella Bengtsson

Acentrella insignificans (McDunnough)

Descriptions/Diagnoses: McDunnough 1926 (adult, as *Baetis insignificans*), Morihara and McCafferty 1979a (larva, as *Baetis insignificans*). Widespread in the western United States from Alaska to California east through the Rocky Mountains to Montana, South Dakota, Colorado, and New Mexico. All species in the genus *Acentrella* are known to anglers as the "tiny blue-winged olives".

Guenther and McCafferty (2008) reported *A. insignificans* from numerous sites in the Black Hills, including Battle Creek, False Butte Creek, and Hot Brook. The USU WCMAFE database includes Black Hills records for this species including Elk Creek and Spring Creek, as well as Bear Butte Creek and the Belle Fourche River not too distant north of the Black Hills.

Anafroptilum Kluge

Anafroptilum album (McDunnough)

Descriptions/Diagnoses: Webb 2002 (larva, adult, as *Centroptilum album*), McDunnough 1926 (adult, as *Centroptilum album*). A widespread species with scattered distribution records from Saskatchewan to Ontario, south to Oregon, Kansas, Tennessee, and Massachusetts. Kluge (2011) established that the genus *Centroptilum* is exclusively an African genus and erected the genus *Anafroptilum* for species to be moved out of *Centroptilum*, but only formally moved *bifurcatum*; Jacobus and Wiersema (2014) clarified the North American situation by formally moving eight of the *Centroptilum* species, including *album*, to *Anafroptilum* and two species to *Neocloeon* Traver. Up to date TSNs do not yet exist for *Anafroptilum* or *A. album*; the latter as *Centroptilum album* is 100878.

Jurgens (1968) reported both *Centroptilum* and *Neocloeon* from several Black Hills streams, including French Creek, Spring Creek, Rapid Creek, Spearfish Creek, Battle Creek, and Castle Creek;

however, with the confusion in the taxonomy, it is difficult to say if these represented *A. album*. Guenther and McCafferty (2008) reported the single Black Hills species-level record for *C. album* in Whitewood Creek.

Baetis Leach

Anglers refer to most species in this genus primarily as "blue-winged olives" (BWO) even though they have neither blue wings nor olive-colored bodies, although some call them simply "baetids" according to the Latin name. Notably, the name "blue-winged olive" was originally coined for a British ephemerellid and has been applied to numerous families of mayflies.

Baetis bicaudatus Dodds

Descriptions/Diagnoses: Dodds 1923 (adult, larva), Morihara and McCafferty 1979a (larva). Widespread western Nearctic and eastern Palaearctic species; occasionally the only mayfly representative in high alpine streams in the Rocky Mountains (McCafferty et al. 1993). One of the species commonly called a "blue-winged olive".

In the literature, *B. bicaudatus* was only known from one locality, Spearfish Creek, in the Black Hills (McCafferty and Kondratieff 1999), although I have seen it in several small streams in the northern Black Hills.

Baetis brunneicolor McDunnough

Descriptions/Diagnoses: Morihara and McCafferty 1979a (larva), McDunnough 1925 (adult). Distribution is primarily northeastern North America, extending south to North Carolina and west to Illinois and Indiana, with some specimens reported from Idaho, Washington, and the Black Hills (Morihara and McCafferty 1979a; Guenther and McCafferty 2008). Called by anglers the "blue-winged rusty dun".

In the Black Hills, *B. brunneicolor* has been reported from French Creek and Grizzly Bear Creek (Guenther and McCafferty 2008).

Baetis flavistriga McDunnough

Descriptions/Diagnoses: Morihara and McCafferty 1979a (larva), McDunnough 1921, 1923 (adult). One of the most abundant *Baetis* species in eastern North America, its distribution appears to limit it to mountainous regions. Another "blue-winged olive".

It has been found at numerous localities in the Black Hills, including Battle Creek, Beaver Creek, Boxelder Creek, Cold Spring Creek, French Creek, Grizzly Bear Creek, Rapid Creek, Spring Creek, and Whitewood Creek (McCafferty *et al.* 1993; Guenther and McCafferty 2008). The USU WCMAFE database includes records from Beaver, Elk, and Spring creeks.

Baetis intercalaris McDunnough

Descriptions/Diagnoses: Morihara and McCafferty 1979a (larva), Waltz et al. 1996 (atypical larval form), McDunnough 1921 (adult). Distributed from Florida and Alabama north to South Dakota, Manitoba, and Quebec. Anglers commonly refer to this species as the "small eastern blue-winged olive".

Guenther and McCafferty (2008) reported *B. intercalaris* from Boxelder Creek and Rapid Creek in the Black Hills. The closest record in the USU WCMAFE database is from the White River in Shannon County, South Dakota.

Baetis magnus **McCafferty and Waltz**

Descriptions/Diagnoses: Morihara and McCafferty 1979a (larva, as *Baetis* sp. B), Durfee and Kondratieff 1993 (adult). The large larvae are found in lower elevation streams (<2,450 m) in Arizona, Colorado, Nebraska, and New Mexico (Durfee and Kondratieff 1993) and South Dakota. The "iron blue quill".

This species was reported in the literature from False Bottom Creek (McCafferty and Kondratieff 1999). Given the distribution that I have encountered, *B. magnus* is more common on the Plains in streams draining from the Black Hills than in the Black Hills proper.

Baetis tricaudatus **Dodds**

Descriptions/Diagnoses: Morihara and McCafferty 1979a (larva), Dodds 1923 (adult). Extremely variable species, probably the most widespread baetid in North America (McCafferty *et al.* 1993). Another "blue-winged olive".

This species is widespread in the Black Hills, including Battle Creek, Bear Butte Creek, Beaver Creek, Boxelder Creek, Cascade Springs, Castle Creek, Highland Creek, Hot Brook, Iron Creek, Jim Creek, Little Spearfish Creek, Rapid Creek, Slate Creek, Spearfish Creek, Spring Creek, and Whitewood Creek (Guenther and McCafferty 2008) and can be very abundant in samples. The USU WCMAFE database includes records from Elk Creek, Castle Creek, Whitewood Creek, Spring Creek, and the Belle Fourche River. I observed this species in nearly every stream I worked on in the Black Hills.

Callibaetis **Eaton**

Callibaetis are generally found in lentic waters, where anglers commonly refer to all species as "speckled duns" or "speckled spinners", since the wings of adults are frequently speckled.

Callibaetis ferrugineus **(Walsh)**

Descriptions/Diagnoses: Check 1982 (larva, adult). Previously known as *C. americanus* Banks in western North America, this species is widespread in the Nearctic from Alaska and northern Canada to California, east to Manitoba and south to Florida in mid- to high elevations (Check 1982, McCafferty and Waltz 1990, McCafferty *et al.* 1993, McCafferty 1996).

Guenther and McCaferty (2008) reported *C. ferrugineus* from Bismark Lake Campground and Sylvan Lake in the Black Hills.

Callibaetis fluctuans **(Walsh)**

Descriptions/diagnoses: Check 1982 (larva, adult), Burks 1953 (adult). Distributed in a transcontinental band from Washington, Oregon, and central California to southeastern Canada and Georgia (Check 1982, McCafferty *et al.* 1993).

Reported from Boxelder Creek and Fall Creek Reservoir in the Black Hills (Guenther and McCafferty 2008).

Callibaetis pallidus **Banks**

Descriptions/Diagnoses: Check 1982 (larva, adult), Needham 1903 (as *Callibaetis skokianus*). Ontario, New York and Pennsylvania west to Saskatchewan and south to Colorado, Arizona, and Kansas (Check 1982, McCafferty *et al.* 1993).

The only known Black Hills record is from Paulson's Pond in Custer County.

Callibaetis pictus (Eaton)

Descriptions/Diagnoses: Check 1982 (larva, adult), Lugo-Ortiz and McCafferty 1996. Widespread in the West (northern Washington and the Black Hills of South Dakota south to Costa Rica) at elevations up to about 3200 m (Check 1982, McCafferty 1990, McCafferty *et al.* 1993, Lugo-Ortiz and McCafferty 1996).

In and near the Black Hills, *C. pictus* is known from Stockade Lake and Hawkwright Creek (Guenther and McCafferty 2008).

Camelobaetididus Demoulin

Camelobaetidius warreni (Traver and Edmunds)

Descriptions/Diagnoses: Lugo-Ortiz and McCafferty 1995b (larva), Traver and Edmunds 1968 (adult and larva, as *Dactylobaetis warreni*). This species is known from Arizona (Lugo-Ortiz and McCafferty 1995b), California, Colorado and South Dakota.

This species was previously reported from Hot Brook in the Black Hills as *C. cepheus* or *Dactylobaetis cepheus* (McCafferty 1990; McCafferty *et al.* 1993).

Diphetor Waltz and McCafferty

Diphetor hageni (Eaton)

Descriptions/Diagnoses: Morihara and McCafferty 1979a (larva, as *Baetis hageni*), Meyer and McCafferty 2001 (adult). Transcontinental species, ranging from British Columbia to New Brunswick and Nova Scotia, south to California, Colorado, Indiana, Missouri, and North Carolina (McCafferty *et al.* 1993, Meyer and McCafferty 2001). In Wisconsin, this species appears to be parthenogenetic (Bergman and Hilsenhoff 1978); it is unknown if populations in the Black Hills are parthenogenetic or not. Like *Baetis*, this species is known as the "blue-winged olive".

Guenther and McCafferty (2008) reported *D. hageni* from Beaver Creek, Boxelder Creek, Flynn Creek, Grace Coolidge Creek (as "Greater Cool Creek"), Jim Creek, Rapid Creek, Slate Creek, South Fork Rapid Creek, Spearfish Creek, and Spruce Gulch Creek in the Black Hills. The USU WCMAFE database includes a record from Spruce Gulch Creek.

Fallceon Waltz and McCafferty

Fallceon quilleri (Dodds)

Descriptions/Diagnoses: Morihara and McCafferty 1979a (larva, as *Baetis quilleri*), Lugo-Ortiz *et al.* 1994 (adult). Widespread in relatively low elevations throughout the southwestern United States, in the Rocky Mountains north to Colorado (McCafferty *et al.* 1993), and central IA (Klubertanz 1995). Another species known as the "blue-winged olive".

In the Black Hills, *F. quilleri* is probably the second most abundance baetid species present, being reported from Beaver Creek, Boxelder Creek, Cherry Creek, Flynn Creek, Hot Brook, Palmer Gulch Creek, Spearfish Creek, and Whitewood Creek (Guenther and McCafferty 2008). The USU WCMAFE database includes several records of this species, including Whitewood Creek, Elk Creek, Spring Creek, Flynn Creek, Cass Pig Creek, Bear Butte Creek, the Cheyenne River, and the Bell Fourche River in South Dakota and Blacktail Creek near Hulett, Wyoming. I have collected several adults of *F. quilleri* on Annie Creek near its confluence with Spearfish Creek.

Iswaeon McCafferty and Webb

Iswaeon anoka (Daggy)

Descriptions/Diagnoses: Morihara and McCafferty 1979b (larva, as *Heterocloeon frivolum*), McCafferty et al. 2006 (adult, in description of subgenus *Iswaeon*). Distributed from Alberta, Idaho, and Oklahoma east to Pennsylvania, Virginia, and North Carolina, with most speciemens collected east of the Mississippi River (McCafferty *et al.* 1993, 2005, McCafferty 2006). Known to anglers as a "tiny blue-winged olive".

Guenther and McCafferty (2008) reported the two known records of this species from the Black Hills in Boxelder Creek and Spring Creek.

Labiobaetis Novikova and Kluge

Labiobaetis propinquus (Walsh)

Descriptions/Diagnoses: Morihara and McCafferty 1979a, b (larva, as *Baetis propinquus*), Morihara and McCafferty 1979b (adult, as *Baetis propinquus*). Distributed from central and western North America from northern Canada and Alaska south into northern Mexico. This species has been variably included in the genera *Baetis*, *Pseudocloeon*, and *Labiobaetis*. Morihara and McCafferty (1979b) included the species in *Baetis* as part of the *Baetis propinquus* group. There are many references to *Pseudocloeon propinquus* in fly-fishing manuals and websites, but Waltz and McCafferty (1985, 1987) later restricted *Pseudocloeon* to include only the original species *P. kraepelini*, known from Java, so this species really should not be included in that genus (McCafferty et al. 2010), although ITIS has not yet (s of 2016) been updated to reflect that change. McCafferty and Waltz (1995) erected *Labiobaetis* and placed *propinquus* in that genus. This species, along with the rest of *Labiobaetis* is commonly known to anglers as a "tiny blue-winged olive".

McCafferty and Kondratieff (1999) reported *L. propinquus* (as *P. propinquum*) from Squaw Creek (= Cleopatra Creek) and Little Spearfish Creek.

Paracloeodes Day

Paracloeodes minutus (Daggy)

Descriptions/Diagnoses: Daggy 1945 (larva, adult, as *Pseudocloeon minutum*). This species has been collected from California, Colorado, Illinois, Indiana, Minnesota, and Iowa, south into Mexico and Costa Rica. Aptly known to anglers as a "tiny blue-winged olive".

Paracloeodes minutus is not common in the Black Hills, being reported from the Cheyenne River south of the Black Hills in Guenther and McCafferty (2008). Near to the Black Hills, the USU WCMAFE database has records from Elk Creek near Elm Springs in Meade County and the White River in Shannon County, South Dakota.

Procloeon Bengtsson

Shearer (2006) reported *Procloeon* sp. from the Black Hills. The closest literature records to the Black Hills of a species in this genus, *Procloeon viridoculare*, are from Cherry Creek in Meade County (Guenther and McCafferty 2008). Similarly, the closest record of this genus to the Black Hills in the USU WCMAFE database is from Black Pipe Creek in Mellette County, a little south of Belvidere, South Dakota. Anglers sometimes call all species in this genus the "tiny sulphur duns".

Caenidae

Only one species of Caenidae has been reported from in or near the Black Hills. Shearer (2006) reported from the Black Hills only at the genus level. It is probable that other species will eventually be collected from either the Black Hills proper, or possibly in areas just outside the Black Hills (e.g., *C. latipennis* Banks, which has been reported throughout the rest of the state of South Dakota). All caenids are known commonly to anglers as "Angler's curses", or less frequently "little white spinners".

Caenis Stephens

Caenis amica Hagen

Descriptions/Diagnoses: Provonsha 1990 (larva, adult). This species is widespread throughout the U.S. and southern Canada.

Guenther and McCafferty (2008) reported *C. amica* from Horsethief Lake in the Black Hills, and Rapid City, while the USU WCMAFE database includes a record from the Cheyenne River near Spencer, Wyoming.

Ephemerellidae

Ephemerellidae is represented in South Dakota by three species, all of which are reported from the Black Hills, although generally uncommon. Fully reliable characters have only recently been discovered to separate larvae of the two species of *Ephemerella* (Jacobus and McCafferty 2003), so they were usually left as *Ephemerella* sp. (Shearer 2006) and I usually left them as the slashed taxon *Ephemerella dorothea/excrucians*.

Drunella Needham

Drunella doddsii (Needham)

Descriptions/Diagnoses: Needham 1927 (as *Ephemerella doddsii*), Allen and Edmunds 1962 (as *Ephemerella doddsi*). Mangum and Winget (1991) provided a fairly thorough review of the ecology of this species via an environmental profile based on data from 813 sites across the western United States. Known to anglers as the "western green drake".

McCafferty and Kondratieff (1999) provided the single Black Hills record for *D. doddsii* in the Black Hills in Spearfish Creek, which is the easternmost record for this species.

Ephemerella Walsh

As I was identifying invertebrates in biomonitoring samples, I usually left *Ephemerella* species at the genus level or as a slashed taxon because reliable characters separating larvae of the two species known from the Black Hills had only recently been identified. Although several species of *Ephemerella* are known by numerous common names, anglers generally refer to both Black Hills species as "pale morning duns".

Ephemerella dorothea Needham

Descriptions/Diagnoses: Jacobus and McCafferty 2003 (larva), McDunnough 1924 (as *Ephemerella infrequens*). The western subspecies *E. d. infrequens* McDunnough is Distributed from British Columbia south to Oregon, and east to South Dakota, Colorado, and New Mexico.

The western subspecies *E. d. infrequens* is the form that was reported from Spearfish in the Black Hills, at the far eastern limit of its range (Guenther and McCafferty (2008). The USU WCMAFE database includes records of this species, as *E. infrequens*, from Castle Creek and Elk Creek.

Ephemerella excrucians Walsh

Descriptions/Diagnoses: Jacobus and McCafferty 2003 (larva), Walsh 1862 (adult). Widespread in the western Nearctic from British Columbia, along the Pacific Coast into Mexico, and east to Wisconsin, Iowa, Colorado, and New Mexico.

Ephemerella excrucians was previously known as *E. inermis* in some early records, which included Beaver Creek, Boxelder Creek, Highland Creek, Jim Creek, Rapid Creek, and Spearfish Creek. Searches for E. inermis in the USU WCMAFE database turn up records from Castle Creek and Elk Creek in South Dakota and White Claw Creek in Wyoming.

Ephemeridae

The burrowing mayflies of the family Ephemeridae are widespread in the U.S. and Canada, with three species reported from South Dakota. One species, *Ephemera simulans*, was reported for South Dakota by McCafferty (1975), and the locality in the Black Hills for that one specimen was later published in Guenther and McCafferty (2008). Shearer (2006) reported larvae of *Ephemera* at the genus level.

Ephemera Linnaeus

Ephemera simulans Walker

Description/diagnosis: Ide 1935 (larva), McCafferty and Edmunds 1973 (adult, as description of subgenus *Ephemera*). Most common and widespread North American species of *Ephemera*, being found across the entire United States, except the extreme southwest and southeast regions (McCafferty 1975, 1994). Known to anglers as the "brown drake".

Guenther and McCafferty (2008) reported *E. simulans* from Boxelder Creek in the Black Hills.

Heptageniidae

Four species of Heptageniidae are known from the Black Hills (Guenther and McCafferty 2008). All four species are uncommonly collected in streams in the Black Hills, although they may be much more abundant elsewhere.

Ecdyonurus Eaton

Ecdyonurus criddlei (McDunnough)

Descriptions/Diagnoses: Bednarik and Edmunds 1980 (larva, as *Heptagenia criddlei*), McCafferty 2004 (adult, as diagnosis of *Ecdyonurus* vs. *Nixe*). This species name was moved from *Nixe* into *Ecydonurus* by McCafferty (2004). This species is widely distributed in the Rocky Mountains from British Columbia and Alberta south to Arizona and New Mexico into Mexico. Anglers know this species as the "little slate-winged dun".

Shearer (2006) reported larvae of *Nixe* (likely *Ecdyonurus criddlei*) from the Black Hills. The Black Hills (French Creek at the Hazelrodt Picnic Grounds) represent the easternmost limits for *Ecdyonurus criddlei* (McCafferty 1990, Guenther and McCafferty 2008).

Epeorus Eaton

Epeorus longimanus (Eaton)

Descriptions/Diagnoses: Edmunds and Allen 1964 (larva, adult). This is the most widely distributed species of *Epeorus*, being found from Alaska and British Columbia south to southern California, central Arizona, and northern New Mexico. Anglers know this species as the "slate brown dun".

The Black Hills represent the easternmost limits for *Epeorus longimanus* in the United States. It has been reported from Grizzly Bear Creek, Iron Creek, Jenny Gulch, Jim Creek, Rapid Creek, and Spearfish Creek (Guenther and McCafferty 2008), and I have collected an adult male of this species on Annie Creek near its confluence with Spearfish Creek. Shearer (2006) reported larvae of *Epeorus* from the Black Hills.

Heptagenia Walsh

Heptagenia elegantula (Eaton)

Descriptions/Diagnoses: Bednarik and Edmunds 1980 (larva), Eaton 1888 (adult, as *Rhithrogena elegantula*). This species also is widespread throughout the Rocky Mountains. It is found frequently in large, silty streams and rivers, and anglers know it as the "pale evening dun".

Although widespread in eastern South Dakota, *H. elegantula* has only been reported from Battle Creek within the Black Hills (Guenther and McCafferty 2008).

Stenacron Jensen

Stenacron interpunctatum (Say)

Descriptions/Diagnoses: Webb 2002 (larva, adult). Distributed in most states and provinces east of the Rocky Mountains. No temperature tolerance was provided for *Stenacron interpunctatum* in Grafe et al. (2002), so the eurythermal warm designation is based on data from Vieira et al. (2006). Anglers call this species (among others) the "light cahill".

Stenacron interpunctatum has been collected in Battle Creek and Rapid Creek in the Black Hills, which constitute the westernmost limits of its range (Guenther and McCafferty 2008).

Leptohyphidae

Two species of Leptohyphidae are known from South Dakota, and both are reported from the Black Hills. Our ecological knowledge of these two species does not discriminate between them, so they can be left at the genus level for biomonitoring identifications.

Tricorythodes Ulmer

When I identified invertebrates in biomonitoring samples, I usually reported *Tricorythodes* at the genus level; some early reports of *T. minutus* should be relegated to the genus level, as species-level identification keys were not available at the time, and it was incorrectly assumed that *T. minutus* (= *T. explicatus*) was the only species in the area. Anglers know all of the *Tricorythodes* as "Tricos".

Shearer (2006) reported *Tricorythodes* larvae from the Black Hills, Jurgens (1968) reported *Tricorythodes* from French Creek, Rapid Creek, Spearfish Creek, and Battle Creek, and the USU WCMAFE database includes records from Elk Creek, Beaver Creek, Spring Creek, and the Belle Fourche River. I have collected larvae of *Tricorythodes* in the Fall River near Hot Springs.

Tricorythodes explicatus (Eaton)

Descriptions/Diagnoses: Allen and Murvosh 1987 (larva); Kilgore and Allen 1973 (larva, as *Tricorythodes minutus*), Baumgartner 2009 (adult). Distributed from Wyoming and Utah south through Arizona, New Mexico, Texas, and Mexico to Guatemala, east to South Dakota, Nebraska, and Missouri. An apparently disjunct population has also been discovered in New England (Baumgartner 2009).

Tricorythodes explicatus is widespread and the more common of the two species of *Tricorythodes* found in the Black Hills; the name *T. minutus* was synonymized into this taxon by Baumgardner (2009). It has been found in Beaver Creek, Boxelder Creek, Fall River, French Creek, Grizzly Bear Creek, and Spring Creek (Guenther and McCafferty 2008, Baumgardner 2009).

Tricorythodes fictus Traver

Descriptions/Diagnoses: Baumgardner et al. 2003 (larva); Traver 1935 (adult). *Tricorythodes fictus* is at the northern limits of its range in the Black Hills, with its distribution extending south through Oklahoma, Texas, and Chihuahua, Mexico.

Tricorythodes fictus has been collected in Spring Creek (Guenther and McCafferty 2008).

Leptophlebiidae

Of the seven species of Leptophlebiidae in South Dakota, all but one are found in the Black Hills. Only one or two records are reported each in Guenther and McCafferty (2008) for *Leptophlebia nebulosa*, *Paraleptophlebia debilis*, and *P. memorialis*. The Black Hills are the easternmost limit for *P. memorialis* and the westernmost limit for *P. mollis*. Note that abdominal gills are easily broken off of larval specimens, necessitating identification only to the family level if they are missing.

Choroterpes Eaton

Shearer (2006) reported the genus *Choroterpes*, for which no records from the Black Hills were reported in Guenther and McCafferty (2008). It is possible that these represented the very similar *Neochoroterpes oklahoma*, which has been reported from the Black Hills.

Leptophlebia Westwood

Leptophlebia nebulosa (Walker)

Descriptions/Diagnoses: Burian 2001 (larva, adult). Distributed in northern North America from the Yukon and Northwest Territories north of the Arctic Circle south to Colorado, Oklahoma, and Virginia. Known by anglers (among othe species) as the "black quill".

Apparently, only two literature records of this species exist for South Dakota: Iron Creek at Spearfish Creek/U.S. Highway 14A (Burian 2001) and the Cheyenne River just south of the Black Hills (McCafferty and Kondratieff 1999).

Neochoroterpes Allen

Neochoroterpes oklahoma (Traver)

Descriptions/Diagnoses: Henry 1993 (larva and adult). This is the most widespread species in the genus, being distributed from South Dakota south to Colorado, Texas, and northern Mexico (Henry 1993). Apparently this species is not common enough to warrant a common name from the angling community.

Widespread in western South Dakota, the literature records for *N. oklahoma* abut the Black Hills (Belle Fourche River, Cheyenne River, Redwater River), but do not include records from within the Black Hills proper.

Paraleptophlebia Lestage

Because gill loss in *Paraleptophlebia* is such a large problem in biomonitoring samples (rough handling during capture with Hess, Surber, or Kick samplers), members of this genus are usually left at the genus level for identifications. I saw it from many streams in the northern Black Hills. Jurgens (1968) reported *Paraleptophlebia* from French Creek, Spring Creek, Battle Creek, Castle Creek, and the Fall River, and Drewes (1984) found *Paraleptophlebia* in Slate Creek. The USU WCMAFE database includes one record of this genus in the Black Hills, from Castle Creek.

Paraleptophlebia adoptiva (McDunnough)

Descriptions/Diagnoses: Burks 1953 (adult), Webb 2002 (larva, adult). Distributed throughout eastern and central North America. Known as the "blue quill" to anglers.

Guenther and McCafferty (2008) reported *P. adoptiva* from Boxelder Creek, Castle Creek, and Rapid Creek in the Black Hills.

Paraleptophlebia debilis (Walker)

Descriptions/Diagnoses: Burks 1953 (adult), Webb 2002 (larva, adult). Most of North America except the southwestern United States. The "mahogany dun" in the angling world.

McCafferty and Kondratieff (1999) reported *P. debilis* from Rapid Creek in the Black Hills.

Paraleptophlebia memorialis (Eaton)

Descriptions/Diagnoses: Kilgore and Allen 1973 (larva), Eaton 1884 (adult). Widely distributed in western North America from British Columbia and Alberta south to Arizona and New Mexico. No common name from the angling community, since this is a rather uncommon species.

McCafferty and Kondratieff (1999) reported *P. memorialis* from Iron Creek, Rapid Creek, and Sunday Gulch in the Black Hills. The Black Hills are the easternmost range limit for this species.

Paraleptophlebia mollis (Eaton)

Descriptions/Diagnoses: Burks 1953 (adult), Needham et al. 1935 (larva). Distributed throughout eastern and central North America. Known to anglers as the "American iron blue quill", the "dark blue quill", or the "Jenny spinner".

Black Hills records for *P. mollis* include Boxelder Creek, Grizzly Bear Creek, Jim Creek, and Rapid Creek (McCafferty 1990). The Black Hills represent the westernmost range limit for this species.

Siphlonuridae

The Black Hills represent the easternmost limits of the ranges of both of the two species of Siphlonuridae known from South Dakota. Shearer (2006) reported larvae of *Siphlonurus* from the Black Hills. Because they prefer margins of streams and other more lentic habitats, siphlonurids are rarely collected in most stream biomonitoring samples.

Siphlonurus Eaton

Both species of *Siphlonurus* are known as "gray drakes" in the angling community.

Siphlonurus columbianus McDunnough

Descriptions/Diagnoses: Allen 1955 (larva, adult), McDunnough 1925 (adult). Distributed from British Columbia and Alberta south to California, Colorado, and South Dakota.

McCafferty (1990) reported this species from Grizzly Bear Creek in the Black Hills.

Siphlonurus occidentalis (Eaton)

Descriptions/Diagnoses: Jensen (1966) cites Traver 1935 and Edmunds 1952 (PhD dissertation) for descriptions. Distributed from British Columbia and Alberta, south to California, Arizona, New Mexico, and South Dakota. McCafferty et al. (1993) suggest that this species is the most common *Siphlonurus* in the western United States.

Guenther and McCafferty (2008) reported this species from Spearfish, Horsethief Lake, Palmer Gulch Creek, and a stream at Sylvan Lake in the Black Hills

.

Ephemeroptera Traits Table

Table 3: Ephemeroptera Traits Table

Species	TSN	Seen	Tolerance Value	FFG	Habit
Ameletidae	568544	+	0	GC	SW
Ameletus sp.	100996	+	0	GC	SW
Baetidae	100755	+	4	GC	SW
Acentrella	100801	+	4	GC	SW
Acentrella insignificans	568572	+	4	GC	SW
Anafroptilum	--	+	2	GC	SW
Anafroptilum album	--	+	2	GC	SW
Baetis	100800	+	5	GC	SW
Baetis bicaudatus	100823	+	5	GC	SW
Baetis brunneicolor	100825	-	5	GC	SW
Baetis flavistriga	100835	+	4	GC	SW
Baetis intercalaris	100808	-	5	OM	SW
Baetis magnus	568577	+	5	GC	SW
Baetis tricaudatus	100817	+	5	GC	SW
Callibaetis	100903	+	9	GC	SW
Callibaetis ferrugineus	100918	/	9	GC	SW
Callibaetis fluctuans	100820	/	9	GC	SW
Callibaetis pallidus	100926	/	9	GC	SW
Callibaetis pictus	100927	/	9	GC	SW
Camelobaetidius	568548	+	5	GC	SW
Camelobaetidius warreni	568592	+	5	GC	SW
Diphetor	568550	+	5	GC	SW
Diphetor hageni	568598	+	5	GC	SW
Fallceon	568551	+	4	GC	SW
Fallceon quilleri	568601	+	5	GC	SW
Iswaeon	776928	-	5	SC	SW
Iswaeon anoka	776948	-	5	SC	SW
Labiobaetis	568552	-	4	SC	SW
Labiobaetis propinquus	568605	-	4	OM	SW
Paracloeodes	100899	+	5	SC	SW
Paracloeodes minutus	100901	+	5	SC	SW
Procloeon	206622	-	5	OM	SW
Caenidae	101467	+	7	GC	--
Caenis	101478	+	7	GC	SP
Caenis amica	101480	+	7	OM	SP
Ephemerellidae	101232	+	1	GC	CN
Drunella	101365	-	0	SC	CN
Drunella doddsii	101368	-	0	SC	CN
Ephemerella	101233	+	1	GC	CN
Ephemerella dorothea infrequens	101272	/	1	GC	CN
Ephemerella excrucians	101276	/	1	GC	CN
Ephemeridae	101525	-	4	GC	BU
Ephemera	101526	-	4	GC	BU
Ephemera simulans	101530		4	GC	BU
Heptageniidae	100504	+	4	SC	CN
Ecdyonurus	697960	-	4	SC	CN
Ecdyonurus criddlei	698149	-	2	SH	CN
Epeorus	100626	+	0	GC	CN
Epeorus longimanus	100637	+	0	SC	CN
Heptagenia	100602	-	4	SC	CN
Heptagenia elegantula	100604	-	4	SC	CN
Stenacron	100713	-	4	GC	CN
Stenacron interpunctatum	100714	-	4	OM	CN

Species	TSN	Seen	Tolerance Value	FFG	Habit
Leptohyphidae	568545	+	4	GC	CN
Tricorythodes	101405	+	5	GC	SP
Tricorythodes explicatus	101419	/	5	GC	SP
Tricorythodes fictus	101420	/	5	GC	SP
Leptophlebiidae	101095	+	2	GC	SW
Choroterpes	101108	-	2	GC	CN
Leptophlebia	101148	-	2	GC	SW
Leptophlebia nebulosa	101165	-	2	GC	SW
Neochoroterpes	568558	+	2	GC	SW
Neochoroterpes oklahoma	609646	+	2	GC	SW
Paraleptophlebia	101187	+	1	GC	SW
Paraleptophlebia adoptiva	101203	/	1	GC	SW
Paraleptophlebia debilis	101193	/	1	GC	SW
Paraleptophlebia memorialis	101214	/	4	GC	SW
Paraleptophlebia mollis	101218	/	1	GC	SW
Siphlonuridae	100951	+	7	GC	SW
Siphlonurus	100953	+	7	GC	SW
Siphlonurus columbianus	100954	/	7	GC	SW
Siphlonurus occidentalis	100955	/	7	GC	SW

Species	Temperature Range	Lotic	Lentic	Current	Substrate	Voltinism
Ameletidae	E: cool	3.0	1.1	1.0	6.2	Uni
Ameletus sp.	E: cool	3.0	1.1	1.0	6.2	Uni
Baetidae	E: warm	--	--	--	--	--
Acentrella	E: warm	4.0	--	2.0	5.0	Multi
Acentrella insignificans	E: warm	4.0	--	2.0	5.0	Multi
Anafroptilum	E: warm	3.5	1.1	1.1	3.9	Multi
Anafroptilum album	E: warm	3.5	1.1	1.1	3.9	Multi
Baetis	E: warm	3.5	1.1	2.0	6.1	Multi
Baetis bicaudatus	S: cold	2.9	1.1	2.7	6.3	Multi
Baetis brunneicolor	E: warm	3.8	3.0	0.5	5.5	Multi
Baetis flavistriga	E: cool	3.5	5.0	2.0	6.1	Multi
Baetis intercalaris	E: cool	3.8	1.1	3.0	5.7	Multi
Baetis magnus	E: warm	2.2	1.1	2.0	6.1	Multi
Baetis tricaudatus	E: warm	3.3	5.0	2.3	6.4	Multi
Callibaetis	E: warm	3.9	1.1	0.4	1.3	Multi
Callibaetis ferrugineus	E: warm	3.0	4.0	0.4	0.0	Multi
Callibaetis fluctuans	E: warm	5.0	1.1	0.4	1.3	Multi
Callibaetis pallidus	E: warm	3.9	1.1	0.4	1.3	Multi
Callibaetis pictus	E: warm	3.7	5.0	0.4	1.3	Multi
Camelobaetidius	E: warm	4.3	--	2.0	4.3	Multi
Camelobaetidius warreni	E: warm	4.0	--	2.0	4.0	Multi
Diphetor	E: warm	3.5	--	--	5.5	Multi
Diphetor hageni	E: warm	3.3	--	--	5.5	Multi
Fallceon	E: warm	3.9	--	--	--	Multi
Fallceon quilleri	E: warm	3.9	--	--	--	Multi
Iswaeon	E: warm	4.0	--	1.7	3.7	Multi
Iswaeon anoka	E: warm	4.0	--	1.7	3.7	Multi
Labiobaetis	E: warm	4.1	1.3	2.0	5.8	Multi
Labiobaetis propinquus	E: warm	4.2	1.3	2.0	5.0	Multi
Paracloeodes	E: warm	4.2	--	2.0	1.0	Multi
Paracloeodes minutus	E: warm	4.5	--	2.0	1.0	Multi
Procloeon	E: cool	3.5	4.0	1.2	4.6	Multi
Caenidae	E: warm	--	--	--	4.0	--
Caenis	E: warm	3.8	1.1	0.8	1.9	Multi
Caenis amica	E: warm	3.9	3.4	0.5	1.7	Multi
Ephemerellidae	--	--	--	--	--	--
Drunella	E: cool	3.6	1.8	2.6	6.0	Uni
Drunella doddsii	E: cool	3.3	1.8	2.3	6.5	Uni
Ephemerella	E: cool	3.5	1.1	2.0	5.2	Uni
Ephemerella dorothea infrequens	E: cool	3.5	1.1	2.0	5.2	Uni
Ephemerella excrucians	E: cool	4.0	5.0	0.5	4.0	Uni
Ephemeridae	E: cool	--	--	--	--	--
Ephemera	E: cool	4.0	1.0	1.3	1.8	Mero
Ephemera simulans	E: cool	4.5	4.8	0.8	2.2	Mero
Heptageniidae	--	--	--	--	--	--
Ecdyonurus	E: warm	3.6	--	1.3	5.7	Uni
Ecdyonurus criddlei	E: warm	3.3	--	1.5	7.0	Uni
Epeorus	E: cool*	3.1	--	2.3	6.6	Uni
Epeorus longimanus	E: cool	2.6	--	2.0	7.0	Uni
Heptagenia	E: warm	4.1	1.1	2.3	5.8	Uni
Heptagenia elegantula	E: warm	4.2	1.1	2.3	4.5	Uni
Stenacron	E: warm*	3.5	1.2	1.6	5.7	Uni
Stenacron interpunctatum	E: warm*	3.7	5.0	1.5	6.5	Uni
Leptohyphidae	E: warm	--	--	--	--	--
Tricorythodes	E: warm	3.9	1.4	0.8	3.7	Multi
Tricorythodes explicatus	E: warm	3.9	1.4	0.8	3.7	Uni
Tricorythodes fictus	E: warm	3.9	1.4	0.8	3.7	Uni

Species	Temperature Range	Lotic	Lentic	Current	Substrate	Voltinism
Leptophlebiidae	E: warm	--	--	--	--	--
Choroterpes	E: warm	3.9	1.2	1.0	2.0	Uni
Leptophlebia	E: warm	3.3	1.0	0.9	3.4	Uni
Leptophlebia nebulosa	E: warm	3.0	1.0	1.3	7.0	Uni
Neochoroterpes	E: warm	3.9	4.0	0.8	5.6	Uni
Neochoroterpes oklahoma	E: warm	4.5	4.0	2.0	7.0	Uni
Paraleptophlebia	E: warm	3.4	1.2	1.6	5.0	Uni
Paraleptophlebia adoptiva	E: warm	3.4	1.2	1.6	5.0	Uni
Paraleptophlebia debilis	E: cool	3.3	1.2	1.0	5.5	Uni
Paraleptophlebia memorialis	E: warm	4.0	1.2	1.5	4.5	Uni
Paraleptophlebia mollis	E: warm	4.0	5.0	1.6	4.6	Uni
Siphlonuridae	--	--	--	--	--	--
Siphlonurus	E: hot*	3.5	1.1	0.5	1.6	Uni
Siphlonurus columbianus	E: hot*	3.5	1.1	0.5	1.6	Uni
Siphlonurus occidentalis	E: hot*	3.3	3.9	0.5	1.6	Uni

Odonata

Introduction

Dragonflies and damselflies are one of the few groups of insects that have attracted mass popular appeal (the others being butterflies and some beetles). The quick, commanding flight of dragonflies and their striking color patterns make them an obvious part of the ecological community. Damselflies, though less well known (and often more drab in coloration), form the rest of the Odonata. The penchant for odonates to prey on pestiferous insects such as mosquitoes has generally endeared many people to them, although some people have a fear of odonates, particularly the large, formidable-looking darners (Aeshnidae).

All odonates are predatory, both in the larval and adult forms. They can indeed make a serious dent in the populations of smaller insects (e.g., mosquitoes, no-see-ums, black flies) around streams and lakes, a welcome thought to recreationists everywhere.

The common names that I use are from Westfall and May (1996) and Needham et al. (2000), and anglers commonly refer to all of these insects simply as "dragonflies".

Overview of the Odonata in the Black Hills

In the United States, the dragonflies and damselflies are well known at the state level (Westfall and May 1996, Needham et al. 2000), and field guides exist for several states or regions. A complete checklist of North American Odonata was also recently released (Paulson and Dunkle 2012), updating records compiled since Westfall and May (1996) and Needham et al. (2000). A guide to the Odonata of the Black Hills has not been written, although the list of species in South Dakota and Wyoming is fairly well known. Most of the distributional data for these states, however, are reported only at the county level (e.g., Bick et al. 1977, Kondratieff 2000a). Molnar and Lavigne (1979, 1994) reviewed the Odonata of Wyoming, updated by data compiled into a "working document" by Sims (http://bugsofboogercounty.files.wordpress.com/2012/08/wyoming-odonate-distribution3.pdf, accessed 5 April 2013).

Kondratieff (2000a) has provided county maps for 76 of the 79 species known from South Dakota through the NPWRC website (http://www.npwrc.usgs.gov/resource/distr/insects/dfly/sd/toc.htm). These data should not be considered to be complete, based on a caveat within the NPWRC website. It reads as follows:

"This Web site is a 'work in progress,' consisting of information on the known distribution of Dragonflies and Damselflies (Odonata) in the United States. Distribution maps and county checklists were created by extracting information on Odonata distribution from publications listed in References. Users of the Web site should recognize that lack of a confirmed sighting in a given county does not necessarily mean that the species is absent from that county.

"Because it is a 'work in progress,' the Dragonflies and Damselflies (Odonata) of the United States Web site is constantly being updated. Additional families, photos, and species accounts will be added as funds and time permit, and distribution maps of species that are already covered are being updated as new county records are established. Distribution maps are currently limited to states of the conterminous United States."

Finally, the Dragonfly Society of the Americas maintains extensive dragonfly and damselfly distribution data and checklists on its website, www.odonatacentral.org. Records are submitted by anyone and verified by experts, so this web site was also consulted to build this list. Many of the records represent county-level distributional data, but some records include more specific collection locality information. I accessed the website periodically through the preparation of this work, and transmissions were interrupted because the website was actually being updated at the time, suggesting that the records are up to date! The website also has excellent photographs of nearly all of the species of Odonata known to occur in the Black Hills (as of 9 August 2011, it did not have images of only 12 species). The images can be accessed online using the following html address and replacing the ##### with the five-digit OdonataCentral image number that I provide in the description for each species:

http://odonatacentral.org/index.php/GalleryAction.getTaxaImages/taxon_id/#####.

Distributional data on odonates in the Black Hills at levels finer than the county level are few. More specific locality information for certain species can be found in Bick and Hornuff (1972), Provonsha and McCafferty (1977), Hummel (1999), Paulson et al. (1999). Additional data are often found in notes published occasionally in the DSA's newsletter, *Argia*. Although many records in the DSA website are at the county level, some have more specific information on collection localities.

Based on the distribution of odonates in Custer (55 species), Fall River (42 species), Lawrence (45 species), and Pennington (46 species) counties in South Dakota and Crook (45 species) and Weston (30 species) counties in Wyoming, approximately 77 species are reported from in or around the Black Hills. (Even though the Black Hills extend into the southwestern corner of Meade County, South Dakota, this county was excluded because the vast majority of the county lies on the Plains.)

The Black Hills area is clearly the most odontalogically diverse region within either South Dakota or Wyoming. Larval odonates are uncommonly encountered in stream biomonitoring programs in the area, since most species live in lentic habitats, so most reports are from spotting or collections of adults.

Species Accounts

Aeshnidae

Six species of darners are reported from the Black Hills, based on Bick and Hornuff (1972), county-level distributions from the NPWRC website, and the DSA website. Shearer (2006) reported larvae of unidentified species of *Aeshna* from Black Hills streams.

Aeshna Fabricius

Aeshna canadensis Walker – Canada Darner

Descriptions/diagnoses: Walker 1958 (larva); Needham et al. 2000 (adult); OdonataCentral image #44939. Transcontinental across Canada, extending south along the Pacific Coast to California, but otherwise no farther south than Montana, Nebraska, Missouri, and West Virginia. South Dakota is not among the states reported for this species' distribution in Needham et al. (2000).

The DSA website reported only one Black Hills record for this species: French Creek Natural Area in Custer State Park.

Aeshna constricta Say – Lance-Tipped Darner

Descriptions/diagnoses: Musser 1962 (larva), Needham et al. 2000 (adult); OdonataCentral image #44941. Transcontinental from British Columbia and Washington state east to the Canadian Maritimes and Virginia.

Paulson et al. (1999) reported this species from Stockade Lake and from Spring Creek at the Willow Springs B&B. The DSA website reported county-level records for Custer and Pennington counties, as well as a record from a small unnamed pond in Lawrence County.

Aeshna interrupta Walker – Variable Darner

Descriptions/diagnoses: Musser 1962 (larva), Needham et al. 2000 (adult); OdonataCentral image #44949. Among its three subspecies, this species is distributed across North America from Alaska, British Columbia, and California east to the Canadian Maritimes, West Virginia, and New York. *Aeshna interrupta interna*, the subspecies found in the Black Hills, is found throughout the Rocky Mountains from Alberta and British Columbia south to New Mexico and California.

Bick and Hornuff (1972) reported *Aeshna interrupta interna* from Custer State Park. The DSA website reported the following Black Hills records for this species: Deerfield Lake, French Creek east of Blue Bell Campground in Custer State Park, and Spearfish Creek. Sims (2012) added a record from Weston County.

Aeshna palmata Hagen – Paddle-Tail Darner

Descriptions/diagnoses: Musser 1962 (larva), Needham et al. 2000 (adult); OdonataCentral image #44958. The Kamchatka Peninsula in Russia and western North America from Alaska and the Yukon south to California, New Mexico, Nebraska, and Iowa.

The DSA website reported only county-level records for this species in Crook, Custer, Lawrence, and Pennington counties. Sims (2012) reported this species from Crook and Weston counties. I have collected adult males of this species on Annie Creek and False Bottom Creek.

Aeshna umbrosa Walker – Shadow Darner

Descriptions/diagnoses: Musser 1962 (larva), Needham et al. 2000 (adult); OdonataCentral image #44969. Found in every Canadian province and territory, south to California, New Mexico, Arkansas, and Georgia. The nominate subspecies *A. u. umbrosa* is found only in states east of the Continental Divide.

The DSA website reported the following Black Hills records for this species: Ditch Creek 1 mile south of Ditch Creek Campground, French Creek east of Blue Bell Campground in Custer State Park, and Little Spearfish Creek.

Anax Leach

Anax junius (Drury) – Common Green Darner

Descriptions/diagnoses: Musser 1962 (larva), Needham et al. 2000 (adult); OdonataCentral image #45009.

Molinar & Lavigne (1979) reported this species from Crook County, Wyoming. In South Dakota, the DSA website reported this species from Custer County, from Mud Lake in the Coxes GPA, and from the Cottonwood Springs Recreation Area.

Rhionaeschna Förster

These species will key to *Aeshna* in most generic identification keys; however, these two species were moved to the genus *Rhionaeschna* by von Ellenreider (2003) in a review of the Neotropical species of *Aeshna*.

Rhionaeschna californica (Calvert) – California Darner

Descriptions/diagnoses: Musser 1962 (larva, as *Aeshna californica)*, Needham et al. 2000 (adult, as *Aeshna californica*). Distributed along the Pacific Coast from British Columbia to Baja California, east to Montana, South Dakota, Colorado, and Arizona. This species was moved to the primarily Neotropical genus *Rhionaeschna* by von Ellenreider (2003).

This species was reported from Crook and Weston counties in Molinar and Lavigne (1979, as *Aeshna californica*). The DSA website reported this species from the Mirror Lake GPA, as well as Crook, Custer, Pennington, and Weston counties.

Rhionaeschna multicolor (Hagen) – Blue-Eyed Darner

Descriptions/diagnoses: Musser 1962 (larva, as *Aeshna multicolor)*, Needham et al. 2000 (adult, as *Aeshna multicolor*); OdonataCentral image #45324. Western North America from British Columbia along the Pacific Coast to Panama, east in the United States to Montana, Iowa, Kansas, and Texas. This species was also moved to *Rhionaeschna* by von Ellenrieder (2003).

Molinar and Lavigne (1979) reported this species from Weston County. Paulson et al. (1999) reported *R. multicolor* (as *Aeshna multicolor*) from Little Spearfish Creek. The DSA website reported county-level records only for this species from Fall River, Lawrence, and Weston counties.

Calopterygidae

Three species of calopterygid damselflies are known from the Black Hills.

Calopteryx Leach

Calopteryx aequabilis Say – River Jewelwing

Description/diagnosis: Needham 1903 (larva); Westfall and May 1996 (adult); OdonataCentral image #42239. Distributed from the Pacific Coast states east to New England, south to Colorado, Nebraska, Indiana, and Pennsylvania.

The nearest localities figured by Johnson (1974) for *C. aequabilis* in South Dakota northeast of the Black Hills, possibly in the Plains region of Meade County, and in the northwestern corner of Nebraska near Chadron. Molinar and Lavigne (1979) recorded this species from Crook County. The DSA website reported the following Black Hills records for this species: Crook County, Custer County, Lawrence County, Pennington County. The NPS reported this species from Wind Cave National Park on their website (http://home.nps.gov/wica/naturescience/dragonflies-and-damselflies.htm).

Calopteryx maculata (Beauvois) – Ebony Jewelwing

Description/diagnosis: Walker 1953 (larva); Westfall and May 1996 (adult); OdonataCentral image #42252. Distributed from the Atlantic seaboard west to Manitoba and the base of the Rocky Mountains in Montana and Colorado, south to Texas and the Gulf Coast.

The nearest locality figured by Johnson (1974) for C. maculata was in south central South Dakota, near Tripp County. The DSA website reported the following Black Hills records for this species: Cold Springs Creek along Crook County Road 207, French Creek Natural Area in Custer State Park, Hot Brook at Chataqua Park in Hot Springs, Newton Fork Spring Creek.

Hetaerina Hagen in Selys

Hetaerina americana (Fabricius) – American Rubyspot

Description/diagnosis: Walker 1953 (larva), Westfall and May 1996 (adult); OdonataCentral image #42272. Distributed in eastern Canada, nearly every one of the contiguous American states, Mexico, and Central America south to Honduras.

Bick and Hornuff (1972) reported *H. americana* from the Belle Fourche River at Devil's Tower National Monument in the Black Hills, a record repeated in Molinar & Lavigne (1979), along with a record from Weston County. The DSA website reported the following Black Hills records for this species: Crook County, Custer County, Hot Brook at Chataqua Park in Hot Springs, Mirror Lake GPA. The NPS reported this species from Wind Cave National Park on their website (http://home.nps.gov/wica/naturescience/dragonflies-and-damselflies.htm).

Coenagrionidae

A total of 23 species of coenagrionid damselflies are known from the Black Hills, based on Bick and Hornuff (1972) and county-level distributions from the NPWRC website. Shearer (2006) reported larvae of unidentified species of *Argia* and *Enallagma* from Black Hills streams. In biomonitoring studies, Coenagrionidae is often left at the family level due to damaged larval specimens; when specimens were in good enough condition to identify to genus, they were individuals of *Argia*.

Amphiagrion Selys

The DSA website reported the following Black Hills records for numerous specimens that were apparently considered to be intermediate between *A. abbreviatum* and *A. saucium* and unassignable to either species: Fall River County, Lawrence County, Ditch Creek 1 mile south of Ditch Creek Campground, French Creek Natural Area in Custer State Park. Even though *A. saucium* has not yet been officially reported from the Black Hills, a species account is included.

Amphiagrion abbreviatum (Selys) – Western Red Damsel

Description/diagnosis: Cook and Antonelli 1969 (larva), Westfall and May 1996 (adult); OdonataCentral image #42664. Distributed in western North America from British Columbia and Saskatchewan south to Baja California and New Mexico.

Based on records in Molinar and Lavigne (1979), the DSA website reported the following Black Hills records for this species: Crook County, Weston County.

Amphiagrion saucium (Burmeister) – Eastern Red Damsel

Description/diagnosis: Walker 1953 (larva), Westfall and May 1996 (adult). Distributed from Ontario, Quebec, and Nova Scotia south along the Atlantic seaboard to Georgia and west to Minnesota,

Indiana, Tennessee, and Mississippi. This species has not yet been reported from the Black Hills, but the DSA website reports that some specimens from the Black Hills are apparently intermediate between *A. abbreviatum* and *A. saucium*.

Argia Rambur

Argia alberta Kennedy – Paiute Dancer

Description/diagnosis: Westfall and May 1996 (adult); OdonataCentral image #42716. Distributed in the western US from California east to South Dakota, Kansas, and Arizona. Westfall and May (1996) stated that northern populations, including those in South Dakota and Wyoming, have generally been found near warm or hot springs, while southern populations have been found near cooler streams.

The DSA website reported the following Black Hills records for this species: Custer, Fall River, and Lawrence counties.

Argia apicalis (Say) – Blue-Fronted Dancer

Description/diagnosis: Walker 1953 (larva), Westfall and May 1996 (adult); OdonataCentral image #42720. Distributed from the Dakotas, Colorado, and New Mexico eastward to Ontario, New England, New York, the Atlantic seaboard, the Gulf Coast, and south to Nuevo Leon, Mexico.

The DSA website reported the following Black Hills records for this species: French Creek Natural Area in Custer State Park. Sims (2012) also reported this species from Crook County.

Argia emma Kennedy – Emma's Dancer

Description/diagnosis: Kennedy 1916 (adult and larva); Westfall and May 1996 (adults). Western North America from the Pacific Coast through the Rocky Mountains to South Dakota and Nebraska. Kennedy (1916) indicated that the larvae are vigorously active in streams in Washington, found on roots and brush in pools and under cobbles in riffles.

Bick and Hornuff (1972) reported *Argia emma* from moderately to swiftly flowing streams in Crook County, Wyoming, near the Black Hills and in Custer State Park. Molinar and Lavigne (1979) likewise reported this species from Crook County. The DSA website reported the following Black Hills records for this species: Pennington County, French Creek Natural Area in Custer State Park, Box Elder Creek. The USU WCMAFE database includes records from Grace Coolidge Creek in Custer County and the Cheyenne River near Oral, South Dakota.

Argia fumipennis (Burmeister) – Variable Dancer

Description/diagnosis: Byers 1930 (larva), Westfall and May 1996 (adult); OdonataCentral image #42747. The northern subspecies, *A. f. violacea* (Hagen) is the subspecies represented in the Black Hills; it is distributed from Arizona and Durango east to southeastern Canada and Alabama.

Bick and Hornuff (1972) reported *A. fumipennis violacea* from the Belle Fourche River in Devil's Tower National Monument. Molinar and Lavigne (1979) also reported this species from Crook County as *A. violacea*. Paulson et al. (1999) reported this species from French Creek in Custer State Park. The DSA website reported the following Black Hills records for this species: Crook County, Custer County, Lawrence County, Deerfield Lake, Hot Brook at Chataqua Park in Hot Springs.

Argia immunda (Hagen) – Kiowa Dancer

Description/diagnosis: Westfall and May 1996 (adult); the larva apparently has not yet been formally described in the literature; OdonataCentral image #42760. Distributed from the south-central

United States from South Dakota south through Mexico to Belize. Usually found near streams, but occasionally along lake shores, Westfall and May (1996) stated that it is found at hot springs in South Dakota.

The DSA website reported the following Black Hills records for this species: Custer County, Hot Brook at Chataqua Park in Hot Springs.

Argia lugens (Hagen) – Sooty Dancer

Description/diagnosis: Needham 1904 (larva), Westfall and May 1996 (adult); OdonataCentral image #42778. Distributed in southwestern United States as far north as Utah and Colorado, south along the Pacific coast states of Mexico to Oaxaca and Veracruz. Note that the distribution in the Black Hills is north of the distribution reported in Westfall and May (1996).

Hummel (1999) reported *A. lugens* from the Black Hills vicinity (Hot Brook in Hot Springs) for the first (and apparently only) time. The DSA website repeated that record for this species: Hot Brook at Chataqua Park in Hot Springs.

Argia moesta (Hagen) – Powdered Dancer

Description/diagnosis: Byers 1930 (larva), Westfall and May 1996 (adult); OdonataCentral image #42782. Distributed from eastern Canada (Ontario to New Brunswick) south along the Atlantic seaboard to Florida and west to California and Baja California, central Mexico, and the Gulf Coast states.

The DSA website reported the following record for this species: Hot Brook at Chataqua Park in Hot Springs. That report from a collection made in 1981 was the only report of *A. moesta* from South Dakota as of 2011.

Argia plana Calvert – Springwater Dancer

Description/diagnosis: Westfall 1990 (larva), Westfall and May 1996 (adult); OdonataCentral image #42794. Distributed in central United States from Wisconsin and South Dakota south through Mexico to Guatemala.

The DSA website reported the following Black Hills records for this species: Custer County, Fall River County, Mirror Lake GPA.

Argia vivida Hagen – Vivid Dancer

Description/diagnosis: Walker 1953 (larva), Westfall and May 1996 (adult); OdonataCentral image #42824. Distributed in western North America from British Columbia and Alberta south to Baja California and as far east as Nebraska and South Dakota. It appears to prefer small spring-fed streams, including hot springs. Kennedy (1916) noted that the nymphs of *A. vivida* are rather sluggish when removed from the water.

Molinar and Lavigne (1979) reported this species from Crook and Weston counties. The DSA website reported the following Black Hills records for this species: Crook County, Lawrence County, Weston County, French Creek east of the Blue Bell Campground in Custer State Park, Hot Brook at Chataqua Park in Hot Springs, Spring Creek at Spring Creek Trailhead below Sheridan Lake.

Coenagrion Kirby

Nearctic larvae of *Coenagrion resolutum* and *Enallagma annexum* are very difficult to separate at the genus level. The two species can usually be separated by the number of antennal segments (6 in *Enallagma*, 7 in *Coenagrion*), although this character can be variable.

Coenagrion resolutum (Hagen) – Taiga Bluet

Description/diagnosis: Walker 1914, Baker and Clifford 1980 (larva); Westfall and May 1996 (adult). Distributed across all of Canada south to California and Arizona in the west and Ohio and Pennsylvania in the east. Westfall and May (1996) suggested that this species may range farther north in Canada than any other damselfly in North America.

Molinar and Lavigne (1979) reported this species from Crook County. The DSA website reported the following Black Hills records for this species: Crook County, dam on Spearfish Creek upstream of Savoy.

Enallagma Charpentier

Although all of the species of *Enallagma* known from the Black Hills have been described in their adult and larval forms, identification of larvae is largely based on gills (which are often lost from specimens collecting in biomonitoring studies) and the cerci of last instars (which are sufficiently mature to determine gender by observing presence or absence of a developing ovipositor). As such, nearly all identifications of *Enallagma* in biomonitoring studies are left at the genus level.

Enallagma anna Williamson – River Bluet

Description/diagnosis: Garrison 1984 (larva), Westfall and May 1996 (adult); OdonataCentral image #42969. Western North America from Alberta to California and New Mexico, eastward to South Dakota and Nebraska. Westfall and May (1996) reported that the larvae appear to be restricted to low streams and rivers in the arid mountains of western North America, sometimes near warm springs.

Collections in and near the Black Hills included Bick and Hornuff (1972), which reported *Enallagma anna* from streams in Crook County, Wyoming, including from a stream fed by hot springs. Molinar and Lavigne (1979) subsequently also reported this species from Crook and Weston counties. Paulson et al. (1999) reported this species from Spring Creek at the Willow Springs B&B. The DSA website reported the following Black Hills records for this species: Crook County, Custer County, Fall River County, Lawrence County, Pennington County, Weston County.

Enallagma annexum (Hagen) – Northern Bluet

Description/diagnosis: Walker 1953 (larva, as *Enallagma cyathigerum*), Westfall and May 1996 (adult, as *Enallagma cyathigerum*). Distributed from Alaska, the Northwest Territories, and Newfoundland, south to Baja California, New Mexico, Nebraska, Indiana, and Virginia. Based on Turgeon et al. (2005), this species is the Nearctic counterpart to the Palaearctic *E. cyathigerum*, with which it used to be considered synonymous, and many North American reports used that name.

Bick and Hornuff (1972) stated that *E. cyathigerum* (= *E. annexum*) was the most abundant damselfly they collected during their collecting trip, being found in all habitats except swiftly flowing streams in both Carbon and Weston counties in Wyoming. They did not specify if the collections were made in or just near the Black Hills. Molinar and Lavigne (1979) subsequently also reported this species (as *E. cyathigerum*) from those counties. The DSA website reported the following Black Hills records: Crook County, Lawrence County, Fall River County, Pennington County, Weston County, French Creek in Custer State Park.

Enallagma antennatum (Say) – Rainbow Bluet

Description/diagnosis: Needham 1903 (larva), Westfall and May 1996 (adult); OdonataCentral image #42970. Eastern North America from Quebec and West Virginia east to Montana and Colorado.Westfall and May (1996) reported that this species is found in vegetation in ponds and slow streams, often near lake inlets and outlets.

Bick and Hornuff (1972) reported *E. antennatum* from the Belle Fourche River in Devil's Tower National Monument. Molinar and Lavigne (1979) reported this species from Crook County. Paulson et al. (1999) reported this species from the Lakota Lake Picnic Area. The DSA website reported the following Black Hills records for this species: Crook County, Custer County, dam on Spearfish Creek upstream of Savoy, Deerfield Lake.

Enallagma basidens Calvert – Double-Striped Bluet

Description/diagnosis: Bird 1931, Huggins 1978 (larva); Westfall and May 1996 (adult); OdonataCentral image #42972. Distributed from Ontario south through the central United States to South Carolina, Georgia, and the Gulf Coast in the east, to Colorado, Arizona, and California in the West, and to the northern tier of Mexican states as far south as San Luis Potosí.

The DSA website reported the following Black Hills records for this species: Cold Brook Recreation Area.

Enallagma boreale Selys – Boreal Bluet

Description/diagnosis: Walker 1944 (larva), Westfall and May 1996 (adult); OdonataCentral image #42973. Distributed transcontinentally from Alaska and northern Canada south to California and New Mexico in the west and Ohio and West Virginia in the east. Westfall and May (1996) indicate that it prefers usually fishless lentic and slow water habitats.

Bick and Hornuff (1972) reported *E. boreale* from ponds in Weston County, Wyoming, in or near the Black Hills. Molinar and Lavigne (1979) also reported it from Crook and Weston counties. The DSA website reported the following Black Hills records for this species: Crook County, Custer County, Lawrence County, Weston County, Cottonwood Springs Recreation Area, Deerfield Lake.

Enallagma carunculatum Morse – Tule Bluet

Description/diagnosis: Walker 1953 (larva), Westfall and May 1996 (adult); OdonataCentral image #42974. Distributed from British Columbia to the Canadian Maritimes, south to Baja California, New Mexico, Kansas, Kentucky, and Maryland. It is common in large, slow rivers and lakes, and less common in small ponds (Westfall and May 1996).

Molinar and Lavigne (1979) reported this species from Weston County. Paulson et al. (1999) reported this species from cattail-bordered ponds at the Pioneer Rest Area. The DSA website reported the following Black Hills records for this species: Custer County, Lawrence County, Pennington County, Weston County, Cottonwood Springs Recreation Area.

Enallagama civile (Hagen) – Familiar Bluet

Description/diagnosis: Walker 1953 (larva), Westfall and May 1996 (adult); OdonataCentral image #42976. Distributed from the southern tier of Canadian provinces south to South America, including the Bahamas and the Greater Antilles, as well as Hawaii; Westfall and May (1996) suggest that this is the most widely distributed damselfly in North America. It is often an early colonizer in ponds and slow streams.

Molinar and Lavigne (1979) reported this species from Crook and Weston counties. The DSA website reported the following Black Hills records for this species: Crook County, Custer County, Pennington County, Weston County, dam on Spearfish Creek upstream of Savoy.

Enallagma clausum Morse – Alkali Bluet

Description/diagnosis: Walker 1944 (larva), Westfall and May 1996 (adult); OdonataCentral image #42977. Distributed from the southern tier of Canadian provinces through the northern Great

Plains to Kansas, Nevada, and California. Westfall and May (1996) state that it is often found in saline and alkaline waters and can occur in freshwater lakes.

Molinar and Lavigne (1979) reported this species from Crook and Weston counties. The DSA website reported the following Black Hills records for this species: Crook County, Weston County, Deerfield Lake.

Enallagma ebrium (Hagen) – Marsh Bluet

Description/diagnosis: Walker 1914 (larva), Westfall and May 1996 (adult); OdonataCentral image #42988. From British Columbia to the Canadian Maritimes, south to Washington, Colorado, Tennessee, and Virginia. Westfall and May (1996) report that it inhabits marshes and ponds.

Molinar and Lavigne (1979) reported this species from Weston County. Bick and Hornuff (1972) reported *E. ebrium* from Custer State Park. The DSA website reported the following Black Hills records for this species: Custer County, Lawrence County, Weston County.

Enallagma exsulans (Hagen) – Stream Bluet

Description/diagnosis: Walker 1953 (larva), Westfall and May 1996 (adult); OdonataCentral image #42990. Distributed from Ontario, Quebec, and the Canadian Maritimes south and west across the United States to the Gulf Coast, Oklahoma, Nebraska, and the Dakotas; south into Tamaulipas and Nuevo Leon, Mexico.

The DSA website reported the following Black Hills records for this species: Deerfield Lake.

Enallagma hageni (Walsh) – Hagen's Bluet

Description/diagnosis: Walker 1914 (larva), Westfall and May 1996 (adult); OdonataCentral image #42992. From British Columbia to the Canadian Maritimes, south to Colorado, Missouri, West Virginia, and Virginia. Westfall and May (1996) report that it inhabits marshes and ponds and more acidic waters than *E. ebrium*, which it closely resembles.

Bick and Hornuff (1972) reported *E. hageni* from a site four miles south of Devil's Tower National Monument. Molinar and Lavigne (1979) reported this species from Crook County. Paulson et al. (1999) reported this species from cattail-bordered ponds at the Pioneer Rest Area. The DSA website reported the following Black Hills records for this species: Crook County, Custer County, Fall River County, Lawrence County, Pennington County.

Enallagma praevarum (Hagen) – Arroyo Bluet

Description/diagnosis: Garrison 1984 (larva), Westfall and May 1996 (adult); OdonataCentral image #43003. Southwestern, from California and South Dakota to Oaxaca and Nuevo Leon, Mexico. Prefers small ponds, but also found on small streams (Westfall and May 1996).

Bick and Hornuff (1972) reported *E. praevarum* from a well-vegetated site in Crook County, Wyoming, not specifying the flow regime of the stream or its proximity to the Black Hills. Molinar and Lavigne (1979) reported this species from Crook County. The DSA website reported the following Black Hills records for this species: Crook County, Fall River County, Lawrence County, Pennington County, French Creek east of Blue Bell Campground in Custer State Park.

Ischnura Charpentier

Although all of the species of *Ischnura* known from the Black Hills have been described in their adult and larval forms, identification of larvae is largely based on gills, which are often lost from specimens collecting in biomonitoring studies. As such, nearly all identifications of *Ischnura* in biomonitoring studies are left at the genus level.

Ischnura cervula Selys – Pacific Forktail

Description/diagnosis: Kennedy 1915 (larva), Westfall and May 1996 (adult); OdonataCentral image #43052. Distributed in western North America from British Columbia and Alberta, south through the Rocky Mountains to New Mexico, Sonora, and Baja California.

The DSA website reported the following Black Hills records for this species: Mud Lake in the Coxes GPA.

Ischnura damula Calvert – Plains Forktail

Description/diagnosis: Walker 1953 (larva), Westfall and May 1996 (adult); OdonataCentral image #43055. Western North America from British Columbia and Saskatchewan to Arizona and Texas. Westfall and May (1996) state that this species prefers ponds.

Bick and Hornuff (1972) reported *I. damula* at a site 10 miles south of Newcastle, Wyoming, near the Black Hills, while Molinar and Lavigne (1979) reported this species from Crook and Weston counties. The DSA website reported the following Black Hills records for this species: Crook County, Custer County, Weston County, Cottonwood Springs Recreation Area, Deerfield Lake, Yates Pond in Cheyenne Crossing.

Ischnura perparva McLachlan – Western Forktail

Description/diagnosis: Kennedy 1915 (larva), Westfall and May 1996 (adult); OdonataCentral image #43084. Garrison (1981) provided additional morphometric data on larvae of this species. Distributed from British Columbia and Manitoba south to California, New Mexico, and Oklahoma. Inhabits a wide range of lotic habitats, where larvae cling to instream structures and vegetation (Westfall and May 1996).

Molinar and Lavigne (1979) reported this species from Crook and Weston counties. Paulson et al. (1999) reported this species from Spring Creek at the Willow Springs B&B. The DSA website reported the following Black Hills records for this species: Crook County, Custer County, Lawrence County, Weston County.

Ischnura verticalis (Say) – Eastern Forktail

Description/diagnosis: Needham 1903 (larva), Westfall and May 1996 (adult); OdonataCentral image #43105. Distributed from Manitoba and the Canadian Maritimes south to Colorado, Texas, Alabama, and Georgia. This species inhabits slow streams, backwaters, and other lentic habitats (Westfall and May 1996).

Bick and Hornuff (1972) reported *I. verticalis* from a site 4 miles south of Devil's Tower National Monument and from a site 10 miles south of Newcastle, Wyoming, in and near the Black Hills. Molinar and Lavigne (1979) likewise reported this species from Crook and Weston counties. The DSA website reported the following Black Hills records for this species: Crook County, Lawrence County, Pennington County, Weston County, French Creek east of Blue Bell Campground in Custer State Park.

Nehalennia Selys

Nehalennia irene (Hagen) – Sedge Sprite

Description/diagnosis: Needham 1903 (larva), Westfall and May 1996 (adult); OdonataCentral image #43211. Distributed from the southern tier of Canadian provinces south to California, Wyoming, Iowa, and Maryland; Westfall and May (1996) include a questionable record from Mississippi. This species inhabits a wide variety of lentic habitats, including marshes and sedge fens (Westfall and May 1996).

Molinar and Lavigne (1979) reported this species from Crook and Weston counties. The DSA website reported the following Black Hills records for this species: Crook County, Custer County, Weston County.

Corduliidae

Three species of corduliid dragonflies are known from the Black Hills, based on Bick and Hornuff (1972) and county-level distributions from the NPWRC website.

Epitheca Burmeister

The genus *Epitheca* is in a state of flux, taxonomically. Some specialists (e.g., Walker 1966) believe that *Epitheca* is a large, diverse genus, and have synonymized *Tetragoneuria* and Epicordulia into Epitheca or used them as subgenera, while others (e.g., Needham et al. 2000) believe *Epitheca* is a separate genus restricted to the Palaearctic Realm and *Tetragoneuria* and *Epicordulia* are valid genera. I have chosen to follow the former classification herein, but I do not have any qualms either way.

Epitheca costalis (Selys) – Slender baskettail or Stripe-Winged Baskettail

Descriptions/diagnoses: Needham et al. 2000 (adult, as *Tetragoneuria costalis*); it appears that the nymph is as yet undescribed; OdonataCentral image #45501. Distributed in the southeastern United States from the Black Hills south to Texas east to the Atlantic seaboard from New Jersey south to Florida. Needham et al. (2000) does not include South Dakota or Wyoming in within the distribution of this species.

Epitheca cynosura (Say) – Common Baskettail

Descriptions/diagnoses: Walker and Corbet 1975 (larva), Needham et al. 2000 (adult, as *Tetragoneuria cynosura*); OdonataCentral image #45502. This species is more widely distributed than *E. costalis*, being found from Colorado, South Dakota, and Minnesota, south to Texas and the entire Gulf Coast, east to Ontario, Quebec, and the Canadian Maritime provinces of New Brunswick and Nova Scotia as well as the Atlantic seaboard.

Epitheca petechialis (Muttkowski) – Dot-Winged Baskettail

Descriptions/diagnoses: Needham et al. 2000 (adult, as *Tetragoneuria petechialis*); it appears tht the nymph is as yet undescribed; OdonataCentral image #45504. This species is known from a rather narrow swath of the United States from the Black Hills south through Colorado and Kansas to Oklahoma, Texas, and New Mexico.

Molinar and Lavigne (1979) reported this species from Crook County. The DSA website reported the following Black Hills records for this species: Custer County, Pennington County, Cook Lake (Crook Co.), Mirror Lake GPA.

Epitheca spinigera (Selys) – Spiny Baskettail

Descriptions/diagnoses: Walker 1913 (larva), Needham et al. 2000 (adult, as *Tetragoneuria spinigera*). This is a northern species that is transcontinental in Canada from British Columbia to Prince Edward Island, south into the United States to California, Montana, the Dakotas, Iowa, Illinois, Indiana, West Virginia, and New Jersey.

The DSA website reported the following Black Hills records for this species: Custer County.

Somatochlora Selys

Shearer (2006) reported larvae of unidentified species of *Somatochlora* from Black Hills streams.

Somatochlora ensigera Martin – Plains Emerald

Descriptions/diagnoses: Walker and Corbet 1975 (larva), Needham et al. 2000 (adult). Distributed from Saskatchewan, Manitoba, and Ontario southwest to Wyoming, Colorado, South Dakota, Iowa, and Ohio.

Molinar and Lavigne (1979) reported this species from Weston County. The DSA website reported the following Black Hills records for this species: Custer County, Weston County.

Somatochlora minor Calvert – Ocellated Emerald

Descriptions/diagnoses: Walker and Corbet 1975 (larva), Needham et al. 2000 (adult). Distributed transcontinentally in Canada from the Yukon Territory and British Columbia to Labrador and the Maritimes, south to Oregon, Colorado, South Dakota, Minnesota, Michigan, and New York.

Bick and Hornuff (1972) reported *Somatochlora minor* from Custer State Park. The DSA website reported the following Black Hills records for this species: French Creek east of Blue Bell Campground in Custer State Park, Lawrence County.

Gomphidae

Five species of clubtail dragonflies are known from the Black Hills, based on Bick and Hornuff (1972) and county-level distributions from the NPWRC website.

Arigomphus Needham

Arigomphus cornutus (Tough) – Horned Clubtail

Descriptions/diagnoses: Walker 1958 (larva), Needham et al. 2000 (adult); no OdonataCentral image available. Distributed from Montana, North Dakota, Manitoba, Ontario, and Quebec south to Colorado, Nebraska, Iowa, Illinois, and Indiana. Prefers lentic habitats.

Molinar and Lavigne (1979) reported this species (as *Gomphus cornutus*) from Crook County. Paulson et al. (1999) reported this species from Stockade Lake, Bismark Lake, and the Lakota Lake Picnic Area. The DSA website reported the following Black Hills records for this species: Crook County, Deerfield Lake, Mirror Lake GPA, French Creek east of Blue Bell Campground in Custer State Park.

Erpetogomphus Selys

Erpetogomphus designatus Hagen in Selys – Eastern Ringtail

Descriptions/diagnoses: Novelo-Gutiérrez 2005 (larva); Needham et al. 2000 (adult); OdonataCentral image #46080. Distributed from Montana, South Dakota, Illinois, Indiana, Ohio, West Virginia, and Virginia south to Arizona, New Mexico, and the Gulf Coast and onward into Mexico.

Jergens (1968) reported *Erpetogomphus* at the genus level from the Fall River; it is likely that this was the species represented. The DSA website reported the following Black Hills records for this species: Fall River County.

Gomphus Leach

Gomphus graslinellus Walsh – Pronghorn Clubtail

Descriptions/diagnoses: Walker 1958 (larva), Needham et al. 2000 (adult); OdonataCentral image #46148. Distributed from British Columbia and Washington east to Ontario, Michigan, Ohio, Kentucky, and Arkansas.

Jergens (1968) reported *Gomphus* at the genus level from French Creek and the Fall River; it is likely that this was the species represented. Bick and Hornuff (1972) reported *G. graslinellus* from Custer State Park, and Bick et al. (1977) added records from Fall River County, as well. Molinar and Lavigne (1979) reported this species from Crook County. The DSA website reported the following Black Hills records for this species: Fall River County, Cook Lake (Crook Co.), Deerfield Lake, French Creek east of Blue Bell Campground in Custer State Park.

Ophiogomphus Selys

Ophiogomphus severus Hagen – Pale Snaketail

Descriptions/diagnoses: Musser 1962 (larva), Needham et al. 2000 (adult); OdonataCentral image #46471. There are two subspecies; O. s. montanus is western, in the main Rocky Mountain cordillera, while the nominate subspecies' distribution extends from Alberta, Oregon, Idaho, and Nevada east to Saskatchewan, South Dakota, Nebraska, Kansas, and Arkansas.

Jergens (1968) reported *Ophiogomphus* at the genus level from French Creek, Spring Creek, Battle Creek, Castle Creek, the South Fork of Rapid Creek, and the Fall River; it is likely that this was the species represented. Molinar and Lavigne (1979) reported this species from Crook and Weston counties. Paulson et al. (1999) reported this species from French Creek in Custer State Park. Shearer (2006) reported larvae of *Ophiogomphus* (probably *O. severus*) from Black Hills streams. I have larval specimens of *O. severus* in my collection from Whitewood Creek and from the Fall River near Hot Springs. The DSA website reported the following Black Hills records for this species: Crook County, Pennington County, Weston County, Box Elder Creek at the Box Elder Forks Campground, French Creek east of Blue Bell Campground in Custer State Park.

The temperature designation in Grafe et al. (2002) for *Ophiogomphus* was eurythermal warm, but the literature statements recorded in Vieira et al. (2006) suggest that this genus could be classified as stenothermal cold.

Stylurus Needham

Stylurus intricatus (Hagen in Selys) – Brimstone Clubtail

Descriptions/diagnoses: Musser 1962 (larva, as *Gomphus intricatus*), Needham et al. 2000 (adult); OdonataCentral image #46747. Distributed from Alberta and Saskatchewan south to California, Arizona, New Mexico, and Texas.

This species has not yet ben reported from the Black Hills, but its widespread western and central Nearctic distribution suggests that it may occur there or in the near vicinity. Molinar and Lavigne (1979) did report this species, as *Gomphus intricatus*, from Sweetwater County, Wyoming, in the southwest portion of the state.

Lestidae

Four species of spread-winged damselflies are known from the Black Hills, based on Bick and Hornuff (1972) and county-level distributions from the NPWRC website.

Archilestes Selys

Archilestes grandis (Rambur) – Great Spreadwing

Descriptions/diagnoses: Needham 1904 (larva), Westfall and May 1996 (adult); OdonataCentral image #43736. Transcontinental from California to Wisconsin to New York, south through Mexico and Central America to Colombia and Venezuela.

Shearer (2006) reported larvae of *Archilestes* from Black Hills streams. The DSA website reported the following Black Hills records for this species: Custer County.

Lestes Leach

Lestes congener Hagen – Spotted Spreadwing

Descriptions/diagnoses: Walker 1914 (larva), Westfall and May 1996 (adult); OdonataCentral image #43800. Transcontinental from British Columbia and the Northwest Territories south to California, New Mexico, Alabama, and Virginia.

The DSA website reported the following Black Hills records for this species: Custer County, Fall River County, and Weston County. Sims (2012) added a record from Crook County.

Lestes disjunctus Selys – Northern Spreadwing

Descriptions/diagnoses: Walker 1914 (larva), Westfall and May 1996 (adult); OdonataCentral image #43804. Between the two subspecies (*L. d. australis* and *L. d. disjunctus*), *L. disjunctus* is found throughout North America, from Alaska to the Canadian Maritimes, south to California and Florida.

Bick and Hornuff (1972) reported *L. disjunctus* from bog ponds in Crook and Weston counties in the Wyoming portion of the Black Hills. Molinar and Lavigne (1979) subsequently reported this species from Crook County. Paulson et al. (1999) reported this species from cattail-bordered ponds at the Pioneer Rest Area. The DSA website reported the following Black Hills records for this species: Crook County, Fall River County, Lawrence County, Pennington County, French Creek east of Blue Bell Campground in Custer State Park.

Lestes dryas Kirby – Emerald Spreadwing

Descriptions/diagnoses: Kennedy 1915 (larva, as *Lestes uncatus*), Westfall and May 1996 (adult); OdonataCentral image #43807. Northern Holarctic distribution, with populations in Asia and Europe, and distributed throughout North America, except south and east of Virginia and Nebraska.

Molinar and Lavigne (1979) reported this species from Weston County. The DSA website reported the following Black Hills records for this species: Crook County, Weston County, Deerfield Lake.

Lestes unguiculatus Hagen – Lyre-Tipped Spreadwing

Descriptions/diagnoses: Walker 1914 (larva), Westfall and May 1996 (adult); OdonataCentral image #43865. Transcontinental from the Pacific Coast (British Columbia to California) east to the Canadian Maritimes and New England.

Bick and Hornuff (1972) reported *L. unguiculatus* from well-vegetated ponds in Crook and Weston counties in the Wyoming portion of the Black Hills. Molinar and Lavigne (1979) subsequently also reported this species from Crook and Weston counties. The DSA website reported the following Black Hills records for this species: Crook County, Fall River County, Lawrence County, Weston County, French Creek Natural Area in Custer State Park.

Libellulidae

A total of 18 species of libellulid dragonflies are known from the Black Hills, based primarily on county-level distributions from the NPWRC website.

Brechmorhoga **Kirby**

Brechmorhoga mendax **(Hagen) – Pale-Faced Clubskimmer**

Descriptions/diagnoses: Musser 1962 (larva), Needham et al. 2000 (adult); OdonataCentral image #46864. A southwestern species ranging from California and Baja California Sur north and east to Utah, South Dakota, Oklahoma, Texas, and Tamaulipas and Nuevo Leon in Mexico.

I have collected larvae of *Brechmorhoga mendax* from the Fall River near Hot Springs and from Battle Creek near Hermosa. The DSA website reported the following Black Hills records for this species: Hot Brook at Chataqua Park in Hot Springs.

Leucorrhinia **Brittinger**

Leucorrhinia hudsonica **(Selys) – Hudsonian Whiteface**

Descriptions/diagnoses: Walker 1916 (larva), Needham et al. 2000 (adult). Transcontinentally distributed from Alaska and British Columbia to the Canadian Maritimes, south to California, Idaho, Wyoming, Minnesota, Michigan, Pennsylvania, and Maryland.

Molinar and Lavigne (1979) reported this species from Weston County. Paulson et al. (1999) reported this species from Bismark Lake. The DSA website reported the following Black Hills records for this species: Custer County, Weston County.

Leucorrhinia intacta **(Hagen) – Dot-Tailed Whiteface**

Descriptions/diagnoses: Walker 1916 (larva), Needham et al. 2000 (adult); OdonataCentral image #47109. Distributed transcontinentally from British Columbia to the Canadian Maritimes, south to California, Utah, Colorado, Kansas, Missouri, Tennessee, and Virginia.

Molinar and Lavigne (1979) reported this species from Crook County. The DSA website reported the following Black Hills records for this species: Crook County, Custer County, Lawrence County, Pennington County, Cottonwood Springs Recreation Area.

Libellula **Linnaeus**

Grafe et al. (2002) characterized *Libellula* as eurythermal hot, but a weighted average of the literature references in Vieira et al. (2006) would characterize the genus as eurythermal warm.

Libellula forensis **Hagen – Eight-Spotted Skimmer**

Descriptions/diagnoses: Musser 1962 (larva), Needham et al. 2000 (adult); OdonataCentral image #47126. Western North America from British Columbia to South Dakota, south to California, Arizona, New Mexico, and Nebraska.

Molinar and Lavigne (1979) reported this species from Weston County. Paulson et al. (1999) reported this species from Stockade Lake. The DSA website reported the following Black Hills records for this species: Custer County, Weston County, Mud Lake in the Coxes GPA, Mirror Lake GPA. Sims (2012) added a record for Crook County.

Libellula luctuosa Burmeister – Widow Skimmer

Descriptions/diagnoses: Walker and Corbet 1975 (larva), Needham et al. 2000 (adult); OdonataCentral image #47131. Distributed from Nova Scotia, south along the Atlantic seaboard to Georgia and the Gulf Coast, and west to the Dakotas, Colorado, and California, extending north into Oregon and Washington.

Paulson et al. (1999) reported this species from the Cheyenne River at Interstate Highway 90. The DSA website reported the following Black Hills records for this species: Custer County, Cottonwood Springs Recreation Area, Mud Lake in the Coxes GPA, Mirror Lake GPA.

Libellula pulchella Drury – Twelve-Spotted Skimmer

Descriptions/diagnoses: Musser 1962 (larva), Needham et al. 2000 (adult); OdonataCentral image #47136. Broadly distributed across the southern tier of Canadian provinces and US states from coast to coast.

Molinar and Lavigne (1979) reported this species from Crook and Weston counties. The DSA website reported the following Black Hills records for this species: Crook County, Lawrence County, Weston County, French Creek east of Blue Bell Campground in Custer State Park, Hot Brook at Chataqua Park in Hot Springs, Spring Creek at Forest Service Road 305. The USU WCMAFE database includes a record from the Cheyenne River near Spencer, Wyoming.

Libellula quadrimaculata Linnaeus – Four-Spotted Skimmer

Descriptions/diagnoses: Musser 1962 (larva), Needham et al. 2000 (adult); OdonataCentral image #47137. A Holarctic species from northern Europe, Asia, and North America as far south as California, Arizona, New Mexico, Texas, Arkansas, Illinois, Indiana, West Virginia, and North Carolina.

Molinar and Lavigne (1979) reported this species from Crook and Weston counties. The DSA website reported the following Black Hills records for this species: Crook County, Custer County, Fall River County, Lawrence County, Weston County, Twin Lakes (Pennington County).

Libellula saturata Uhler – Flame Skimmer

Descriptions/diagnoses: Musser 1962 (larva), Needham et al. 2000 (adult); OdonataCentral image #47138. A southern species from central Mexico north to Oregon, Idaho, Montana, and South Dakota.

Molinar and Lavigne (1979) reported this species from Crook County. The DSA website reported the following Black Hills records for this species: Hot Brook at Chataqua Park in Hot Springs. I have also collected adults of this distinctive species in Hot Springs, where it was quite abundant along the Fall River through town.

Perithemis Hagen

Perithemis tenera (Say) – Eastern Amberwing

Descriptions/diagnoses: Walker and Corbet 1975 (larva), Needham et al. 2000 (adult); OdonataCentral image #47478. Ontario and Maine south across the United States to Virginia, Texas, New Mexico, and Arizona, south into Mexico to Durango.

The DSA website reported the following Black Hills records for this species: Pennington County.

Plathemis Hagen

Plathemis lydia (Drury) – Common Whitetail

Descriptions/diagnoses: Walker and Corbet 1975 (larva), Needham et al. 2000 (adult); OdonataCentral image #47488. Distributed transcontinentally from British Columbia to the Maritimes, south through all the contiguous United States to Nuevo Leon, Mexico.

The DSA website reported the following Black Hills records for this species: Custer County, Hot Brook at Chataqua Park in Hot Springs, Puddle Dam #1. Molinar and Lavigne (1979) reported this species, as *Libellula lydia*, from Campbell County, Wyoming, just west of Crook and Weston counties.

Sympetrum Newman

Sympetrum corruptum (Hagen) – Variegated Meadowhawk

Descriptions/diagnoses: Walker 1917, Musser 1962 (larva); Needham et al. 2000 (adult); OdonataCentral image #47560. A Holarctic species, with distribution in northern Asia and the the Russian Far East, migratory populations along the Pacific Coast from British Columbia south to Honduras, and a transcontinental distribution to the Canadian Maritimes, the Atlantic seaboard, Florida, and the Gulf Coast.

Molinar and Lavigne (1979) reported this species from Crook and Weston counties. The DSA website reported the following Black Hills records for this species: Crook County, Lawrence County, Weston County, Cottonwood Springs Recreation Area.

Sympetrum costiferum (Hagen) – Saffron-Winged Meadowhawk

Descriptions/diagnoses: Walker 1917, Musser 1962 (larva); Needham et al. 2000 (adult); OdonataCentral image #47561. Transcontinentally distributed from British Columbia to the Maritimes, south to California, Utah, New Mexico, Texas, Missouri, Ohio, and Pennsylvania, north into the Northwest Territories.

The DSA website reported the following Black Hills records for this species: Fall River County, Pennington County. Sims (2012) added a record of this species from Crook County.

Sympetrum danae (Sulzer) – Black Meadowhawk

Descriptions/diagnoses: Musser 1962 (larva, as *S. danae* and *S. internum*), Needham et al. 2000 (adult); OdonataCentral image #47564. Transcontinental from Alaska, British Columbia, and the Pacific Coast states east to the Maritimes, New York, Missouri, and New Mexico.

The DSA website reported the following Black Hills records for this species: Lawrence County, Pennington County.

Sympetrum internum Montgomery – Cherry-Faced Meadowhawk

Descriptions/diagnoses: Musser 1962 (larva, as *S. danae* and *S. internum*), Needham et al. 2000 (adult); OdonataCentral image #47583. Transcontinental from Alaska, British Columbia, and the Pacific Coast states east to the Maritimes, New England, Ohio, Missouri, and New Mexico.

Molinar and Lavigne (1979) reported this species from Crook County. The DSA website reported the following Black Hills records for this species: Crook County, Custer County, Fall River County, Lawrence County, Pennington County.

Sympetrum madidum (Hagen) – Red-Veined Meadowhawk

Descriptions/diagnoses: Cannings 1981 (larva), Needham et al. 2000 (adult). Distributed from the Yukon and Northwest Territories south to California, Utah, Colorado, Nebraska, and Missouri.

Molinar and Lavigne (1979) reported this species from Crook County. Subsequently, the DSA website reported only that record as well.

Sympetrum obtrusum (Hagen) – White-Faced Meadowhawk

Descriptions/diagnoses: Walker 1917 (larva), Needham et al. 2000 (adult); OdonataCentral image #47594. Another transcontinental species in Canada from British Columbia to the Maritimes, south in the United States to California, Utah, Colorado, Kansas, Iowa, Kentucky, and North Carolina.

Molinar and Lavigne (1979) reported this species from Crook County. Paulson et al. (1999) reported this species from cattail-bordered ponds at the Pioneer Rest Area. The DSA website reported the following Black Hills records for this species: Crook County, Fall River County, Pennington County, Bismark Lake in Custer County.

Sympetrum pallipes (Hagen) – Striped Meadowhawk

Descriptions/diagnoses: Walker 1917 (larva), Needham et al. 2000 (adult); OdonataCentral image #47597. Western North America from British Columbia south along the coast to California, and east to Saskatchewan, the Dakotas, Nebraska, and Texas. Notes on collection locations for this species on the DSA website noted that the Fall River site was small, warm, and spring-fed, with a murky/silty-gray coloration.

Molinar and Lavigne (1979) reported this species from Crook and Weston counties. The DSA website reported the following Black Hills records for this species: Crook County, Lawrence County, Weston County, French Creek Natural Area in Custer State Park, Fall River at Allen Ranch Campground, Smith Dam #1.

Sympetrum rubicundulum (Say) – Ruby Meadowhawk

Descriptions/diagnoses: Walker 1917, Musser 1962 (larva); Needham et al. 2000 (adult); OdonataCentral image #47603. Eastern North America from the Canadian Maritimes south along the Atlantic seaboard to Georgia, and west to Manitoba, the Dakotas, Colorado, and Alabama.

The DSA website reported the following Black Hills records for this species: Fall River County, Pennington County, French Creek Natural Area in Custer State Park. I have collected an adult male of this species on the north side of Rapid City.

Sympetrum semicinctum (Say) – Band-Winged Meadowhawk

Descriptions/diagnoses: Walker 1917 (larva), Needham et al. 2000 (adult); OdonataCentral image #47606. *Sympetrum occidentale* Bartenev and its varieties are considered here to be a western subspecies of *S. semicinctum*, although I am not going to make any formal taxonomic synonymy here. With records from *S. occidentale*, this is a transcontinental species from British Columbia to the Maritimes, south through the United States to California, Arizona, New Mexico, Kansas, Missouri, Alabama, Tennessee, and North Carolina. *Sympetrum occidentale*, in its stricter sense, is western with a distribution eastward to Minnesota, Missouri, and Oklahoma. The form (or subspecies) named *fasciatum* Walker is the form that is found in the Black Hills.

Molinar and Lavigne (1979) reported this species, as *Sympetrum occidentale*, from Weston County. Paulson et al. (1999) reported this species, also as *S. occidentale*, from the Cheyenne River at I-90. The DSA website reported the following Black Hills records for this species: Fall River County,

Lawrence County, Pennington County, Weston County, Lake Ranch (Crook Co.). Sims (2012) added an additional record for Crook County.

Tramea Hagen

Tramea lacerata Hagen – Black Saddlebags

Descriptions/diagnoses: Bick 1951 (early instar larva), Musser 1962 (late instar larva), Needham et al. 2000 (adult); OdonataCentral image #47658. Transcontinental in the United States with a handful of states not reporting this species, British Columbia to Quebec in Canada, south into Mexico, the Caribbean islands, including Cuba, and Hawaii.

Paulson et al. (1999) reported this species from the Cheyenne River at I-90. The DSA website reported the following Black Hills records for this species: Custer County, Fall River County, Pennington County.

Odonata Traits Table

Table 4:   Odonata Traits Table

Species	TSN	Seen	Tolerance Value	FFG	Habit
Aeshnidae	101596	-	3	PR	CB
Aeshna	101602	-	5	PR	CB
Aeshna canadensis	185977	-	5	PR	CB
Aeshna constricta	101609	-	5	PR	CB
Aeshna interrupta	593419	-	5	PR	CB
Aeshna palmata	592674	-	5	PR	CB
Aeshna umbrosa	101605	-	5	PR	CB
Anax	101597	-	8	PR	CB
Anax junius	101598	-	8	PR	CB
Rhionaeschna	721973	-	5	PR	CB
Rhionaeschna californica	722023	-	5	PR	CB
Rhionaeschna multicolor	722025	-	5	PR	CB
Calopterygidae	102043	-	--	PR	CB
Calopteryx	102052	-	6	PR	CB
Calopteryx aequabilis	102056	-	6	PR	CB
Calopteryx maculata	102055	-	6	PR	CB
Hetaerina	102048	-	5	PR	CB
Hetaerina americana	102050	-	5	PR	CB
Coenagrionidae	102077	+	9	PR	CB
Amphiagrion	102093	-	5	PR	CB
Amphiagrion abbreviatum	102094	-	5	PR	CB
Amphiagrion saucium	102095	-	5	PR	CB
Argia	102139	+	7	PR	CN
Argia alberta	102153	/	7	PR	CN
Argia apicalis	102140	/	7	PR	CN
Argia emma	102142	/	7	PR	CN
Argia fumipennis	102143	/	7	PR	CN
Argia immunda	592408	/	7	PR	CN
Argia lugens	592426	/	7	PR	CN
Argia moesta	102146	/	7	PR	CN
Argia plana	102151	/	7	PR	CN
Argia vivida	102150	/	7	PR	CN
Coenagrion	102155	-	9	PR	CB
Coenagrion resolutum	102156	-	9	PR	CB
Enallagma	102102	-	9	PR	CB
Enallagma anna	592485	-	9	PR	CB
Enallagma annexum	722162	-	9	PR	CB
Enallagma antennatum	102103	-	9	PR	CB
Enallagma basidens	102125	-	9	PR	CB
Enallagma boreale	102121	-	9	PR	CB
Enallagma carunculatum	102123	-	9	PR	CB
Enallagma civile	102122	-	9	PR	CB
Enallagma clausum	102132	-	9	PR	CB
Enallagma ebrium	102128	-	9	PR	CB
Enallagma exsulans	102112	-	9	PR	CB
Enallagma hageni	102129	-	9	PR	CB
Enallagma praevarum	102127	-	9	PR	CB

Species	TSN	Seen	Tolerance Value	FFG	Habit
Ischnura	102078	-	9	PR	CB
Ischnura cervula	102080	-	9	PR	CB
Ischnura damula	102086	-	9	PR	CB
Ischnura perparva	102087	-	9	PR	CB
Ischnura verticalis	102079	-	9	PR	CB
Nehallenia	102135	-	9	PR	CB
Nehalennia irene	102137	-	9	PR	CB
Corduliidae	102020	-	2	PR	--
Epitheca	102035	-	2	PR	CB
Epitheca costalis	593049	-	2	PR	CB
Epitheca cynosura	185986	-	2	PR	CB
Epitheca petechialis	593050	-	2	PR	CB
Epitheca spinigera	185987	-	2	PR	CB
Somatochlora	101947	-	9	PR	SP
Somatochlora ensigera	101965	-	9	PR	SP
Somatochlora minor	101958	-	9	PR	SP
Gomphidae	101664	+	1	PR	BU
Arigomphus	101770	-	1	PR	BU
Arigomphus cornutus	101772	-	1	PR	BU
Erpetogomphus	101725	-	4	PR	BU
Erpetogomphus designatus	101726	-	4	PR	BU
Gomphus	101665	-	1	PR	BU
Gomphus graslinellus	101700	-	1	PR	BU
Ophiogomphus	101738	+	1	PR	BU
Ophiogomphus severus	101741	+	1	PR	BU
Stylurus	206626	-	1	PR	BU
Stylurus intricatus	593020	-	1	PR	BU
Lestidae	102058	-	9	PR	CB
Archilestes	102059	-	9	PR	CB
Archilestes grandis	102060	-	9	PR	CB
Lestes	102061	-	9	PR	CB
Lestes congener	102062	-	9	PR	CB
Lestes disjunctus	102063	-	9	PR	CB
Lestes dryas	102066	-	9	PR	CB
Lestes unguiculatus	102068	-	9	PR	CB
Libellulidae	101797	-	9	PR	SP
Brechmorhoga	101834	-	9	PR	SP
Brechmorhoga mendax	101835	-	9	PR	SP
Leucorrhinia	101885	-	9	PR	CB
Leucorrhinia hudsonica	101887	-	9	PR	CB
Leucorrhinia intacta	101888	-	9	PR	CB
Libellula	101893	-	9	PR	SP
Libellula forensis	101911	-	9	PR	SP
Libellula luctuosa	101894	-	9	PR	SP
Libellula pulchella	101895	-	9	PR	SP
Libellula quadrimaculata	101896	-	9	PR	SP
Libellula saturata	101916	-	9	PR	SP
Perithemis	101803	-	9	PR	SP
Perithemis tenera	101804	-	9	PR	SP
Plathemis	101808	-	9	PR	SP

Species	TSN	Seen	Tolerance Value	FFG	Habit
Plathemis lydia	101809	-	9	PR	SP
Sympetrum	101976	-	9	PR	SP
Sympetrum corruptum	101978	-	9	PR	SP
Sympetrum costiferum	101980	-	9	PR	SP
Sympetrum danae	101991	-	9	PR	SP
Sympetrum internum	101982	-	9	PR	SP
Sympetrum madidum	101993	-	9	PR	SP
Sympetrum obtrusum	101981	-	9	PR	SP
Sympetrum pallipes	593348	-	9	PR	SP
Sympetrum rubicundulum	101983	-	9	PR	SP
Sympetrum semicinctum	101990	-	9	PR	SP
Tramea	101818	-	9	PR	SP
Tramea lacerata	101822	-	9	PR	SP

Species	Temperature Range	Lotic	Lentic	Current	Substrate	Voltinism
Aeshnidae	--	--	--	--	--	--
Aeshna	E: warm	2.7	1.1	0.8	7.0	Mero
Aeshna canadensis	E: warm	2.7	1.1	0.8	7.0	Mero
Aeshna constricta	E: warm	2.7	1.1	0.8	7.0	Mero
Aeshna interrupta	E: warm	2.7	1.1	0.8	7.0	Mero
Aeshna palmata	E: warm	2.7	1.1	0.8	7.0	Mero
Aeshna umbrosa	E: warm	2.0	3.0	1.0	7.0	Mero
Anax	--	2.8	1.1	1.0	7.0	Uni
Anax junius	--	2.6	4.0	1.0	7.0	Uni
Rhionaeschna	E: warm	2.7	1.1	0.8	7.0	Mero
Rhionaeschna califomica	E: warm	2.7	1.1	0.8	7.0	Mero
Rhionaeschna multicolor	E: warm	2.7	1.1	0.8	7.0	Mero
Calopterygidae	E: cool	--	--	--	--	--
Calopteryx	E: cool	3.6	1.0	2.0	2.6	Mero
Calopteryx aequabilis	E: cool	3.8	1.0	2.0	0.0	Mero
Calopteryx maculata	E: cool	2.0	1.0	2.0	2.6	Mero
Hetaerina	E: cool	3.8	4.0	2.0	2.5	Mero
Hetaerina americana	E: cool	3.0	4.0	2.0	2.5	Mero
Coenagrionidae	E: hot	--	--	--	--	--
Amphiagrion	E: hot	3.6	1.0	--	--	--
Amphiagrion abbreviatum	E: hot	3.6	1.0	--	--	--
Amphiagrion saucium	E: hot	3.6	1.0	--	--	--
Argia	E: warm	3.6	1.0	1.8	5.7	Uni
Argia alberta	E: warm	3.6	1.0	1.8	5.7	Uni
Argia apicalis	E: warm	4.0	3.7	1.8	7.0	Uni
Argia emma	E: warm	3.0	4.0	1.8	7.0	Uni
Argia fumipennis	E: warm	3.5	4.0	1.8	7.0	Uni
Argia immunda	E: warm	3.6	1.0	1.8	5.7	Uni
Argia lugens	E: warm	3.6	1.0	1.8	5.7	Uni
Argia moesta	E: warm	3.8	4.0	1.8	7.0	Uni
Argia plana	E: warm	3.6	1.0	1.8	5.7	Uni
Argia vivida	E: warm	3.0	4.0	1.8	7.0	Uni
Coenagrion	E: hot	4.0	3.0	--	--	Uni
Coenagrion resolutum	E: hot	4.0	3.0	--	--	Uni
Enallagma	E: hot	3.7	1.0	1.0	4.0	Uni
Enallagma anna	E: hot	3.7	1.0	1.0	4.0	Uni
Enallagma annexum	E: hot	3.7	1.0	1.0	4.0	Uni
Enallagma antennatum	E: hot	3.7	1.0	1.0	4.0	Uni
Enallagma basidens	E: hot	3.7	1.0	1.0	4.0	Uni
Enallagma boreale	E: hot	3.7	1.0	1.0	4.0	Uni
Enallagma carunculatum	E: hot	5.0	5.0	1.0	4.0	Uni
Enallagma civile	E: hot	3.0	3.7	1.0	1.0	Uni
Enallagma clausum	E: hot	3.7	1.0	1.0	4.0	Uni
Enallagma ebrium	E: hot	3.7	1.0	1.0	4.0	Uni
Enallagma exsulans	E: hot	3.7	1.0	1.0	4.0	Uni
Enallagma hageni	E: hot	3.7	3.0	1.0	4.0	Uni
Enallagma praevarum	E: hot	3.7	1.0	1.0	4.0	Uni
Ischnura	E: hot	3.8	1.0	1.0	2.5	Uni
Ischnura cervula	E: hot	3.8	1.0	1.0	2.5	Uni
Ischnura damula	E: hot	3.8	1.0	1.0	2.5	Uni

Species	Temperature Range	Lotic	Lentic	Current	Substrate	Voltinism
Ischnura perparva	E: hot	3.8	1.0	1.0	2.5	Uni
Ischnura verticalis	E: hot	3.8	1.0	1.0	2.5	Uni
Nehallenia	E: hot	5.0	1.1	0.5	n/a	n/a
Nehalennia irene	E: hot	5.0	1.1	0.5	n/a	n/a
Corduliidae	--	--	--	--	--	--
Epitheca	--	4.0	1.1	1.0	--	Mero
Epitheca costalis	--	4.0	1.1	1.0	--	Mero
Epitheca cynosura	--	4.0	4.0	1.0	--	Mero
Epitheca petechialis	--	4.0	1.1	1.0	--	Mero
Epitheca spinigera	--	4.0	1.1	1.0	--	Mero
Somatochlora	--	3.4	1.1	--	0.3	Mero
Somatochlora ensigera	--	3.4	1.1	--	0.3	Mero
Somatochlora minor		3.4	1.1		0.3	Mero
Gomphidae	E: hot	--	--	--	0.5	--
Arigomphus	E: hot	4.0	1.2	0.5	0.4	Mero
Arigomphus cornutus	E: hot	4.0	1.2	0.5	0.4	Mero
Erpetogomphus	E: hot	4.0	1.2	1.0	0.7	Mero
Erpetogomphus designatus	E: hot	4.0	1.2	1.0	0.7	Mero
Gomphus	E: hot	4.3	1.1	1.0	0.5	Mero
Gomphus graslinellus	E: hot	4.3	1.1	1.0	0.5	Mero
Ophiogomphus	E: warm	3.4	5.0	1.0	1.4	Mero
Ophiogomphus severus	E: warm	3.0	5.0	1.0	4.0	Mero
Stylurus	E: hot	4.0	4.0	--	0.1	Mero
Stylurus intricatus	E: hot	4.0	4.0	n/a	0.0	Mero
Lestidae	--	--	--	--	--	Uni
Archilestes	--	3.4	1.1	1.0	--	Uni
Archilestes grandis	--	3.0	3.4	1.0	--	Uni
Lestes	E: hot	4.0	1.1	--	--	Uni
Lestes congener	E: hot	4.0	1.1	--	--	Uni
Lestes disjunctus	E: hot	4.0	1.1	--	--	Uni
Lestes dryas	E: hot	4.0	1.1	--	--	Uni
Lestes unguiculatus	E: hot	4.0	1.1	--	--	Uni
Libellulidae	E: hot	--	--	--	--	--
Brechmorhoga	E: hot	3.7	--	--	1.3	--
Brechmorhoga mendax	E: hot	4.0	--	--	0.0	--
Leucorrhinia	E: hot	4.0	1.1	--	--	--
Leucorrhinia hudsonica	E: hot	4.0	1.1	--	--	--
Leucorrhinia intacta	E: hot	4.0	1.1	--	--	--
Libellula	E: hot	3.4	1.0	--	0.4	Uni
Libellula forensis	E: hot	3.4	1.0	--	0.4	Uni
Libellula luctuosa	E: hot	3.4	1.0	--	0.4	Uni
Libellula pulchella	E: hot	3.0	3.0	--	0.5	Uni
Libellula quadrimaculata	E: hot	3.4	1.0	--	0.4	Uni
Libellula saturata	E: hot	3.4	1.0	--	0.4	Uni
Perithemis	E: hot	4.0	1.1	--	0.0	Uni
Perithemis tenera	E: hot	4.0	3.2	--	0.0	Uni
Plathemis	E: hot	--	--	--	--	--
Plathemis lydia	E: hot	1.0	3.0	--	0.0	--
Sympetrum	E: hot	--	1.0	--	--	--
Sympetrum corruptum	E: hot	--	1.0	--	--	--

Species	Temperature Range	Lotic	Lentic	Current	Substrate	Voltinism
Sympetrum costiferum	E: hot	--	1.0	--	--	--
Sympetrum danae	E: hot	--	1.0	--	--	--
Sympetrum internum	E: hot	--	3.0	--	--	--
Sympetrum madidum	E: hot	--	1.0	--	--	--
Sympetrum obtrusum	E: hot	--	1.0	--	--	--
Sympetrum pallipes	E: hot	--	1.0	--	--	--
Sympetrum rubicundulum	E: hot	--	1.0	--	--	--
Sympetrum semicinctum	E: hot	--	1.0	--	--	--
Tramea	E: hot	4.0	1.1	0.5	0.0	--
Tramea lacerata	E: hot	4.0	1.1	0.5	0.0	--

Semi-Aquatic Orthoptera

Introduction

Nearly all of the approximately 1,200 species of North American Orthoptera reside in terrestrial habitats, and even of those that could be considered to be semi-aquatic, their association with water is usually that they are found feeding exclusively or primarily on aquatic or semi-aquatic plants (above the water line) or inhabit burrows in wet soil along water bodies. Those that jump into a stream when startled, only to be eaten by a hungry brook trout, do not count! Outside of North America, a few Orthoptera are known that regularly swim in the water to reach aquatic macrophytes, which they use for food or oviposition.

Bland (2008) described the biology of a few families of Orthoptera with semiaquatic representatives; although the families occur in the Black Hills, most of the semiaquatic representatives do not. These families include the Acrididae, Tetrigidae, Tridactylidae, Tettigoniidae, Gryllidae, and Gryllotalpidae.

The most aquatic of these groups is the Tridactylidae, or pygmy mole grasshoppers. These insects have flattened, moveable plates on their hind tibiae for swimming and skating on the water surface. Most species of Tridactylidae burrow in sandy banks.

Overview of the Aquatic Orthoptera in the Black Hills

There are no known checklists of aquatic Orthoptera for either South Dakota or the Black Hills, and I have not seen reports of semi-aquatic Orthoptera from biomonitoring efforts in the Black Hills. The only report of one of these semi-aquatic Orthoptera that I can find is a species of *Tetrix* I collected on the banks of Whitewood Creek just north of the Black Hills.

Furthermore, because the Orthoptera are semi-aquatic rather than fully aquatic, species trait information relating to their function within streams is non-existent. Therefore, a traits table has not been constructed for the semi-aquatic Orthoptera.

Species Accounts

Tetrigidae

Tetrix **Latreille**

As mentioned above, I have collected one specimen of a species of *Tetrix* on Whitewood Creek just north of the Black Hills.

Orthoptera Traits Table

Table 5: Orthoptera Traits Table

Species	TSN	Seen	Tolerance Value	FFG	Habit
Tetrigidae	102162	-	--	SH	SP
Tetrix	102171	-	--	SH	SP

Species	Temperature Range	Lotic	Lentic	Current	Substrate	Voltinism
Tetrigidae	--	--	--	--	--	--
Tetrix	--	--	--	--	--	--

Plecoptera

Introduction

The stoneflies are another group, along with the mayflies and caddisflies, that are almost definitive of aquatic entomology. Both nymphs and adults are used as models for fly-fishing patterns, and stoneflies can comprise a large portion of the diets of insectivorous fish in streams. Unlike mayflies and caddisflies, (generally!), stonefly taxa are far less common on the Plains than in mountainous regions, so as streams exit the Black Hills, their stonefly fauna tends to decrease considerably from what it was in the Hills. This isn't a reflection of poor water quality, although stoneflies are the "P" in the EPT taxa metric and are usually good indicators of water quality. Rather, stoneflies just generally prefer faster flowing water and cooler temperatures, and the higher dissolved oxygen concentrations that accompany them, that occur much more frequently in the mountains. That isn't to say that stoneflies do not occur in streams on the Plains or in the transition from mountains to plains; in fact, there are some species that occur only in the Plains.

Many stoneflies are predaceous, although some groups are shredders and become much more abundant in headwaters reaches, in concert with the Stream Continuum Concept (Vannote et al. 1980).

Overview of the Plecoptera in the Black Hills

The most comprehensive treatment of the stoneflies of the Black Hills is Huntsman et al. (1999), in which 27 species in 22 genera and 6 families were reported. It is interesting to note that in 1994, Huntsman's Ph.D. advisor, Dr. Richard W. Baumann, stated that "the most poorly known states are North Dakota, Iowa and eastern South Dakota" in his plea for specimens that could help Huntsman in his research on the Plecoptera of the Black Hills (Baumann 1994). Stewart and Stark (2006) reported state-level distributions for every species of North American stonefly, while Kondratieff and Baumann (2000) reported them at the county level through the NPWRC website, although the data are incomplete. No data from Kondratieff and Baumann (2000) or Stewart and Stark (2002) added substantially to that of Huntsman et al. (1999).

Species Accounts

Capniidae

Capniidae are represented in the Black Hills by six species. In most biomonitoring efforts in the Black Hills capniid larvae have been left at the family level, and the angling community refers to all capniids as "little snowflies". This is due to the general difficulty many taxonomists have with identification of capniids and the generally immature stage of the specimens at time of collection, which is generally in late summer (August and September). However, *Paracapnia angulata* has been identified in biomonitoring samples.

Capnia Pictet

The genus *Capnia* is, at present, an "artificial assemblage of taxa", very much in need of revision and resulting reassignment of most of its species to existing or new genera (Murányi et al. 2014). So far, only one species group (the *decepta* group) has been reassigned, to *Arsapnia* Banks, which does not occur in the Black Hills. The two Black Hills species of *Capnia*, as currently understood, are *C. confusa* and *C. gracilaria*, in the *vernalis* and *gracilaria* species groups, respectively. Based on the structure of the adult male terminalia, Murányi et al. (2014) stated that these two species groups probably belong to the genus *Mesocapnia* Raušer, so they will likely be reassigned in the future.

Capnia confusa Claassen

Descriptions/diagnoses: Dosdall & Lehmkuhl 1979 (larva), Nelson & Baumann 1989 (adult). "Probably the most widespread member of the genus in North America" (Nelson and Baumann 1989), *C. confusa* is distributed from Alaska south to California and east to New Mexico, South Dakota, and Manitoba.

Nelson and Baumann (1989) reported this species from Lawrence County, and Huntsman et al. (1999) reported this species from Deer Creek, East Spearfish Creek, Grace Coolidge Creek, Iron Creek, Little Spearfish Creek, Rapid Creek, South Fork Castle Creek, Spearfish Creek, Sunday Gulch, and Whitewood Creek in South Dakota.

Capnia gracilaria Claassen

Descriptions/diagnoses: Ricker 1943 (larva), Needham and Claassen 1925, Nelson and Baumann 1989 (adult). Distributed from Alaska and British Columbia east to Manitoba, south to Baja California, New Mexico, and South Dakota. The Black Hills are the easternmost limit for *C. gracilaria*.

Nelson and Baumann (1989) reported this species from Lawrence County, and Huntsman et al. (1999) reported this species from Burnt Fork, Deer Creek, East Spearfish Creek, Elk Creek, Elkhorn Spring, Iron Creek, Little Spearfish Creek, Pine Creek, Spearfish Creek, Spokane Creek, Sunday Gulch, Willow Creek, and a tributary to Bear Butte Creek.

Eucapnopsis Okamoto

Eucapnopsis brevicauda Claassen

Descriptions/diagnoses: Ricker 1943, Stewart and Stark 2002 (larva), Needham and Claassen 1925 (adult, as *Capnia brevicauda*). Western Nearctic from Alaska south to California and east to New Mexico, South Dakota, and Montana.

Huntsman et al. (1999) reported this species from Iron Creek, Pine Creek, Rapid Creek, Spearfish Creek, Sunday Gulch, Whitewood Creek, and Willow Creek in South Dakota.

Isocapnia Banks

Isocapnia integra Hanson

Descriptions/diagnoses: Zenger and Baumann 2004 (adult), larva is undescribed.

The immature stages of *Isocapnia* are rarely collected, probably because of their hyporheic habitat (Stewart and Stark 2002). At the time of writing of Huntsman et al. (1999), the specific identity of an interesting, brachypterous *Isocapnia* female had not yet been determined, being limited by the sex of the specimen. Zenger and Baumann (2004) later reported this specimen from Spearfish Creek above its confluence with the Redwater River as *I. integra*.

Paracapnia Hanson

Paracapnia angulata Hanson

Descriptions/diagnoses: Hanson 1961 (adult); Stewart and Stark 2002 (larva). Huntsman et al. (1999) describe the distribution as primarily from Quebec south to Tennessee, but with isolated populations in boreal streams in Saskatchewan and Manitoba, the Black Hills of South Dakota and Wyoming, the Central Rocky Mountains in Wyoming and Colorado, and the Ozark Mountains in Oklahoma.

Huntsman et al. (1999) reported this species from Battle Creek, Bear Butte Creek, Boxelder Creek, Burnt Fork, Elk Creek, Grizzly Bear Creek, Hay Creek, Horse Creek, Iron Creek, Jim Creek, Little Elk Creek, Middle Fork Boxelder Creek, Palmer Creek, Rapid Creek, Spring Creek, and Sunday Gulch in South Dakota, as well as Beaver Creek in Wyoming.

Utacapnia Gaufin

Utacapnia lemoniana (Nebeker and Gaufin)

Descriptions/diagnoses: Nebeker and Gaufin 1965 (adult, larva); Stewart and Stark 2002 (larva). Central Rocky Mountains from Idaho and Nevada east to South Dakota and Colorado. Stewart and Stark (2002) also list Alaska in the distribution. The Black Hills are the easternmost limit for *U. lemoniana*.

Huntsman et al. (1999) reported this species from Burnt Fork, Mercedes Gulch, North Fork Castle Creek, Willow Creek, and a tributary to Bear Butte Creek in South Dakota.

Chloroperlidae

The Black Hills represent the easternmost limits for the range of all five species of Chloroperlidae found there. The temperature preference for all chloroperlids is eurythermal warm, based on Grafe et al. (2002); however, preference based on data compiled for Vieira et al. (2006) suggest that all species of Chloroperlidae in the Black Hills should be classified as eurythermal cool. Most of the Chloroperlidae are known as "sallflies" to anglers.

Alloperla Banks

Jurgens (1968) identified *Alloperla* at the genus level from French Creek, Spearfish Creek, and Castle Creek, and, likewise, Drewes (1984) identified Alloperla at the genus level from Slate Creek. These are the only records of this genus from the Black Hills. The general distribution of this genus (many species east of the Mississippi River and a few species west of the main Rocly Mountain cordillera) causes me to wonder if these were in fact *Alloperla* or another chloroperlid genus.

Paraperla Banks

Paraperla frontalis (Banks)

Descriptions/diagnosis: Stewart and Stark 2002 (larva), Needham and Claassen 1925 (adult). Distributed from Alaska and the Yukon south to California, New Mexico, and South Dakota.

Huntsman et al. (1999) reported this species from East Spearfish Creek, Grace Coolidge Creek, Grizzly Bear Creek, Iron Creek, Pine Creek, Rapid Creek, Spearfish Creek, Sunday Gulch, and Whitewood Creek.

Suwallia Ricker

Suwallia lineosa (Banks)

Descriptions/diagnosis: Dosdall and Lehmkuhl 1979 (larva, parts only), Needham and Claassen 1925 (adult, as *Alloperla lineosa*). Distributed from Alaska, the Yukon, and Saskatchewan south to Oregon, Wyoming, and South Dakota. .

Huntsman et al. (1999) reported this species from Little Spearfish Creek at the Timon Campground and Spearfish Creek at the Botany Bay Picnic Area and at the Spearfish Water Diversion.

Sweltsa Ricker

Shearer (2006) reported larvae of unidentified species of *Sweltsa* from Black Hills streams. The USU WCMAFE database includes a record from Spruce Gulch Creek. I saw this species in most streams from the northern Black Hills in biomonitoring samples.

Sweltsa borealis (Banks)

Descriptions/diagnosis: Claassen 1931 (larva), Needham and Claassen 1925 (adult, as *Alloperla borealis*). Found in the Rocky Mountains from Alaska and the Yukon south to California, New Mexico, and South Dakota.

Huntsman et al. (1999) reported this species from Beaver Creek, Iron Creek, Pine Creek, Spearfish Creek, and a tributary to Bear Butte Creek.

Sweltsa coloradensis (Banks)

Descriptions/diagnosis: Surdick 1985 (larva), Needham and Claassen 1925 (adult, as *Alloperla coloradensis*). Distributed from the Yukon south through British Columbia to California, Arizona, New Mexico, and South Dakota.

Huntsman et al. (1999) reported this species from streams throughout the Black Hills. Streams included Beaver Creek, Boxelder Creek, Deer Creek, East Spearfish Creek, Grace Coolidge Creek, Grizzly Bear Creek, Iron Creek, Jenny Gulch, Little Spearfish Creek, Little Squaw Creek, Palmer Creek, Pine Creek, Rapid Creek, South Fork Boxelder Creek, South Fork Rapid Creek, Spearfish Creek, Spokane Creek, Sunday Gulch, Whitetail Creek, Whitewood Creek, and Willow Creek in South Dakota.

Triznaka Ricker

Triznaka pintada (Ricker)

Descriptions/diagnosis: Baumann and Kondratieff 2008 (adult); Stewart and Stark 2002 (larvae). Distributed in the contiguous United States from the Pacific coast states east to Idaho, South Dakota, Colorado, and New Mexico.

Huntsman et al. (1999) reported this species from streams throughout the Black Hills. Streams included Beaver Creek, Boxelder Creek, Burnt Fork, Flynn Creek, Grace Coolidge Creek, Grizzly Bear Creek, Iron Creek, Jenny Gulch, Jim Creek, Little Elk Creek, Little Spearfish Creek, Little Squaw Creek, Middle Fork Boxelder Creek, Newton Fork, Palmer Creek, Pine Creek, South Fork Boxelder Creek, Spearfish Creek, Spokane Creek, Spring Creek, Sunday Gulch, Whitewood Creek, Willow Creek, and tributaries to Sylvan Lake, Bear Butte Creek, Beaver Creek, and Whitewood Creek in South Dakota, as well as Blacktail Creek, Cold Creek, Houston Creek(?), and Whitelaw Creek in Wyoming.

Leuctridae

Only one species of Leuctridae is found in the Black Hills, at the easternmost limit of its range.

Paraleuctra Hanson

Paraleuctra vershina Gaufin & Ricker

Descriptions/diagnosis: Stewart and Stark 2002, Stewart and Harper 1996 (larva); Stark and Kyzar 2000 (adult). Rocky Mountains from Alaska south to California, New Mexico, and South Dakota. Most leuctrids are known as "needleflies" by anglers, but this species (along with some nemourids) is commonly called the "tiny winter black".

Huntsman et al. (1999) reported this species from Bear Butte Creek(?), East Spearfish Creek, Elkhorn Spring, Grace Coolidge Creek, Grizzly Bear Creek, Horse Creek, Iron Creek, Palmer Creek, Pine Creek, Spearfish Creek, Sunday Gulch, Whitetail Creek, Whitewood Creek, and Willow Creek in South Dakota.

Nemouridae

Nemouridae are represented by five species in the Black Hills. Along with the leuctrid *Paraleuctra vershina*, nemourids are also known as "tiny winter blacks" by anglers.

Amphinemura Ris

Amphinemura banksi Baumann and Gaufin

Descriptions/diagnoses: Stewart and Stark 2002 (larva), Needham and Claassen 1925 (adult, as *Nemoura venusta*). Distributed throughout the central and southern Rocky Mountains, the Black Hills is the easternmost range limit for this species (Huntsman et al. 1999).

Huntsman et al. (1999) reported the collection of *A. banksi* throughout the Black Hills, including Beaver Creek, Burnt Fork, Castle Creek, Deer Creek, Ditch Creek, East Spearfish Creek, Elk Creek, Flynn Creek, Hay Creek, Iron Creek, Jenny Gulch, Little Spearfish Creek, Newton Fork Castle Creek, Palmer Creek, Pine Creek, Rapid Creek, South Fork Boxelder Creek, South Fork Rapid Creek, Spearfish Creek, Spring Creek, Sunday Gulch, Willow Creek, a stream entering Sylvan Lake, and a tributary to Beaver Creek in South Dakota, as well as Beaver Creek, Blacktail Creek, Cold Creek, Cold Springs Creek, Lytle Creek, and Whitelaw Creek in Wyoming. The USU WCMAFE database includes records (at the genus level) from Spearfish Creek, Whitewood Creek, and Spruce Gulch Creek.

Malenka Ricker

Malenka coloradensis (Banks)

Descriptions/diagnoses: Claassen 1931, Shepard and Stewart 1983 (larva), Needham and Claassen 1925 (adult, as *Nemoura coloradensis*). Huntsman et al. (1999) reported that the Black Hills represents the northeastern range limit for this generally southern Rocky Mountain species (Arizona north to Idaho and Montana).

Drewes (1984) reported *Malenka* at the genus level from Slate Creek. Huntsman et al. (1999) reported the collection of *M. coloradensis* throughout the Black Hills, including Elkhorn Spring, French Creek(?), Iron Creek, Little Spearfish Creek, South Fork Rapid Creek, Spearfish Creek, Whitewood Creek, a stream entering Sylvan Lake, and tributaries to Bear Butte Creek and Whitewood Creek in South Dakota, as well as Blacktail Creek in Wyoming. The USU WCMAFE database includes a record (at the genus level) from Spruce Gulch Creek. Shearer (2006) reported larvae of *Malenka* from Black Hills streams.

Nemoura Latreille

Nemoura trispinosa Claassen

Descriptions/diagnoses: Stewart and Stark 2002 (larva), Claassen 1923 (adult). Known from Manitoba and the Wyoming Black Hills, east to Illinois, New York, and Labrador (Huntsman et al. 1999).

Huntsman et al. (1999) reported *N. trispinosa* from Sylvan Lake, Burnt Fork, Spearfish Creek, Whitewood Creek(?), and tributaries to Bear Butte Creek and Whitewood Creek in South Dakota, as well as Blacktail Creek and Whitelaw Creek in Wyoming.

Prostoia Ricker

Prostoia besametsa (Ricker)

Descriptions/diagnoses: Baumann 1975, Stewart and Stark 1988 (larva), Ricker 1952 (adult). Western Nearctic distribution from Alaska south to California, New Mexico, and Nebraska. The Black Hills and the Nebraska Sand Hills are the eastern limits of this species (Grubbs et al. 2014).

Huntsman et al. (1999) reported *P. besametsa* from Battle Creek, Beaver Creek, Boxelder Creek, East Spearfish Creek, Grizzly Bear Creek, Hay Creek, Iron Creek, Little Spearfish Creek, South Fork Boxelder Creek, South Fork Rapid Creek, Spearfish Creek, Sunday Gulch, Whitetail Creek, Whitewood Creek, and a tributary to Bear Butte Creek in South Dakota, as well as Beaver Creek, North Redwater Creek, and Whitelaw Creek in Wyoming.

Zapada Ricker

Zapada cinctipes (Banks)

Descriptions/diagnoses: Dosdall and Lehmkuhl 1979 (larva), Needham and Claassen 1925 (adult, *Nemoura cinctipes*). Huntsman et al. (1999) reported that *Z. cinctipes* is one of the most common stonefly species in the Black Hills. It is distributed from Alaska east to Manitoba and south to California, New Mexico, and South Dakota.

Huntsman et al. (1999) reported *Z. cinctipes* from throughout the Black Hills. Streams included Battle Creek, Bear Butte Creek, Beaver Creek, Boxelder Creek, Burnt Fork, Castle Creek, Deer Creek, Ditch Creek, East Spearfish Creek, Elk Creek, Estes Creek, False Bottom Creek, Grizzly Bear Creek, Hay Creek, Iron Creek, Jim Creek, Little Elk Creek, Little Spearfish Creek, Meadow Creek, Middle Fork Boxelder Creek, Newton Fork, North Boxelder Creek, North Fork Castle Creek, Palmer Creek, Pine Creek, Rapid Creek, Slate Creek, South Fork Boxelder Creek, South Fork Castle Creek, South Fork Rapid Creek, Spearfish Creek, Spokane Creek, Spring Creek, Strawberry Creek(?), Sunday Gulch, Whitetail Creek, Whitewood Creek, and Willow Creek in South Dakota, as well as Beaver Creek, Cold Creek, Lytle Creek, and North Redwater Creek in Wyoming. The USU WCMAFE database includes records from Castle Creek and Whitewood Creek. I saw this species in nearly every biomonitored stream in the northern Black Hills.

Perlidae

Four species of Perlidae occur in or around the Black Hills. *Acroneuria abnormis* and *Perlesta decipiens* were actually found in the Belle Fourche River and Cheyenne River, respectively. Of the other two species, *Claassenia sabulosa* is uncommon, and *Hesperoperla pacifica*, at the eastern limit of its distribution, is common in all sizes of streams in the Black Hills. It is likely that the stoneflies identified as *Arcynopteryx* by Jurgens (1968) represented either *Claassenia sabulosa* or *Hesperoperla pacifica*, in which case they were reported from French Creek, Spring Creek, Battle Creek, and the South Fork of Rapid Creek. Shearer (2006) reported larvae of both *Claassenia* and *Hesperoperla* from Black Hills streams at the genus level. Anglers commonly call all perlids the "golden stones".

Acroneuria Pictet

Acroneuria abnormis (Newman)

Descriptions/diagnoses: Stark and Gaufin 1976, Dosdall and Lehmkuhl 1979 (larva), Needham and Claassen 1925 (adult). Distributed from Alberta and Montana south to New Mexico and Louisiana, east to the Atlantic seaboard.

Jurgens (1968) reported *Acroneuria*, at the genus level, from Spearfish Creek, the South Fork of Rapid Creek, Battle Creek, and Castle Creek; likewise, Drewes (1984) reported *Acroneuria* at the genus level from Slate Creek. Huntsman et al. (1999) reported *A. abnormis* from the Belle Fourche River in both Wyoming and South Dakota.

Claassenia Wu

Claassenia sabulosa (Banks)

Descriptions/diagnoses: Stewart and Stark 2002 (larvae), Needham and Claassen 1925 (adult, as *Perla sabulosa*). Distributed from British Columbia south along the Pacific coast to California and east to Quebec, South Dakota, Colorado, and New Mexico.

Huntsman et al. (1999) reported this species from Boxelder Creek, Rapid Creek, Spearfish Creek, and Whitewood Creek. The USU WCMAFE database has a record from Grace Coolidge Creek.

Hesperoperla Banks

Hesperoperla pacifica (Banks)

Descriptions/diagnoses: Stewart and Stark 2002 (larva), Needham and Claassen 1925 (adult, as *Alloperla pacifica*). Distributed from Alaska south along the Pacific coast to California, east to Saskatchewan, South Dakota, Colorado, and New Mexico.

Huntsman et al. (1999) reported this species from throughout the Black Hills. Streams included Battle Creek, Beaver Creek, Boxelder Creek, Castle Creek, Ditch Creek, Grizzly Bear Creek, Iron Creek, East Spearfish Creek, Elk Creek, Hay Creek, Jim Creek, Little Elk Creek, Little Spearfish Creek, Little Squaw Creek, Newton Fork, Rapid Creek, Slate Creek, South Fork Boxelder Creek, South Fork Castle Creek, South Fork Rapid Creek, Spearfish Creek, Sunday Gulch, West Strawberry Creek(?), Whitetail Creek, Whitewood Creek, and tributaries to Bear Butte Creek and Whitewood Creek in South Dakota, as well as Beaver Creek, Cold Springs Creek, Lytle Creek, Sand Creek, and Whitelaw Creek in Wyoming. The USU WCMAFE database has records from Castle Creek and Spruce Gulch Creek.

Perlesta Banks

Perlesta decipiens (Walsh)

Descriptions/diagnoses: Stark et al. 1989, 1998 (larva); DeWalt et al. 2001 (adult). Distributed in the central United States from North Dakota, Wisconsin, and Pennsylvania south to Colorado, Texas, Arkansas, and Virginia. Sometimes known to anglers as the "widespread stone".

Huntsman et al. (1999) reported this species from the Belle Fourche River at Belle Fourche and from the Cheyenne River near Angostura Reservoir, while the USU WCMAFE database includes records from the Belle Fourche River near Belle Fourche and Elk Creek near Elm Springs, all of which are outside of the Black Hills, proper.

Perlodidae

The Perlodidae are represented by six species in the Black Hills. Huntsman et al. (1999) point out some interesting biological observations on these stoneflies, such as the tendency for *Isogenoides elongates* and *Isoperla longiseta* to inhabit the larger rivers in the Black Hills, although this is not supported by the data from Vieira et al. (2006).

Isogenoides Klapálek

Isogenoides elongatus (Hagen)

Descriptions/diagnoses: Ricker 1952, Baumann et al. 1977 (larvae, selected parts), Needham and Claassen 1925 (adult, as *Isogenus elongatus*). Distributed from British Columbia, Alberta, and Manitoba south to Washington, Arizona, New Mexico, and South Dakota. Huntsman et al. (1999) reported that this species is typically found in larger creeks and rivers. Anglers call members of this genus the "springflies".

Huntsman et al. (1999) reported *I. elongatus* from Rapid Creek and Spearfish Creek.

Isoperla Banks

Jurgens (1968) reported *Isoperla* at the genus level from Spring Creek, Spearfish Creek, French Creek, Battle Creek, Castle Creek, the South Fork of Rapid Creek, and the Fall River. Drewes (1984) found *Isoperla* in Slate Creek, reporting their presence at the genus level. Shearer (2006) reported larvae of unidentified species of *Isoperla* from Black Hills streams. Anglers refer to members of this genus by several names including "yellow sally", "stripetails", and "yellow stones".

Even though the USU WCMAFE database includes records of many species of *Isoperla*, the only records from the Black Hills were at the genus level: Castle Creek and Elk Creek.

Isoperla longiseta Banks

Descriptions/diagnoses: Szczytko and Stewart 1979 (larva, selected parts), Needham and Claassen 1925 (adult). Distributed from British Columbia east to Manitoba south to New Mexico, Iowa, and Illinois. Huntsman et al. (1999) reported that this species is typically found in larger creeks and rivers.

Huntsman et al. (1999) reported *I. longiseta* from the Belle Fourche River at Belle Fourche and Newell, and the Cheyenne River(?) in Oral, South Dakota.

Isoperla phalerata (Smith)

Descriptions/diagnoses: Szczytko and Stewart 1979, Stark et al. 1988 (larva, selected parts), Stewart and Stark 2002 (larva), Needham and Claassen 1925 (adult, as *Perla phalerata*), Szczytko and

Kondratieff 2015 (adult). Central Rocky Mountain distribution from Oregon and Idaho east to New Mexico, Colorado, and South Dakota. Huntsman et al. (1999) pointed out that this species is considered to be uncommon throughout its range, but particularly common in the Black Hills; they suggest that this might be due to reduced competition from other similarly sized stoneflies.

Huntsman et al. (1999) and Szczytko and Kondratieff (2015) reported this species from streams throughout the Black Hills. Streams included Battle Creek, Bear Butte Creek, Beaver Creek, Boxelder Creek, Castle Creek, Cold Springs Creek, Deer Creek, East Spearfish Creek, Elk Creek, Grace Coolidge Creek, Grizzly Bear Creek, Iron Creek, Jenny Gulch, Little Spearfish Creek, Little Squaw Creek, Meadow Creek, Middle Fork Boxelder Creek, Newton Fork, North Fork Castle Creek, Palmer Creek, Rapid Creek, Redwater River, South Fork Boxelder Creek, South Fork Rapid Creek, Spearfish Creek, Spring Creek, Sunday Gulch, and Whitewood Creek in South Dakota, as well as Beaver Creek, Cold Creek, Cold Springs Creek, Redwater Creek, and Sand Creek in Wyoming.

Isoperla quinquepunctata (Banks)

Descriptions/diagnoses: Frison 1942 (as *I. patricia*), Jewett 1959, Szczytko and Stewart 1979 (larva, selected parts), Needham and Claassen 1925 (adult). Distributed from British Columbia east to Saskatchewan, south to Baja California, New Mexico, and Nebraska. Anglers commonly call this species the "little yellow stonefly".

Szczytko and Stewart (1978) synonymized *I. patricia* Frison with *I. quinquepunctata*; *I. patricia* was originally described from Spearfish. Huntsman et al. (1999) reported this species from streams throughout the Black Hills. Streams included Battle Creek, Beaver Creek, Belle Fourche River, Boxelder Creek, Deer Creek, Elk Creek, Flynn Creek, French Creek, Grace Coolidge Creek, Grizzly Bear Creek, Hay Creek, Iron Creek, Jenny Gulch, Jim Creek, Little Elk Creek, Little Spearfish Creek, Little Squaw Creek, Meadow Creek, Newton Fork, North Fork Castle Creek, Palmer Creek, Rapid Creek, Redwater River, South Fork Boxelder Creek, Spearfish Creek, Spokane Creek, Spring Creek, Sunday Gulch, West Strawberry Creek(?), Whitewood Creek, and tributaries to Bear Butte Creek and Beaver Creek in South Dakota, as well as Beaver Creek, Cold Creek, Cold Springs Creek, Lame Jones Creek, North Redwater Creek, Redwater Creek, Sand Creek, and Whitelaw Creek in Wyoming.

Isoperla transmarina (Newman)

Descriptions/diagnoses: Newman 1838 (adult); Szczytko and Harden and Mickel 1952 (larva). Northern transcontinental distribution from British Columbia to the Canadian Maritimes, south to the Black Hills in South Dakota and Wyoming, Minnesota, Michigan, West Virginia, and North Carolina.

Huntsman et al. (1999) reported *I. transmarina* from Beaver Creek above Cook Lake in Wyoming and Rapid Creek at the inlet to Pactola Reservoir in South Dakota.

Skwala Ricker

Skwala americana (Klapálek)

Descriptions/diagnoses: Stewart and Stark 2002 (larva), Needham and Claassen 1925 (adult, as *Perlodes americana*). British Columbia east to Manitoba, south to California, Arizona, New Mexico, and South Dakota. Known to anglers as the "large springfly".

Huntsman et al. (1999) reported this species from Battle Creek, Boxelder Creek, Burnt Fork, Castle Creek, Elk Creek, Grizzly Bear Creek, Horse Creek, Iron Creek, North Boxelder Creek, North Fork Castle Creek, Rapid Creek, South Fork Boxelder Creek, Spearfish Creek, Spring Creek, Sunday Gulch, West Strawberry Creek(?), Whitetail Creek, Whitewood Creek, and a tributary to Elk Creek in South Dakota, as well as Beaver Creek and Whitelaw Creek in Wyoming. The USU WCMAFE database includes records from Spruce Gulch Creek.

Plecoptera Traits Table

Table 6:   Plecoptera Traits Table

Species	TSN	Seen	Tolerance Value	FFG	Habit
Capniidae	102643	+	1	SH	SP
Capnia	102688	-	1	SH	SP
Capnia confusa	102702	-	1	SH	SP
Capnia gracilaria	102712	-	1	SH	SP
Eucapnopsis	102785	-	1	SH	SP
Eucapnopsis brevicauda	102786	-	1	SH	SP
Isocapnia	102740	-	1	SH	SP
Isocapnia integra	102745	-	1	SH	SP
Paracapnia	102804	+	1	SH	SP
Paracapnia angulata	102805	+	1	SH	SP
Utacapnia	102752	-	1	SH	SP
Utacapnia lemoniana	102754	-	1	SH	SP
Chloroperlidae	103202	+	1	PR	CN
Paraperla	103233	-	1	PR	CN
Paraperla frontalis	103234	-	1	PR	CN
Suwallia	103254	+	1	PR	CN
Suwallia lineosa	103259	+	1	PR	CN
Sweltsa	103273	+	1	PR	CN
Sweltsa borealis	103281	/	1	PR	CN
Sweltsa coloradensis	103283	/	1	PR	CN
Triznaka	103308	-	1	PR	CN
Triznaka pintada	103309	-	1	PR	CN
Leuctridae	102840	+	0	SH	SP
Paraleuctra	102887	+	0	SH	SP
Paraleuctra vershina	102899	+	0	SH	SP
Nemouridae	102517	+	2	SH	SP
Amphinemoura	102540	+	2	SH	SP
Amphinemoura banksi	102546	+	2	SH	SP
Malenka	102567	+	2	SH	SP
Malenka coloradensis	102575	+	2	SH	SP
Nemoura	102526	-	2	SH	SP
Nemoura trispinosa	102535	-	2	SH	SP
Prostoia	102584	+	2	SH	SP
Prostoia besametsa	102585	+	2	SH	SP
Zapada	102591	+	2	SH	SP
Zapada cinctipes	102594	+	2	SH	SP
Perlidae	102914	+	1	PR	CN
Acroneuria	102917	-	1	PR	CN
Acroneuria abnormis	102919	-	1	PR	CN
Claassenia	102930	+	3	PR	CN
Claassenia sabulosa	102932	+	3	PR	CN
Hesperoperla	102971	+	1	PR	CN
Hesperoperla pacifica	102972	+	1	PR	CN
Perlesta	103251	+	5	PR	CN
Perlesta decipiens	609909	+	5	PR	CN
Perlodidae	102994	+	2	PR	CN
Isogenoides	103124	-	2	PR	CN

Species	TSN	Seen	Tolerance Value	FFG	Habit
Isogenoides elongatus	103130	-	2	PR	CN
Isoperla	102995	+	2	PR	CN
Isoperla longiseta	103030	/	2	PR	CN
Isoperla phalerata	103027	/	2	PR	CN
Isoperla quinquepunctata	103045	/	2	PR	CN
Isoperla transmarina	103036	/	2	PR	CN
Skwala	103102	+	2	PR	CN
Skwala americana	568735	+	2	PR	CN

Species	Temperature Range	Lotic	Lentic	Current	Substrate	Voltinism
Capniidae	S: cold	--	--	--	5.5	--
Capnia	S: cold	3.6	1.1	3.0	6.0	Uni
Capnia confusa	S: cold	3.6	1.1	3.0	6.0	Uni
Capnia gracilaria	S: cold	3.6	1.1	3.0	6.0	Uni
Eucapnopsis	S: cold	3.6	5.0	--	5.5	Uni
Eucapnopsis brevicauda	S: cold	3.6	5.0	--	5.5	Uni
Isocapnia	S: cold	3.8	1.4	--	5.0	Uni
Isocapnia integra	S: cold	3.8	1.4	--	5.0	Uni
Paracapnia	S: cold	3.6	1.2	2.0	4.0	Uni
Paracapnia angulata	S: cold	3.4	5.0	2.0	4.0	Uni
Utacapnia	S: cold	4.0	1.7	--	5.5	Uni
Utacapnia lemoniana	S: cold	4.0	1.7	--	5.5	Uni
Chloroperlidae	E: warm	--	--	--	6.0	--
Paraperla	E: warm	4.0	1.7	2.0	7.0	Uni
Paraperla frontalis	E: warm	4.0	1.7	2.0	7.0	Uni
Suwallia	E: warm	3.8	1.2	2.0	6.4	Uni
Suwallia lineosa	E: warm	3.8	1.2	2.0	6.4	Uni
Sweltsa	E: warm	3.6	1.1	2.0	5.8	Mero
Sweltsa borealis	E: warm	3.6	1.1	2.0	5.8	Mero
Sweltsa coloradensis	E: warm	3.8	5.0	2.0	5.5	Mero
Triznaka	E: warm	3.9	5.0	--	5.1	Uni
Triznaka pintada	E: warm	3.9	5.0	--	5.1	Uni
Leuctridae	S: cold	--	--	--	4.0	--
Paraleuctra	S: cold	3.5	1.3	--	7.0	Uni
Paraleuctra vershina	S: cold	3.5	1.3	--	7.0	Uni
Nemouridae	E: warm	--	--	--	7.0	--
Amphinemoura	E: warm	3.2	1.7	2.3	5.5	Uni
Amphinemoura banksi	E: warm	3.2	1.7	2.3	5.5	Uni
Malenka	E: warm	3.3	5.0	--	7.0	Uni
Malenka coloradensis	E: warm	3.3	5.0	--	7.0	Uni
Nemoura	E: warm	3.2	1.3	2.5	5.2	Uni
Nemoura trispinosa	E: warm	3.0	5.0	2.0	4.8	Uni
Prostoia	S: cold	3.7	5.0	2.0	5.5	Uni
Prostoia besametsa	S: cold	3.9	5.0	2.0	5.5	Uni
Zapada	E: cool	3.7	1.0	--	7.8	Mero
Zapada cinctipes	E: cool	3.7	3.5	--	7.8	Mero
Perlidae	E: warm	--	--	--	--	--
Acroneuria	E: warm	4.1	1.2	2.3	4.7	Uni
Acroneuria abnormis	E: warm	4.3	--	2.5	5.0	Uni
Claassenia	E: cool	4.0	5.0	2.7	7.0	Mero
Claassenia sabulosa	E: cool	4.0	5.0	2.7	7.0	Mero
Hesperoperla	E: cool	3.6	5.0	2.5	6.0	Mero
Hesperoperla pacifica	E: cool	3.6	5.0	2.5	6.0	Mero
Perlesta	E: warm	3.7	--	1.7	4.3	Uni
Perlesta decipiens	E: warm	3.8	--	2.0	3.5	Uni
Perlodidae	E: cool	--	--	--	8.5	--
Isogenoides	E: cool	4.0	1.7	2.3	5.8	
Isogenoides elongatus	E: cool	3.9	1.7	2.3	5.8	Uni
Isoperla	E: cool	3.8	1.1	2.1	4.8	Uni
Isoperla longiseta	E: cool	3.8	1.1	2.1	4.8	Uni

Species	Temperature Range	Lotic	Lentic	Current	Substrate	Voltinism
Isoperla phalerata	E: cool	3.8	1.1	2.1	4.8	Uni
Isoperla quinquepunctata	E: cool	4.1	1.1	2.1	4.8	Uni
Isoperla transmarina	E: cool	4.2	1.1	2.1	4.8	Uni
Skwala	E: warm	4.1	1.7	2.5	6.0	Uni
Skwala americana	E: warm	4.2	1.7	2.5	6.0	Uni

Aquatic Hemiptera

Introduction

The true bugs are represented in the aquatic realm includes several families in the Heteroptera and excludes the clades that used to be called the "Homoptera" (comprising the Cicadomorpha, Fulgoromorpha, and Sternorrhyncha, sensu Song et al. 2012). A few species in the Hemiptera (particularly in the Belostomatidae) are among the largest insects to be found in the water. While most terrestrial hemipterans are phytophagous, most aquatic forms are predaceous, feeding on other invertebrates and even a few small vertebrates (Schumann et al. 2012). Their piercing-sucking mouthparts are able to stab into prey, inject saliva containing digestive enzymes such as proteases and phospholipase, and extract the pre-digested and liquefied prey's body contents. The bites of aquatic Hemiptera are often quite painful, compared with those of many terrestrial forms, such as cimicids and triatomine reduviids, which have developed relatively stealthy feeding habits on humans.

A few families skate on the surface of the water rather than swimming or clinging to the bottom or other submerged structures. These species are either predaceous, feeding on stranded organisms that get caught in the surface tension of the water, or scavengers. This ability has been extensively studied, particularly the morphological features, such as hydrofuge hairs on the tarsi, allowing such activity.

Overview of the Aquatic Hemiptera in the Black Hills

The Hemiptera, as a group, have not been treated at the state level in South Dakota since Harris (1937, 1943), and have never been treated specifically for the Black Hills only. Henry and Froeschner (1988) also reported many species at the state level only for both South Dakota and Wyoming. Revisions of various taxonomic groups are the basis for most of the information we have on the hemipterans in the Black Hills.

Anglers commonly refer to these insects by the common names of the families and generally do not have specialized angling names for each species, genus, or family.

Species Accounts

Belostomatidae

The Belostomatidae are known in the Black Hills from very few collections. Tolerance values were not provided for these taxa in Grafe (2002), and there was no published temperature range for *B fluminea*.

Belostoma Latreille

Belostoma fluminea Say

Descriptions/Diagnoses: Lauck 1964 (adult). Distributed from the Atlantic coast (Quebec and New England south to Florida), west to Manitoba, Colorado, and Arizona.

The USU WCMAFE database has a record of this species (at the genus level) from Spruce Gulch Creek. I have specimens of *B. fluminea* that I collected from False Bottom Creek, Whitewood Creek, and Sturgis, as well as from a site on the Belle Fourche River five miles south of Newell on the Plains north of the Black Hills.

Lethocerus Mayr

Lethocerus americanus (Leidy)

Descriptions/Diagnoses: Goodwyn 2006 (adult). Menke (1963) described the distribution of *L. americanus* as "coast to coast in the United States." The accompanying map showed a broad band from Washington, Oregon, and southern California east to Texas, Missouri, and the Atlantic seaboard from Virginia to Maine. In Canada, the shading representing its distribution extended across the southern tier of Canadian provinces, with scattered records from Saskatchewan to Newfoundland.

I collected a male adult of *L. americanus* using electrofishing gear on Whitewood Creek just north of the Black Hills; I somehow managed to avoid being bitten even though the bug got lost in the lining of my jacket and crawled all over the inside of my jacket during the vigorous electrofishing activities.

Corixidae

A total of 21 species are known from the state of South Dakota, of which 7 have published localities in and near the Black Hills (Hungerford 1948). No records of Corixidae in Hungerford (1948) from Wyoming were from the northeastern portion of the state near the Black Hills. It is likely that additional species occur in the Black Hills. Most corixids are phytophagous, but a few are predaceous; I have never known a corixid to bite a human, though.

For biomonitoring purposes, adult corixids should probably be identified to the genus level, since intrageneric variation in traits has not yet been identified for any genus of Corixidae. Immatures should be left at the family level, since characters have not been identified to reliably separate the immature stages even to genus level. Many of the published records are from cities and towns instead of specific streams because corixids are often collected at lights.

Cenocorixa Hungerford

Cenocorixa utahensis (Hungerford)

Description/Diagnosis: Hungerford 1948. Distributed from British Columbia, Alberta, and Manitoba south to California, Arizona, New Mexico, and Texas.

Hungerford (1948) reported this species from Piedmont and Rapid City.

Corisella Lundblad

Corisella tarsalis (Fieber)

Description/Diagnosis: Hungerford 1948. Distributed from California, Montana, South Dakota, Manitoba, Wisconsin, and New York south to Texas and west-central Mexico.

Hungerford (1948) reported this species from Nisland and Rapid City, north and east of the Black Hills, respectively.

Hesperocorixa Kirkaldy

Hesperocorixa laevigata (Uhler)

Description/Diagnosis: Hungerford 1948. Distributed from British Columbia, Alberta, and Manitoba south to California, Arizona, New Mexico, Texas, Mississippi, North Carolina, and Hidalgo, Mexico.

Hungerford (1948) reported this species from Newell and Wasta, both several miles from the Black Hills, but indicative that it might occur within the Black Hills.

Hesperocorixa vulgaris (Hungerford)

Description/Diagnosis: Hungerford 1948. Widely distributed in the Nearctic, from the Northwest Territories, British Columbia, Washington, and California east to Quebec, the Canadian Maritimes, New England, and the Atlantic Coast south to Georgia, Mississippi, Texas, and Colorado.

Hungerford (1948) reported this species from Wasta, 61 km east of Rapid City.

Sigara Fabricius

The USU WCMAFE database has records of this genus collected at the Cheyenne River near Spencer, Wyoming.

Sigara alternata (Say)

Description/Diagnosis: Hungerford 1948. Transcontentally distributed from the Northwest Territories and British Columbia south to California and east to the Atlantic seabord as far south as North Carolina.

Hungerford (1948) reported this species from Piedmont, Pringle, and Wasta.

Sigara grossolineata Hungerford

Description/Diagnosis: Hungerford 1948. Distributed from Saskatchewan, Montana, Utah, and California east to Quebec, New England, Pennsylvania, Ohio, and Kansas.

Hungerford (1948) reported this species from Hot Springs, Piedmont, Rapid City, and the "State Game Lodge."

Sigara lineata Förster

Description/Diagnosis: Hungerford 1948. The first North American corixid to be described, this species is distributed from Saskatchewan to Ontario, south to South Dakota, Illinois, and Ohio.

Shearer (2006) reported *Sigara lineata* from the Black Hills without designating a particular locality.

Trichocorixa **Kirkaldy**

Trichocorixa verticalis **(Fieber)**

Description/Diagnosis: Sailer in Hungerford 1948. There are numerous subspecies that have been named within T. verticalis, each with its own nearly discrete distribution. The distribution of the species includes the Gulf Coast and Atlantic Coast from southern Texas to Maine, the Caribbean islands, and Bermuda (*T. v. verticalis*); the Atlantic Coast from South Carolina to Maine (*T. v. sellaris*), a north-south band from western Hudson Bay south to Arizona, New Mexico, and Texas (*T. v. interiores*); central Quebec near the St. Lawrence River (*T. v. fenestrata*); central California (*T. v. californica*); and southern California (*T. v. saltoni*).

Sailer in Hungerford (1948) reported the subspecies *T. v. interiores* from Piedmont, South Dakota.

Gelastocoridae

Gelastocoridae, or toad bugs, are semi-aquatic and are found in places such as stream margins. They are undoubtedly present in the Black Hills, particularly *Gelastocoris oculatus*, although I have not seen any literature references documenting their presence. If it is found that they occur in the Black Hills, immature stages are described in Brown, L. N., and J. E. McPherson. 1994. Life history and laboratory rearing of *Gelastocorus oculatus oculatus* (Fabricius) (Hemiptera: Gelastocoridae) with descriptions of immature stages. *Proceedings of the Entomological Society of Washington* 95: 516-526.

Gerridae

Gerridae, or water striders, are present in the Black Hills, although I have not found primary literature sources documenting their presence. The species listed here are possible, given their known distributions or collection in biomonitoring studies. Gerrids are not frequently encountered in biomonitoring samples because they are extremely wary of humans and can quickly glide away on the water surface. Additionally, most biomonitoring samples are not designed for effective capture of gerrids, given the restricted area for sampling. Nevertheless, they occasionally show up in biomonitoring samples from smaller streams because there is less area to which to escape the sampler.

Descriptions have rarely been provided for immature gerrids in the published literature, so my references to descriptions and diagnoses refer only to adults.

Aquarius **Schellenberg**

Aquarius remigis **(Say)**

Description/Diagnosis: Drake and Harris 1934 (as *Gerris remigis*). Even though the actual boundaries of the distribution are not known, Drake and Harris (1934) reported that *A. remigis* (Say) occurs in all of the contiguous 48 states, Canada, Mexico, and Guatemala; undoubtedly, this could include the Black Hills in South Dakota and Wyoming.

Gerris **Fabricius**

Shearer (2006) reported *Gerris* from the Black Hills.

Gerris buenoi **Kirkaldy**

Description/Diagnosis: Drake and Harris 1934. Drake and Harris (1934) indicate that this species is transcontinental in northern United States and southern Canada, and Stonedahl and Lattin (1982) indicated that it has been collected as far south as California and Colorado.

Gerris comatus Drake and Hottes

Description/Diagnosis: Drake and Harris 1934. Drake and Harris (1934) described the distribution of this species from the Atlantic seaboard (New York, New Jersey, and Maryland) west to Montana, South Dakota, and Colorado. It was not stated if the South Dakota records were in or near the Black Hills.

Gerris gillettei Lethierry and Severin

Description/Diagnosis: Drake and Harris 1934. Drake and Harris (1934) and Stonedahl and Lattin (1982) indicate that this species is distributed from Washington, Oregon, and California east to Montana, Wyoming, Colorado, and Texas. It was not stated how close the Wyoming and Montana localities were to the Black Hills.

Gerris incognitus Drake and Hottes

Description/Diagnosis: Drake and Harris 1934. Stonedahl and Lattin (1982) stated that the distribution of this species ranged from British Columbia south to California and east to Montana, Wyoming, and Colorado, with a record from Quebec. It was not stated how close the Wyoming and Montana localities were to the Black Hills.

Gerris incurvatus Drake and Hottes

Description/Diagnosis: Drake and Harris 1934. Drake and Harris (1934) indicated that the distribution of this species ranged from British Columbia south to California and east to Montana, Wyoming, and Illinois. It was not stated how close the Wyoming and Montana localities were to the Black Hills.

Gerris marginatus Say

Description/Diagnosis: Drake and Harris 1934. Drake and Harris (1934) state that G. marginatus is "probably the commonest and most widely distributed member of the genus in North America, save perhaps *remigis* [now in *Aquarius*]." The distribution included the contiguous 48 states, Canada, Mexico, and Brazil.

Limnoporus Stål

Limnoporus notabilis (Drake and Hottes)

Description/Diagnosis: Drake and Harris 1934 (as *Gerris notabilis*). Drake and Harris (1934) reported specimens from a geographic range as British Columbia south to California, east to Alberta, Montana, South Dakota, Iowa, Colorado, and Arizona. It is unknown if any of the South Dakota and Wyoming records are from the Black Hills.

Trepobates Uhler

Trepobates is considered eurythermal warm according to Grafe et al. (2002), based on the family Gerridae, but would be considered to be eurythermal cool according to a weighted average of literature statements in Vieira et al. (2006).

Hydrometridae

Hydrometridae may or may not occur in the Black Hills. I have not seen any specimens in numerous years of biomonitoring within the Black Hills, and I have not found any published records of this family's presence there.

Naucoridae

The Naucoridae are represented in South Dakota by one species, *Ambrysus mormon*. During fish sampling using electrofishing techniques, I have frequently encountered this species. Individuals can be netted with the fish and end up in the sorting containers. Care must be taken when handling fish so as not to get a naucorid accidentally pinched between the fingers, since they bite readily. The bite feels very much like getting stabbed by a sewing needle.

Ambrysus Stål

Ambrysus mormon Montandon

This species primarily inhabits plains streams, but can be found in the Black Hills at lower elevations. Tinerella and DeLorme (2005) reported collection of *A. mormon* in hot springs, ponds, and cold water springs in southwestern South Dakota. Shearer (2006) reported *Ambrysus* from the Black Hills at the genus level, and the USU WCMAFE database includes a record of *Ambrysus*, at the genus level, from Cass Pig Creek.

This species is easily collected by dragging a sweep net through emergent vegation on the margins. I do not have a reference for current preference available. No records for voltinism were available, either, but it is likely that *A. mormon* is univoltine.

Nepidae

Although there do not appear to be any published records of Nepidae in the Black Hills, the distribution maps in Sites and Polhemus (1994) indicate a single species with a distribution that would include the Black Hills. I have not seen any biomonitoring samples with nepids present.

Ranatra Fabricius

Ranatra fusca Palisot de Beauvois

Descriptions/Diagnosis: Packauskas and McPherson 1986 (larva), Sites and Polhemus 1994 (adult). This species is distributed across the southern border of Canada coast to coast, Washington, Oregon, California, Idaho, and Montana, and North Dakota south to Colorado and Kansas, east to the Atlantic coast as far south as North Carolina. There do not appear to be any specific Black Hills records of *R. fusca* in the published literature.

In my personal collection, I have two specimens of *Ranatra fusca* from the Black Hills, one that I collected on Whitewood Creek just north of the Black Hills and a second that I collected at a hotel porch light in Sturgis.

Notonectidae

Notonectidae undoubtedly occur in the Black Hills, but I have neither literature nor biomonitoring records documenting their presence. In his revision of the genus *Buenoa*, Truxal (1953) reported three species (*B. confusa*, *B. margaritacea*, and *B. macrotibialis*) from South Dakota, but provided no records specifically from the Black Hills. Blatchley (1926) included South Dakota in the range for *B. margaritacea*. Hungerford (1933) did not even have records of *Notonecta* from South

Dakota or Wyoming. The nearest records of any notonectid species in the USU WCMAFE database to the Black Hills are from the Thunder Basin National Grasslands in Wyoming. Interestingly, BugGuide doesn't even have any photographic records from South Dakota submitted as of November 2014. These insects prefer habitats that are not usually sampled in stream biomonitoring programs, so they are rarely collected in those efforts. This is a wide open field for anyone wanting to spend some time in the pools, ponds, and lakes of the Black Hills.

Saldidae

Saldidae occur in the Black Hills but I could find no published records. I have seen Saldidae reported at the family level from biomonitoring samples collected in the Black Hills.

Veliidae

Veliidae occur in the Black Hills, but I do not have species-level identifications. No checklists currently could be found that detail the presence of individual species in the Black Hills; however, Smith and Polhemus (1978) provide general distributional data for the Nearctic species, so I am including those species for which the Black Hills falls within their described distribution.

As with the Gerridae, veliids are not frequently encountered in biomonitoring samples because they are extremely wary of humans and can quickly glide away on the water surface, and biomonitoring sample protocols are not designed for effective capture of veliids.

Microvelia Westwood

The USU WCMAFE database includes one record of *Microvelia*, at the genus level, from Spruce Gulch Creek.

Microvelia albonotata Champion

Descriptions/Diagnoses: Smith 1980. Smith and Polhemus (1978) describe the distribution as "Canada, U.S. east of Rocky Mtns., Mexico to Peru, Caribbean."

Microvelia americana (Uhler)

Descriptions/Diagnoses: Smith 1980. Distributed generally east of the Mississippi River, north into Ontario and Quebec, and west to Iowa, Nebraska, Kansas, and Texas.

Microvelia buenoi Drake

Descriptions/Diagnoses: Smith 1980. Distributed from Alaska and the Northwest Territories south to California and the northern half of the contiguous United States.

Microvelia hinei Drake

Descriptions/Diagnoses: Smith 1980. Drake and Hussey (1955) and Smith and Polhemus (1978) both describe the distribution as "Canada to Argentina."

Microvelia pulchella Westwood

Descriptions/Diagnoses: Smith 1980. The nominate subspecies, according to Drake and Hussey (1955) is distributed in the Neotropics, including the Caribbean Islands, north to Arizona, Texas, Mississippi, and Florida, while the *incerta* subspecies was reported from Canada through the

United States to the Neotropics; therefore, Smith and Polhemus (1978) describe the distribution as "Canada to Argentina, Caribbean."

Rhagovelia **Mayr**

The USU WCMAFE database includes one record of Rhagovelia, at the genus level, from Grace Coolidge Creek.

Rhagovelia distincta **Champion**

Descriptions/Diagnoses: Gould 1931 (adult). Smith and Polhemus (1978) describe the distribution as "Western U.S., Mexico, Middle America." With numerous subspecies recognized, this species is found from Arizona to Wyoming, east to Indiana to Texas, and south into Mexico.

Rhagovelia oriander **Parshley**

Descriptions/Diagnoses: Gould 1931. Smith and Polhemus (1978) describe the distribution as "Midwestern U.S.", while Gould (1931) specifically reported South Dakota, Ohio, Iowa, Minnesota, and Kansas, all at the state level.

Aquatic Hemiptera Traits Table

Table 7:   Aquatic Hemiptera Traits Table

Species	TSN	Seen	Tolerance Value	FFG	Habit
Belostomatidae	103683	+	--	PR	CB
Belostoma	103684	+	--	PR	CB
Belostoma fluminea	103689	+	--	PR	CB
Lethocerus	103699	-	--	PR	CB
Lethocerus americanus	103709	-	--	PR	CB
Corixidae	103364	+	10	PI	SW
Cenocorixa	103501	-	10	PI	SW
Cenocorixa utahensis	103502	-	10	PI	SW
Corisella	103484	-	10	PR	SW
Corisella tarsalis	103486	-	10	PR	SW
Hesperocorixa	103444	+	10	PI	SW
Hesperocorixa laevigata	103452	/	10	PI	SW
Hesperocorixa vulgaris	103450	/	10	PI	SW
Sigara	103369	+	10	PI	SW
Sigara alternata	103370	/	10	PI	SW
Sigara grossolineata	103382	/	10	PI	SW
Sigara lineata	103402	/	10	PI	SW
Trichocorixa	103423	-	10	PR	SW
Trichocorixa verticalis	103431	-	10	PR	SW
Gelastocoridae	103768	-	--	PR	CB
Gelastocoris	103769	-	--	PR	SK
Gelastocoris oculatus	103772	-	--	PR	SK
Gerridae	103801	+	5	PR	SK
Aquarius	717547	+	5	PR	SK
Aquarius remigis	717592	+	5	PR	SK
Gerris	103829	+	5	PR	SK
Gerris buenoi	103842	/	5	PR	SK
Gerris comatus	103834	/	5	PR	SK
Gerris gillettei	103852	/	5	PR	SK
Gerris incognatus	103843	/	5	PR	SK
Gerris incurvatus	103844	/	5	PR	SK
Gerris marginatus	103840	/	5	PR	SK
Limnoporus	103872	-	5	PR	SK
Limnoporus notabilis	103880	-	5	PR	SK
Trepobates	103811	+	10	PR	CB
Naucoridae	103613	+	5	PR	CN
Ambrysus	103614	+	5	PR	CN
Ambrysus mormon	103626	+	5	PR	CN
Nepidae	103747	-	--	PR	CB
Ranatra	103748	-	--	PR	CB
Ranatra fusca	103755	-	--	PR	CB
Notonectidae	103557	-	--	PR	CB
Saldidae	104063	+	10	PR	CB
Veliidae	103885	+	--	PR	SK
Microvelia	103900	+	--	PR	SK
Microvelia albonotata	103913	/	--	PR	SK
Microvelia americana	103902	/	--	PR	SK

Species	TSN	Seen	Tolerance Value	FFG	Habit
Microvelia buenoi	103914	/	--	PR	SK
Microvelia hinei	103908	/	--	PR	SK
Microvelia pulchella	103910	/	--	PR	SK
Rhagovelia	103886	+	--	PR	SK
Rhagovelia distincta	103895	/	--	PR	SK
Rhagovelia oriander	103896	/	--	PR	SK

Species	Temperature Range	Lotic	Lentic	Current	Substrate	Voltinism
Belostomatidae	--	--	--	--	--	--
Belostoma	--	3.4	1.3	1.0	1.0	Multi
Belostoma fluminea	--	3.0	4.0	1.0	1.0	Multi
Lethocerus	E: warm	3.4	1.0	1.0	1.0	Multi
Lethocerus americanus	E: warm	3.4	1.0	1.0	1.0	Multi
Corixidae	E: warm	--	--	--	--	Multi
Cenocorixa	E: warm	5.0	1.2	0.5	1.0	Multi
Cenocorixa utahensis	E: warm	5.0	1.2	0.5	1.0	Multi
Corisella	E: warm	5.0	1.3	0.5	--	Multi
Corisella tarsalis	E: warm	5.0	1.3	0.5	--	Multi
Hesperocorixa	E: warm	4.0	1.2	0.5	--	Multi
Hesperocorixa laevigata	E: warm	4.0	1.2	0.5	--	Multi
Hesperocorixa vulgaris	E: warm	4.0	1.2	0.5	--	Multi
Sigara	E: warm	4.0	1.6	0.7	--	Multi
Sigara alternata	E: warm	4.0	1.6	0.7	--	Multi
Sigara grossolineata	E: warm	4.0	1.6	0.7	--	Multi
Sigara lineata	E: warm	4.0	1.6	0.7	--	Multi
Trichocorixa	E: warm	5.0	1.1	0.5	--	Multi
Trichocorixa verticalis	E: warm	5.0	1.1	0.5	--	Multi
Gelastocoridae	E: warm	--	--	--	--	--
Gelastocoris	E: warm	3.3	1.2	--	0.4	--
Gelastocoris oculatus	E: warm	3.3	1.2	--	0.4	--
Gerridae	E:warm	--	--	--	--	--
Aquarius	E:warm	4.0	4.0	--	--	Multi
Aquarius remigis	E:warm	4.0	4.0	--	--	Multi
Gerris	E:warm	3.5	1.1	0.7	--	Uni
Gerris buenoi	E:warm	3.0	4.0	0.5	--	Uni
Gerris comatus	E:warm	3.5	1.1	0.7	--	Uni
Gerris gillettei	E:warm	3.5	1.1	0.7	--	Uni
Gerris incognatus	E:warm	3.5	1.1	0.7	--	Uni
Gerris incurvatus	E:warm	3.5	1.1	0.7	--	Uni
Gerris marginatus	E:warm	3.4	3.5	1.5	--	Uni
Limnoporus	E:warm	4.0	1.2	--	--	Uni
Limnoporus notabilis	E:warm	4.0	1.2	--	--	Uni
Trepobates	E:warm	3.8	1.0	0.5	--	Uni
Naucoridae	E: warm	--	--	--	5.0	--
Ambrysus	E: warm	4.0	4.0	--	0.0	--
Ambrysus mormon	E: warm	4.0	4.0	--	0.0	--
Nepidae	--	--	--	--	--	--
Ranatra	--	4.0	1.4	1.0	0.0	Multi
Ranatra fusca	--	4.0	1.4	1.0	0.0	Multi
Notonectidae	E: warm	--	--	--	--	--
Saldidae	E: warm	--	--	--	--	--
Veliidae	E:warm	--	--	--	--	Multi
Microvelia	E:warm	3.7	1.1	0.4	6.0	Multi
Microvelia albonotata	E:warm	3.7	1.1	0.4	6.0	Multi
Microvelia americana	E:warm	3.7	1.1	0.4	6.0	Multi
Microvelia buenoi	E:warm	3.7	1.1	0.4	6.0	Multi
Microvelia hinei	E:warm	3.7	1.1	0.4	6.0	Multi
Microvelia pulchella	E:warm	3.0	3.0	0.4	6.0	Multi

Species	Temperature Range	Lotic	Lentic	Current	Substrate	Voltinism
Rhagovelia	E:warm	3.7	--	2.5	10.0	Multi
Rhagovelia distincta	E:warm	3.7	--	2.5	10.0	Multi
Rhagovelia oriander	E:warm	3.7	--	2.5	10.0	Multi

Megaloptera

Introduction

Overview of the Megaloptera in the Black Hills

The Megaloptera have been poorly studied in the Black Hills, specifically, and in South Dakota and Wyoming, in general. Corydalids have not been reported from Wyoming, and the only corydalid species reported from South Dakota is *Chauliodes rastricornis* (Johnson et al. 1997; Penny et al. 1997). This species is collected throughout the Plains ecoregions and may occur along the foothills, but it is unlikely that it occurs at higher elevations within the Black Hills. I have not found any records of the presence of dobsonflies in or near the Black Hills.

Even more interestingly, there have been no reports in the peer-reviewed, published literature to document the presence of alderflies (Sialidae) in the state of South Dakota. Based on the distribution information in Whiting (1981), every neighboring state has at least one, and usually multiple, species of *Sialis*, so it is likely that South Dakota, and possibly the Black Hills, have multiple species.

Species Accounts

Corydalidae

Chauliodes Latreille

Chauliodes rastricornis Rambur

As stated above, the only corydalid species reported from South Dakota is *Chauliodes rastricornis* (Penny et al. 1997). This species is collected throughout the Plains ecoregions and may occur along the flanks of the Black Hills, but it is unlikely that it occurs at higher elevations within the Black Hills. I have not found any records of the presence of dobsonflies in or near the Black Hills.

Sialidae

Sialidae occur in the Black Hills, but as I mentioned earlier, there have been no reports in the peer-reviewed, published literature to document their presence in the state of South Dakota. I have encountered larvae quite frequently in biomonitoring samples, and Shearer (2006) listed Sialis larvae from his samples. Every neighboring state has multiple species of *Sialis* (Whiting 1981), so it is likely that multiple species may live in the Black Hills.

Sialis Latreille

Identification keys do not exist to separate larvae of the species of *Sialis*, so all identifications of larval individuals from biomonitoring samples have been left at the genus level. The USU WCMAFE database includes records of Sialis, at the genus level, from Hat Creek, Spring Creek, and Spruce Gulch Creek, South Dakota. Whiting (1981) reported *Sialis cornuta* Ross and *Sialis velata* Ross from Crook County, Wyoming, so these species are likely candidates for occurring in the Black Hills proper. Whiting (1981) also reported *S. hamata* Ross from Wyoming, but in the far northwest corner of the state.

Sialis cornuta Ross

Description/diagnosis: Ross 1937 (adult). This species is primarily distributed in the West from Washington, Alberta, and Montana, south to Oregon, Utah, and Wyoming (Whiting 1981). The sole Wyoming record is in Crook County.

Sialis velata Ross

Description/diagnosis: Ross 1937 (adult). This species is widespread in southern Canada from British Columbia to Quebec and in the United States from Idaho to New England, south to Utah, Colorado, Texas, Missouri, Tennessee, and Virginia (Whiting 1981). Although S. velata has not been reported in the published literature from South Dakota, it likely occurs there. Wyoming records are only at the county level and include Carbon, Crook, and Laramie counties. Locklin et al. (2006) reported on the life history of this species in Texas.

Megaloptera Traits Table

Table 8:   Megaloptera Traits Table

Species	TSN	Seen	Tolerance Value	FFG	Habit
Corydalidae	115023	-	0	PR	CN
Chauliodes	115024	-	0	PR	CN
Chauliodes rastricornis	115025	-	0	PR	CN
Sialidae	115001	+	--	PR	BU
Sialis	115002	+	4	PR	BU
Sialis cornuta	115013	/	4	PR	BU
Sialis velata	115011	/	4	PR	BU

Species	Temperature Range	Lotic	Lentic	Current	Substrate	Voltinism
Corydalidae	E: warm	--	--	--	--	--
Chauliodes	E: warm	3.0	1.3	--	0.0	--
Chauliodes rastricornis	E: warm	3.0	1.3	--	0.0	--
Sialidae	E: warm	--	--	--	--	Uni
Sialis	E: warm	3.9	1.2	--	2.7	Uni
Sialis cornuta	E: warm	3.9	1.2	--	2.7	Uni
Sialis velata	E: warm	3.9	1.2	--	2.7	Uni

Aquatic Neuroptera

Introduction

Most Neuroptera are terrestrial, including lacewings and ant-lions, which would probably be the most familiar. There are several other groups, some of which are quite rare or uncommon, and some of which are used as biological control agents in pest management programs. But, there is one family (Sisyridae) that has moved into the aquatic realm, and they are called the spongillaflies.

The bizarre larvae of the spongillaflies have long, spindly legs, flexible, threadlike mouthparts, and seven pairs of jointed, moveable tracheal gills ventrolaterally on the abdomen. The larvae are piercing predators of freshwater sponges and some bryozoans, hence their functional feeding group is considered to be piercers. Stoaks et al. (1983) provided several nice observations on interactions between spongillaflies and sponges in North Dakota.

Larvae leave the water to pupate, generally not far from the water but in protected places such as under rocks or tree bark. The adults are crepuscular or nocturnal and are primarily scavengers on invertebrate carrion but are also known to hunt small invertebrates actively and feed on pollen and honeydew. Adults are sometimes found at lights.

Overview of the Aquatic Neuroptera in the Black Hills

The latest review of aquatic Neuroptera does not include any records from South Dakota, Wyoming, or the Black Hills (Bowles 2006). Stoaks et al. (1983) mapped collection localities in North Dakota, demonstrating them to be mostly in the eastern part of that state. Given the widespread distribution of *Climacia areolaris* (Hagen) and *Sisyra vicaria* (Walker) in the eastern United States, they may eventually be found in South Dakota and Wyoming, if not even in the Black Hills. These two species are discussed below.

Species Accounts

Sisyridae

Climacia McLachlan

Grafe et al. (2002) did not provide a temperature preference for any aquatic Neuroptera; the eurythermal cool designation for *Climacia* and *C. areolaris* is based on data from Vieira et al. (2006).

Climacia areolaris (Hagen)

Descriptions/diagnosis: Bowles 2006 (adults, with larval characters discussed in keys). This species is distributed from North Dakota, Colorado, and New Mexico east to the Atlantic coast from Quebec south to Florida, possibly south into Mexico. As mentioned above, it has not yet been reported from the Black Hills, but they are on the edge of the reported distribution, so may occur there.

Sisyra Burmeister

Sisyra vicaria (Walker)

Descriptions/Diagnosis: Bowles 2006 (adults, with larval characters discussed in keys). This species is distributed from British Columbia to Quebec, south to California, Arizona, Texas, Louisiana, Alabama, Mississippi, and Florida.

Aquatic Neuroptera Traits Table

Table 9:   Neuroptera Traits Table

Species	TSN	Seen	Tolerance Value	FFG	Habit
Sisyridae	115085	-	--	--	--
Climacia	115086	-	--	PR	CB
Climacia areolaris	115087	-	--	PR	CB
Sisyra	115090	-	--	PI	CB
Sisyra vicaria	115091	-	--	PI	CB

Species	Temperature Range	Lotic	Lentic	Current	Substrate	Voltinism
Sisyridae	--	--	--	--	--	--
Climacia	E: cool	4.0	1.3	--	--	Multi
Climacia areolaris	E: cool	4.0	4.5	--	--	Multi
Sisyra	--	--	3.0	--	--	Uni
Sisyra vicaria	--	--	3.0	--	--	Uni

Aquatic Coleoptera

Introduction

It has been postulated that one of every four animal species on this planet is a beetle. This is not true of the aquatic animals, where Coleoptera is actually a rather minor order. That, of course, is not to say that they are not important, since they are among the most conspicuous and recognizable elements of the aquatic fauna. After all, the lay person may not have a clue what a mayfly is, but aquatic beetles are so generic in appearance that nearly everyone can at least recognize tht it is a beetle. (This is in comparison to a mayfly or stonefly nymph, which are recognizable to entomologists, fly fishers, and naturalists.)

That said, the beetles do play quite a role in the aquatic ecology. Some of the dytiscids are among the largest aquatic insects and are extremely predatory. Other families have small memebers and are yet important shredders or scrapers. Still others mine into living plant tissues and are rarely seen, even by aquatic entomologists.

Overview of the Aquatic Coleoptera in the Black Hills

The best checklist and source of distributional information on Black Hills and South Dakota beetles is in Kirk and Balsbaugh (1975). It provides localized information on nearly 2,000 beetle species in South Dakota, including many aquatic beetles. I found many of the following records among the data in Kirk and Balsbaugh (1975).

There are no characters that reliably separate terrestrial from aquatic and semiaquatic Coleoptera. The presence of gills or natatory (swimming) hairs is a dead give-away, but many of the aquatic species do not have such features. Even more frustrating is the fact that some families have both aquatic and terrestrial members. Thus, it becomes incumbent upon the person trying to identify an aquatic beetle to become familiar with the forms that occur in the water, so as to begin to identify material in hand.

Species Accounts

Chrysomelidae

In North America, the Chrysomelidae are represented by few aquatic species which are generally restricted to emergent aquatic plants. The most common aquatic chrysomelids are in the subfamily Donaciinae, which has 6 species from 2 genera in South Dakota. Shearer (2006) reported pupae of Chrysomelidae in samples from the Black Hills.

Marx (1957) reviewed the North American *Donacia (Donacia)*, here represented by the genus *Donacia*. Despite having over 8,700 records from across North America, there was not one record from South Dakota, and very few from Wyoming, let alone any from the Black Hills. Clark et al. (2008) reviewed the Donaciinae of Utah, including state-level distributions in the discussions for each species; South Dakota and/or Wyoming were included for several species. The species listed below have not been reported specifically from the Black Hills but are included because their general distribution suggests they probably occur there and the information in Clark et al. (2008) demonstrates that they do occur in either South Dakota or Wyoming.

Larvae of aquatic Chrysomelidae have rarely been described; therefore, I refer only to literature describing adults in the references to descriptions and diagnoses, except for *Donacia subtilis*, for which I did find a description of the larva.

Several spcecies of Chrysomelidae have been introduced to the United States for biological control of pest weed species, including aquatic plants. Since there are very few records of invasive aquatic plant species in the Black Hills (http://www/eddmaps.org/), it is unlikely that there have been intensive biological control efforts using these natural enemies in the Black Hills. Native beetles on native aquatic plants could certainly be collected. Hopefully, field notes might indicate the presence of aquatic plants that might host some species. A literature review of insect herbivores of aquatic and wetland plants in the United States was prepared by Harms and Grodowitz (2009) and could be useful for future researchers looking for chrysomelids on aquatic plants in the Black Hills.

As these insects prefer habitats that are not often sampled as part of biomonitoring efforts, I have only seen one aquatic chrysomelid in the Black Hills. Given that most chrysomelids are terrestrial, a good standard operating procedure regarding chrysomelids found in aquatic samples is to treat it as terrestrial if it does not exactly match the descriptions / diagnoses / drawings of known aquatic species. Additionally, species trait data are poorly known and sparse.

Donacia Fabricius

Donacia hirticollis Kirby

Description/diagnosis: Marx 1957. Distributed transcontinentally from the Yukon south to central California, east to the Atlantic Coast from Newfoundland south to New Jersey. Marx (1957) did not include any data from South Dakota and the nearest locality to the Black Hills (figured on a map) is from the Medicine Bow Mountains in south central Wyoming; however, Clark et al. (2008) reported this species from South Dakota and Wyoming at the state level.

Donacia magnifica LeConte

Description/diagnosis: Marx 1957. British Columbia across the southern Canadian provinces to Quebec and south to California, Utah, Colorado, North Dakota, Michigan, and Maine. Marx (1957) did not include any data from South Dakota and the nearest locality to the Black Hills is from the Medicine Bow Mountains in south central Wyoming. Clark et al. (2008) also reported this species from the state of Wyoming.

Donacia subtilis **Kunze**

Description/diagnosis: Marx 1957 (adult); MacGillivray 1965 (larva). Transcontinental distribution from British Columbia and California to the Atlantic Coast from Prince Edward Island south to North Carolina and Georgia; reported most frequently east of the Great Lakes. The nearest locality to the Black Hills reported in Marx (1957) is in north central Colorado, near Greeley, but Clark et al. (2008) reported this species from the state of South Dakota.

Donacia tuberculifrons **Schaeffer**

Description/diagnosis: Marx 1957. Distributed from Manitoba, Minnesota, and Nebraska east to Quebec, Vermont, Connecticut, New Jersey, and Virginia. The nearest localitiy to the Black Hills reported in Marx (1957) is in north central Nebraska. Clark et al. (2008) reported this species from the state of South Dakota.

Plateumaris **Thomson**

Plateumaris germari **(Mannerheim)**

Description/diagnosis: Askevold 1991. Distributed from Alaska, British Columbia, Washington, Oregon, and California east to the Canadian Maritimes, New England, New Jersey, and Pennsylvania. Clark et al. (2008) reported this species from both South Dakota and Wyoming at the state level.

Plateumaris nitida **(Germar)**

Description/diagnosis: Askevold 1991. Distributed from British Columbia south along the Pacific Coast to California, east to the Canadian Maritimes, New England, Virginia, West Virginia, Nebraska, and Colorado. Clark et al. (2008) reported this species from both South Dakota and Wyoming at the state level.

Plateumaris pusilla **(Say)**

Description/diagnosis: Askevold 1991. Distributed from British Columbia south along the Pacific Coast to California, east to the Canadian Maritimes, New England, Pennsylvania, New Jersey, Colorado, and Texas. Clark et al. (2008) reported this species from Wyoming at the state level.

Plateumaris robusta **(Schaeffer)**

Description/diagnosis: Askevold 1991. Distributed from British Columbia and Washington east to Quebec, Michigan, Nebraska, and New Mexico. Clark et al. (2008) reported this species from both South Dakota and Wyoming at the state level.

Curculionidae

I have not seen records or even specimens of aquatic representatives of this distinctive family from the Black Hills. As with the Chrysomelidae, these beetles will pretty much only be present if the emergent aquatic plants that they use as food plants are also present. This habitat is not commonly sampled in biomonitoring efforts, so I have not seen any records of this family in biomonitoring samples. Given that most curculionids are terrestrial, a good standard operating procedure regarding curculionids found in aquatic samples is to treat them as terrestrial if they do not exactly match the descriptions / diagnoses / drawings of known aquatic species. Larvae are very uncommonly encountered, since they are leaf miners and would normally be collected only by dissecting the plants. Hopefully, field notes might indicate the presence of aquatic plants that might host some species.

Several spcecies of Curculionidae have been introduced to the United States for biological control of pest weed species, including aquatic plants. Since there are very few records of invasive aquatic plant species in the Black Hills (http://www/eddmaps.org/), it is unlikely that there have been intensive biological control efforts using these natural enemies in the Black Hills. Native weevils on native aquatic plants could certainly be collected. Shearer (2006) reported adult Curculionidae from aquatic biomonitoring samples collected in the Black Hills.

Additionally, as with the Chrysomelidae, species trait data are poorly known and sparse.

Based on their distribution otherwise in North America, distribution of their host plants, or general invasive status, several genera and species of aquatic Curculionidae are likely to occur in or around the Black Hills. In my opinion, genera would include *Auleutes* Dietz (8 species in North America, with one Holarctic), *Bagous* Germar (33 species in North America), *Listronotus* Jekel (more than 80 species in North America), *Lixus* Fabricius (approximately 70 species in North America), *Notiodes* Schönherr (14 species in North America), and *Rhinoncus* Schönherr (7 species in North America).

Species which I think are likely in the Black Hills include *Euhrychiopsis lecontei* (Dietz) (the milfoil weevil) and *Phytobius leucogaster* (Marsham) (the interoduced European water milfoil weevil). One native species, *Stenopelmus rufinasus* Gyllenhal, is distributed from transcontinentally from California and Oregon to New Jersey and Florida, with BugGuide records in Iowa. It has been spread globally by accident and for biological control of mosquitoferns (*Azolla* spp.) A literature review of insect herbivores of aquatic and wetland plants in the United States was prepared by Harms and Grodowitz (2009) and could be useful for future researchers looking for curculionids on aquatic plants in the Black Hills.

Dryopidae

Only one species of Dryopidae, *Helichus striatus*, has been reported from South Dakota (Brown 1976). Kirk and Balsbaugh (1975) provided records from Lawrence, Pennington, and Fall River counties. Shearer (2006) reported adults of *Helichus* from the Black Hills. There are no published references for current preferences for *H. striatus*. These dryopids comprise a rather unique group in which the larvae are terrestrial (or at most semiaquatic), while the adults are obligatorily aquatic (Ulrich 1986).

Helichus Erichson

Helichus striatus LeConte

Description/diagnosis: Musgrave 1935 (adult); terrestrial larva not yet described. Distributed transcontinentally from South Carolina to Quebec, west to California and British Columbia.

Dytiscidae

Larson et al. (2000) reported 71 species of Dytiscidae from South Dakota and 84 species from Wyoming (Larson et al. 2000); however, nearly all of the records are at the state level and do not distinguish the Black Hills. Kirk and Balsbaugh (1975) reported 24 species from the Black Hills, although some are dubious records, as discussed below.

Dytiscids are generally collected in small numbers; although one biomonitoring sample using EMAP methods from Cleopatra Creek had 250 *Hydroporus* individuals, all other samples had less than 60 individuals/sample for any dytiscid taxon.

Because species-level identification for many dytiscid genera is based on male genitalic characters and is therefore tedious and time consuming, and species-level keys generally do not exist to the larva, many production taxonomy laboratories have elected to identify dytiscids to the genus level, even if the genus could be monospecific in the Black Hills region. In several genera, females cannot be identified to species.

Acilius Leach

Acilius semisulcatus Aubé

Description/diagnosis: Larson et al. 2000 (adult). Larson et al. (2000) indicated that there are questions as to the valid description of the larvae, surmising that Watt (1970) is likely correct. This species has the widest distribution of any *Acilius* species, being distributed from Alaska to British Columbia's Vancouver Island transcontinentally eastward to Labrador, Newfoundland, the Canadian Maritimes, and the U.S. Atlantic coast as far south as Wyoming, South Dakota, Iowa, Illinois, Indiana, Ohio, and Virginia. Larson et al. (2000) reported that this species has broad ecological requirements and can be found in many types of water bodies, particularly vernal pools.

Kirk and Balsbaugh (1975) reported this species from Custer and Pennington counties in South Dakota.

Agabus Leach

In addition to the species listed here, Kirk and Balsbaugh (1975) reported *A. paludosus* (Fabricius) from Pennington County in the Black Hills; however, this is a Palaearctic species and not likely to be found in the Black Hills. Unfortunately, I do not know what species they may have been referring to when using the name *A. paludosus*.

Shearer (2006) reported the genus *Agabus* from streams in the Black Hills, and I have specimens of unidentified *Agabus* from Deadwood and from Battle Creek at Keystone in my personal collection.

Agabus canadensis Fall

Description/diagnosis: Larson et al. 2000 (adult); larva apparently undescribed. Distributed from Yukon Territory, British Columbia, and Washington south and east to Colorado, South Dakota, Wisconsin, and Manitoba. Characteristic of grassland ponds (Larson et al. 2000)

Kirk and Balsbaugh (1975) reported this species from Pennington County, South Dakota.

Agabus disintegratus (Crotch)

Description/diagnosis: Barman et al. 2000 (larva), Larson et al. 2000 (adult). Distributed from Oregon and California east to Texas, Alabama, Virginia, New Jersey, and Connecticut, north to Idaho, South Dakota, Wisconsin, and Ontario. This is the only *Agabus* species in the Black Hills that has longitudinal stripes on the elytra, and it is characteristic of ephemeral ponds (Larson et al. 2000).

Kirk and Balsbaugh (1975) reported this species from Pennington County, South Dakota.

Agabus griseipennis LeConte

Description/diagnosis: Larson et al. 2000 (adult); larva apparently undescribed. Distributed from British Columbia, Alberta, and Saskatchewan south to California, Arizona, New Mexico, and Nebraska. Apparently prefers lentic habitats and very slow lotic habitats, including intermittent streams and shallow lakes (Larson et al. 2000).

Kirk and Balsbaugh (1975) reported this species from Pennington County, South Dakota.

Agabus punctatus **Melsheimer**

Description/diagnosis: Larson et al. 2000 (adult); larva apparently undescribed. Distributed in the U.S. from Massachusetts to South Carolina, west to Kansas and eastern Texas (Larson et al. 2000). Known from small vernal pools (Larson et al. 2000).

Kirk and Balsbaugh (1975) reported this species from Lawrence County, South Dakota.

Boreonectes **Angus**

This genus was recently described for the old *Stictotarsus griseostriatus* (De Geer) group of sibling species (Angus 2010), and many previous records refer to the old genus name, or sometimes the even older genus name *Hydroporus*. The temperature preference in Grafe et al. (2002) for *Stictotarsus* is eurythermal warm, but an average preference based on the literature-based data in Vieira et al. (2006) is stenothermal cold. These preferences would apply to *Boreonectes*, as well. Furthermore, *Boreonectes* and the transferred species had not yet been entered into ITIS as of 2018.

Boreonectes griseostriatus **(DeGeer)**

Description/diagnosis: Larson et al. 2000 (adult, as *Stictotarsus griseostriatus*); larva has not been described, at least from North American material – Bertrand's (1928) description using European material may be of another species (Larson et al. 2000). Holarctic distribution; in North America, it extends from Alaska to Newfoundland and the Canadian Maritimes, south to California, Arizona, New Mexico, Nebraska, Iowa, Tennessee, and North Carolina. This species appears to prefer barren ponds and is a colonizing species (Larson et al. 2000). The TSN for this species as *Stictotarsus griseostriatus* is 728483.

Kirk and Balsbaugh (1975) reported this species from Lawrence County, South Dakota, as *Hydroporus coloradensis*.

Boreonectes striatellus **(LeConte)**

Description/diagnosis: Larson et al. 2000 (adult, as *Stictotarsus striatellus*); larva has not been described. A western species, *B. striatellus* is distributed from the Yukon Territory, Northwest Territories, and Manitoba south through the western U.S. into Mexico; no further east than South Dakota, Colorado, and Texas. This species has very broad ecological requirements in terms of water body sizes, but generally prefers more lentic habitats (Larson et al. 2000). The TSN for this species as *Stictotarsus striatellus* is 568829.

Colymbetes **Clairville**

The temperature preference in Grafe et al. (2002) for *Colymbetes* is eurythermal warm, but an average preference based on the literature-based data in Vieira et al. (2006) is stenothermal cold.

Colymbetes exaratus **LeConte**

Description/diagnosis: Larson et al. 2000 (adult); larvae are as yet undescribed. Distributed from Yukon and British Columbia south and east to Wyoming, Nebraska, Missouri, Illinois, Wisconsin, and Manitoba. Larson et al. (2000) indicate that this species is found in dense emergent vegetation along grassland and parkland ponds and marshes.

Kirk and Balsbaugh (1975) reported this species from the Black Hills but did not specify a county.

Dytiscus **Linnaeus**

Dytiscus marginicollis **LeConte**

Description/diagnosis: Roughley 1990, Larson et al. 2000 (adult); I can find no formal description of the larva. Distributed in western North America from southern British Columbia, Alberta, Saskatchewan, and Manitoba south to Baja California, Arizona, and New Mexico, and into mainland Mexico as far as Durango. Characteristically a lentic species.

I have an adult specimen of this species which I collected in Spearfish at a hotel light on 8 October 1996, representing a **new state record** for this species.

Hydroporus **Clairville**

Hydroporus fuscipennis **Schaum**

Description/diagnosis: Alarie 1991 (larva), Larson et al. 2000 (adult). Holarctic distribution; in North America, it is distributed transcontinentally from Alaska, British Columbia, and Washington to Newfoundland and the St. Lawrence River, south into the contiguous U.S. to Utah, Colorado, North Dakota, Wisconsin, and New Hampshire. Prefers grassland pools that are well-vegetated (Larson et al. 2000).

Kirk and Balsbaugh (1975) reported this species from Pennington County, South Dakota.

Hydroporus notabilis **LeConte**

Description/diagnosis: Larson et al. 2000 (adult); larva apparently undescribed. Distirbution is Holarctic; in North America, distribution extends from Alaska, British Columbia, and Oregon east to Labrador and the Canadian Maritimes, south to Utah, Colorado, Kansas, Iowa, Illinois, Michigan, Pennsylvania, and Rhode Island. Larson et al. (2000) indicate that this species prefers small, nonvegetated pools.

Kirk and Balsbaugh (1975) reported this species from Pennington County, South Dakota.

Hydroporus occidentalis **Sharp**

Description/diagnosis: Larson et al. 2000 (adult); larva apparently undescribed. Distributed from the Yukon Territory south to Washington, Utah, Colorado, Alberta, and Saskatchewan. Larson et al. (2000) report that this species is common in small, silty alpine/subalpine pools, along the margins of lakes, and in small cold springs and seeps.

Kirk and Balsbaugh (1975) reported this species from Lawrence and Pennington counties, South Dakota.

Hygrotus **Stephens**

A total of 31 species of *Hygrotus* occur in the Nearctic, of which 18 species have been reported from South Dakota and/or Wyoming (Larson et al. 2000, listed below); however, no species have been specifically reported from the Black Hills.

The following 18 species of *Hygrotus* have been reported from South Dakota and/or Wyoming: *H. acaroides* (LeConte), *H. bruesi* (Fall), *H. compar* (Fall), *H. dissimilis* Gemminger & Harold, *H. impressopunctatus* (Schaller), *H. infuscatus* (Sharp), *H. marklini* (Gyllenhal), *H. masculinus* (Crotch), *H. nubilus* (LeConte), *H. patruelis* (LeConte), *H. picatus* (Kirby), *H. salinarius* (Wallis), *H. sayi* Balfour-Browne, *H. sellatus* (LeConte), *H. suturalis* (LeConte), *H. tumidiventris* (Fall), *H. turbidus* (LeConte), and *H. unguicularis* (Crotch).

Laccophilus **Leach**

Laccophilus fasciatus terminalis **Sharp**

Description/diagnosis: Larva undescribed, although the larva of another subspecies, *L. f. rufa* Melsheimer was described by Sizer et al. 1998, Zimmerman 1970 (adult). This subspecies has a wide distribution from Oregon, Utah, Wyoming, and North Dakota, south through California, Arizona, New Mexico, and Texas to central Mexico. Often a pioneer species in new, temporary ponds.

Zimmerman (1970) reported specimens of this subspecies from Hill City in Pennington County, South Dakota, and Kirk and Balsbaugh (1975) later also reported this subspecies from the Black Hills. The *rufa* subspecies occurs in far eastern South Dakota.

Laccophilus maculosus decipiens **LeConte**

Description/diagnosis: Barman 1972 (larva), Zimmerman 1970 (adult). This subspecies is distributed from British Columbia, Washington, Oregon, and California east to Saskatchewan, the Dakotas, Nebraska, Colorado, and Texas. Prefers shallow regions of permanent pools and ponds in forested and grassland areas (Larson et al. 2000).

Zimmerman (1970) reported specimens of this subspecies from Devil's Tower National Monument in Crook County, Wyoming, as well as intergrades with the nominate subspecies, *L. m. maculosus*, from Custer State Park in Custer County and in Pennington County, South Dakota. Specimens that are fully the nominal subspecies have not yet been reported from the Black Hills, although they have been reported as close as Rosebud in Todd County, South Dakota. Kirk and Balsbaugh (1975) reported this subspecies from the Black Hills.

Liodessus **Guignot**

Liodessus affinis **(Say)**

Description/diagnosis: Watts 1970 (larva); Larson et al. 2000 (adult). Eastern species, from Ontario to Newfoundland, south to Iowa, Illinois, Indiana, Ohio, and Virginia.

A distribution including South Dakota is not represented for this species in the maps in Larson et al. (2000); however, Kirk and Balsbaugh (1975) reported *Liodessus affinis* from Fall River, Lawrence, and Pennington counties in South Dakota.

Liodessus fuscatus **(Crotch)**

Description/diagnosis: Larson et al. 2000 (adult); larva is undescribed. Another eastern species with a distribution from Saskatchewan east to Labrador and the Canadian Maritimes, south to Texas, Wisconsin, Illinois, Alabama, Georgia, and Florida. Habitat is mossy mats along ponds, fens, and bogs (Larson et al. 2000).

A distribution including South Dakota is not represented for this species in the maps in Larson et al. (2000); however, Kirk and Balsbaugh (1975) reported *Liodessus fuscatus* from Pennington County in South Dakota.

Neoporus **Guignot**

Neoporus dimidiatus **(Gemminger & Harold)**

Description/diagnosis: Larson et al. 2000 (adult); larva is undescribed. Distributed in the western Great Plains from Alberta, Saskatchewan, and Manitoba south to New Mexico and Texas, west into Arizona; east of the Mississippi River, it has been reported from Georgia, Virginia, New England,

Ontario, Quebec, and the Canadian Maritimes. Larson et al. (2000) reported that this species occurs among emergent sedges and reeds in spring-fed pools, beaver ponds, and protected shorelines of large rivers and lakes.

Kirk and Balsbaugh (1975) reported this species from Custer, Fall River, and Pennington counties in the Black Hills as its synonym, *Hydroporus solitarius*.

Neoporus undulatus (Say)

Description/diagnosis: Matta and Peterson 1985, Alarie 1991 (larva, as *Hydroporus undulatus*), Larson et al. 2000 (adult). Distributed transcontinentally from British Columbia to Newfoundland and the Canadian Maritimes, south to Oregon, South Dakota, Arkansas, Alabama, Georgia, and North Carolina. This species prefers emergent vegetation in permanent lentic habitats, such as beaver ponds, marshy streams, and small lakes (Larson et al. 2000).

Kirk and Balsbaugh (1975) reported this species from Lawrence and Pennington counties in the Black Hills as *Hydroporus consimilis* and as *H. undulatus*.

Platambus Thomson

Platambus semivittatus (LeConte)

Description/diagnosis: Barman et al. 1999 (larva, as *Agabus semivittatus*), Larson et al. 2000 (adult, as *Agabus semivittatus*). Distributed transcontinentally from California and Nevada east to Texas, Mississippi, Tennessee, Virginia, Massachusetts, Quebec, Ontario, and Minnesota, south into Mexico. This species prefers streams, seeps, and spring-fed ponds.

Kirk and Balsbaugh (1975) reported this species from Lawrence County, South Dakota, as *Agabus semivittatus*.

Rhantus Dejean

Rhantus gutticollis (Say)

Description/diagnosis: Larson et al. 2000 (adult); larva has not been described. Distributed from British Columbia, Montana, South Dakota, and Wisconsin, south through California, Arizona, New Mexico, and Texas in to Mexico and Central America. Appears to have broad ecological requirements among small streams and ponds (Larson et al. 2000).

Kirk and Balsbaugh (1975) reported this species from Fall River and Pennington counties in the Black Hills (as *Rhantus hoppingi*), while Larson et al. (2000) specifically mentioned the Black Hills of South Dakota in the distribution of this species.

Rhantus sericans Sharp

Description/diagnosis: Larson et al. 2000 (adult); larva has not been described. Distributed from Alaska to Quebec, south to California, New Mexico, Nebraska, Missouri, and Wisconsin. A pioneering species that cis common in most types of lentic habitats in grassland and parkland regions (Larson et al. 2000).

Kirk and Balsbaugh (1975) reported this species from the Black Hills (no county specified) as *Rhantus notatus*. ITIS indicated that *R. notatus* is a synonym of *R. frontalis* (Marsham); however, Larson et al. (2000) consider *R. frontalis* as a solely Palaearctic species, and the specimens attributed to *R. frontalis* in North America to be *R. sericans*.

Thermonectus Dejean

Thermonectus nigrofasciatus ornaticollis (Aubé)

Description/diagnosis: Hilsenhoff 1993 (larva, characters in key); Larson et al. 2000 (adult, characters in key). This subspecies is distributed in "eastern United States, north to Wisconsin and North Dakota, west to Texas, Oklahoma, and Kansas" (Larson et al. 2000). Lentic habitats.

Kirk and Balsbaugh (1975) reported this species from the Black Hills, as *Thermonectes [sic] ornaticollis*.

Uvarus Guignot

Uvarus granarius (Aubé)

Description/diagnosis: Matta 1983 (larva); Larson et al. 2000 (adult). Distributed in eastern North America from Manitoba, Ontario, Quebec, and New Brunswick south to Texas, Iowa, Tennessee, and Florida. Occurs in vegetation mats along the shorelines of small pools (Larson et al. 2000).

A distribution including South Dakota is not represented for this species in the maps in Larson et al. (2000); however, Kirk and Balsbaugh (1975) reported *Uvarus granarius* from Pennington County in South Dakota.

Elmidae

The Elmidae are represented by at least eleven species in the Black Hills (six of which were reported by Kirk and Balsbaugh 1975). Multiple species of *Stenelmis* occur in the state of South Dakota, but only larvae of this genus have been reported from the Black Hills (Shearer 2006). Most riffle beetles are almost exclusively encountered in streams, so lentic habitat preferences do not exist except for *Dubiraphia quadrinotata* and *Stenelmis* sp. At the family level, Elmidae are considered eurythermal warm according to the data in Grafe et al. (2002), but a weighted average of the data in Vieira et al. (2006) would classify Elmidae as eurythermal cool.

Cleptelmis Sanderson

Cleptelmis is represented in the Nearctic by a single species, *C. addenda*. The synonymy of *C. ornata* with *C. addenda* was made according to an obscure paper published in 1989 by Bill Shepard in *Water Beetles of China*, volume 2, that I was not made aware of until 2014. The two species were previously separated by the presence of humeral and apical red spots on the elytra (*C. ornata*) and the lack thereof (*C. addenda*), but the genitalia are identical.

Cleptelmis addenda (Fall)

Descriptions/Diagnoses: Sanderson 1953-1954 (larva, in genus description), Brown 1976 (adult; characters in keys). Distributed from British Columbia south to California and east to Montana, South Dakota, Colorado, and New Mexico. The temperature designation for *C. addenda* using data from Grafe et al. (2002) would be eurythermal cool, but the temperature designation using data from Vieira et al. (2006) would be stenothermal cold.

Shearer (2006) reported both adults and larvae of *Cleptelmis* (presumably *C. addenda*) from the Black Hills, but the state's data that he used were only identified to the genus level.

Dubiraphia Sanderson

Dubiraphia quadrinotata (Say)

Descriptions/Diagnoses: Sanderson 1953-1954 (larva, in genus description), Brown 1976 (larva, adult; characters in keys). Shearer (2006) reported adults and larvae of *Dubiraphia* from the Black Hills, but did not provide a species designation; it is likely that they were *D. quadrinotata*. The temperature designation for *D. quadrinotata* using data from Grafe et al. (2002) would be eurythermal warm, but the temperature designation using data from Vieira et al. (2006) would be eurythermal cool.
I have collected larvae of *Dubiraphia* from the Fall River near Hot Springs.

Heterlimnius Hinton

Heterlimnius corpulentus (LeConte)

Descriptions/Diagnoses: Brown 1976 (larva, adult; characters in keys). Distributed from British Columbia and Montana south to California, Arizona, and New Mexico. Shearer (2006) reported adults and larvae of *Heterlimnius* from the Black Hills.
Heterlimnius corpulentus is one of the most common riffle beetles I saw in the Black Hills.

Lara LeConte

Lara avara LeConte

Descriptions/Diagnoses: Brown 1976 (larva, adult; characters in keys). Brown (1976) reports this species as occurring in rapid mountain and foothill streams from British Columbia to southern California, eastward to Idaho, Wyoming, Utah, and Colorado. The nominate subspecies is rarely collected in the Black Hills, with the other subspecies (*L. a. amplipennis* Darlington) restricted to British Columbia and Washington (Brown 1976).
Lara avara has not previously been reported from South Dakota, but I have seen it in a biomonitoring sample from the Black Hills.

Microcylloepus Hinton

Microcylloepus pusillus (LeConte)

Descriptions/Diagnoses: Brown 1976 (larva, adult; characters in keys). This species has numerous subspecies, most of which are very geographically restricted. The nominate subspecies, however, is widely distributed from California and Montana east to the Atlantic Coast from Florida to Maine, northward into Ontario.
Microcylloepus pusillus is more common in streams as they exit the Black Hills than in the mountainous area of the Black Hills. Shearer (2006) reported adults and larvae of *Microcylloepus* from the Black Hills, and I have collected larvae of *Microcylloepus* from Battle Creek near Hermosa; these are likely *M. pusillus*. The temperature designation for *M. pusillus* using data from Grafe et al. (2002) would be eurythermal warm, but the temperature designation using data from Vieira et al. (2006) would be eurythermal cool.

Narpus Casey

Narpus concolor (LeConte)

Descriptions/Diagnoses: Brown 1976 (larva, adult; characters in keys). Distributed in western North America from British Columbia and Alberta south to California, Arizona, and New Mexico.

Optioservus **Sanderson**

Based on the distribution maps in White (1978), only *O. divergens* occurs in the Black Hills; however, a couple other species have distributions that appear to approach the Black Hills and could feasibly occur there. I have seen a couple specimens of *Optioservus* with a light humeral maculation, which could not be assigned to the immaculate *O. divergens*, but were not easily assignable to any other species either. In almost all cases, the specimens were female, so male genitalia could not be examined. The name that fit those specimens best was *O. seriatus*.

Other species with distributions approaching the Black Hills are *O. quadrimaculatus* (Horn) which is common throughout the northern Rocky Mountains as far south as central California, southern Utah, and southern Colorado (White 1978), and *O. castanipennis* (Fall), which is distributed in the central Rocky Mountains from southern Wyoming to northern New Mexico, west to northern Nevada (De Jong et al. 2013).

Shearer (2006) reported both adults and larvae of *Optioservus* from the Black Hills, not assigning a species name to any of the material available. Likewise, Jurgens (1968) reported *Optioservus* only at the genus level, from French Creek, Spring Creek, and Battle Creek.

Optioservus divergens **(LeConte)**

Descriptions/Diagnoses: White 1978. Distribution ranges from southern California, Arizona, and New Mexico north to Oregon, British Columbia, Idaho, Wyoming, and the Black Hills of South Dakota. This is the only species of *Optioservus* previously reported from South Dakota and the Black Hills.

By far, the most common species of *Optioservus* in the Black Hills is *O. divergens*; in some sites, it is a dominant taxon, numerically, in the entire invertebrate community.

Optioservus seriatus **(LeConte)**

Descriptions/Diagnoses: White 1978. The distribution maps in White (1978) show *O. seriatus* as occurring primarily in northern California, but with scattered reports from Oregon and Washington southeast through Idaho and Wyoming to central Colorado. In the maps, the nearest localities to the Black Hills are in central and southeastern Wyoming (near Casper and Wheatland).

In biomonitoring samples from the Black Hills, I have rarely encountered specimens of *Optioservus* which could not be assigned to *O. divergens* due to the presence of a humeral maculation. Based primarily on the color and relative density of hairs on the pronotal disk and abdominal sternites, these female specimens were tentatively assigned to *O. seriatus* or *O. quadrimaculatus* (Horn). After examination of both of those species from within their known ranges, I am pretty sure, but not certain, that these are *O. seriatus*. Either way, these specimens would represent a new state record, but confirmation will have to wait until males are collected and genitalia can be examined.

Stenelmis **Dufour**

Stenelmis generally prefers larger streams in the western United States, so its presence in Black Hills streams is rare and surprising. Shearer (2006) reported larvae of *Stenelmis* from streams in the Black Hills, and I have collected larvae of *Stenelmis* from the Fall River near Hot Springs. I do not have any idea what species of *Stenelmis* could be represented in these samples.

Zaitzevia Champion

Zaitzevia parvula (Horn)

Descriptions/Diagnoses: Pennak 1953 (larva), Brown 1976 (larva, adult; characters in keys). This species is widely distributed in western North America from British Columbia and Alberta south to California, Arizona, New Mexico, and South Dakota.

Jurgens (1968) reported *Zaitzevia* at the genus level from Battle Creek and the Fall River. Shearer (2006) reported both adults and larvae of *Zaitzevia* from streams in the Black Hills. The temperature designation for *Z. parvula* using data from Grafe et al. (2002) would be eurythermal warm, but the temperature designation using data from Vieira et al. (2006) would be eurythermal cool.

Gyrinidae

Kirk and Balsbaugh (1975) reported two species of whirligig beetles from the Black Hills. Oygur and Wolfe (1991) reported two additional species from in or near the Black Hills, based on distribution maps. I have not seen Gyrinidae from biomonitoring samples in the Black Hills, which is not surprinsing since they would not be common in those kinds of habitats. Based on our limited ecological information on gyrinids, the identification of any whirligig beetles found during biomonitoring could be left at the family level.

Gyrinus Geoffroy

Gyrinus affinis Aubé

Descriptions/diagnoses: Oygur and Wolfe 1991. Distributed from Alaska, the Yukon, and the Nothwest Territories transcontinentally to Newfoundland, south to British Columbia, Wyoming, Kansas, Iowa, Illinois, Indiana, Ohio, Pennsylvania, and New Jersey.

Gyrinus aquiris LeConte

Descriptions/diagnoses: Oygur and Wolfe 1991. Distributed from Manitoba and South Dakota east to Nova Scotia, New Hampshire, and New Jersey, as far south as Ohio. Oygur and Wolfe's (1991) maps do not have a record for this species in the Black Hills; rather, the dot is in far northeastern South Dakota.

Gyrinus maculiventris LeConte

Descriptions/diagnoses: Oygur and Wolfe 1991. Distributed from the Northwest Territories south through western Canada (east to far western Quebec) to California, Colorado, Kansas, Illinois, Ohio, West Virginia, and Maryland. The Black Hills record is a dot on the map in South Dakota, perhaps a bit northeast of the Black Hills proper.

Gyrinus sayi Aubé

Descriptions/diagnoses: Oygur and Wolfe 1991. Distributed from the Northwest Territories south to Washington, and a relatively narrow transcontinental band to Labardor, Newfoundland, and the US coast as far south as Maryland and Delaware. The Black Hills record is located by a dot in far northeastern Wyoming on the map.

Haliplidae

Black Hills records for six species of haliplid beetles were found in Kirk and Bilsbaugh (1975). These insects are uncommon in riffles and I never saw them from biomonitoring samples in the Black Hills; however, they are frequently found in stream margins and slower water such as pools. Description/diagnosis citations refer to adults, since the larvae are unknown for most species.

Haliplus Latreille

Haliplus immaculicollis Harris

Description/diagnosis: Durfee et al. 2005. Distributed from Alaska, the Yukon Territory, the Northwest Territories, Quebec, and the Canadian Maritimes south to Oregon, Colorado, Texas, Indiana, Ohio, and Virginia.

Haliplus pantherinus Aubé

Description/diagnosis: Hilsenhoff and Brigham 1978; Majka et al. 2009 (characters in keys). Distributed from the Canadian Maritimes and New England west to Wisconsin and south to Oklahoma.

Haliplus fulvus Fabricius

Description/diagnosis: Durfee et al. 2005. A Holarctic species, distributed from British Columbia, south to California, and east to the Northwest Territories, Nova Scotia, Massachusetts, Wisconsin, and Colorado. The Kirk and Bilsbaugh (1975) records are based on *Haliplus subguttatus* Crotch, which was synonymized under *H. fulvus* by Vondel (1991), along with *H. salinarius* Wallis; these two species were the previous names for these North American records. This change was not yet reflected in ITIS as of 2018, so a TSN does not exist for this species.

Haliplus triopsis Say

Description/diagnosis: Durfee et al. 2005. Distributed from the Atlantic Coast from Quebec and Maine south to Florida, west to Ontario, Wisconsin, South Dakota, Colorado, and New Mexico.

Peltodytes Régimbart

Peltodytes callosus (LeConte)

Description/diagnosis: Durfee et al. 2005. Distributed from the Pacific Coast from British Columbia south to Baja California and east to Alberta, Minnesota, Kansas, and New Mexico.

Peltodytes edentulus (LeConte)

Description/diagnosis: Durfee et al. 2005. Distributed in a transcontintal band from British Columbia and Washington east to Ontario, Massachusetts, Pennsylvania, and Maryland, south to Colorado, Kansas, Illinois, and Indiana.

Helophoridae

The family Helophoridae is represented in North America by a single genus, *Helophorus*. Only two species of Helophoridae have been reported from the Black Hills (Kirk and Bilsbaugh 1975), one of which, *Helophorus granularis* (Linnaeus), is Palaearctic in distribution. I do not know which

Helophorus species was referred to as *H. granularis*, and increased collecting could result in several new distributional records for this family.

Shearer (2006) reported adults of *Helophorus* from the Black Hills. I collected a larval *Helophorus* from Whitewood Creek just north of the Black Hills.

Helophorus Fabricius

Helophorus linearoides d'Orchymont

Description/diagnosis: Smetana 1988 (adult); larva is undescribed. Distributed from southeastern British Columbia to Manitoba, south to California, Arizona, and Iowa.

Heteroceridae

I anticipate that this family would occur in the Black Hills, but I have no records of it being there.

Hydraenidae

Numerous species of Hydraenidae have been reported from Wyoming (12 species) and South Dakota (3 species), but only two of these have been reported from the Black Hills, both in the genus *Ochthebius* (Perkins 1980). Based on the distribution records on maps in Perkins (1980), the rest are in areas of those states quite distant (far west or far east) from the Black Hills. It is likely that additional collecting efforts will result in new records of hydraenids from this region.

Although hydraenids are not common to the habitats sampled in most biomonitoring programs, I have encountered two genera of hydraenids (*Hydraena* and *Ochthebius*) in biomonitoring samples but the specimens were not identified to species.

Hydraena Kugelann

Although the published literature does not report the genus *Hydraena* from the Black Hills, Shearer (2006) reported adults of *Hydraena* from state data on biomonitoring samples in the Black Hills.

Ochthebius Leach

Ochthebius lineatus LeConte

Descriptions/diagnosis: Perkins 1980, which goes to great lengths to describe the intraspecific variation in this species – a map shows the geographic distribution of 16 different variants of the aedeagal apex across its range. Perkins (1980) described this species as the most frequently collected North American *Ochthebius*, with a distribution extending from British Columbia and the Pacific Coast states, east to Ontario and Wisconsin, south through Iowa, Kansas, eastern Oklahoma, eastern Texas, Mexico, and Central America to Colombia.

Perkins (1980) reported specimens collected in Hot Springs, 6 miles south of Hot Springs, Angostura Dam, Smithwick, Spearfish Creek 3 miles north of Spearfish, Cheyenne Crossing, and the Belle Fourche River at Devils Tower National Monument.

Ochthebius tubus Perkins

Descriptions/diagnosis: Perkins 1980. According to the maps in Perkins (1980), the distribution of this species consists of widely scattered records from northern Montana to Oklahoma to northeast Mexico and the Baja Peninsula, with a concentration of collecting localities along coastal California.

The only record of this species from the Black Hills region is from the Belle Fourche River at Devils Tower National Monument (Perkins 1980).

Hydrochidae

The only published record of Hydrochidae from the Black Hills is in Kirk and Bilsbaugh (1975), which reported *Hydrochus rufipes* as a hydrophilid.

Hydrochus Leach

Hydrochus rufipes Melsheimer

Description/diagnosis: Smetana 1988 (adult); larvae have not been described. Distributed in the eastern United States from New York to Florida, westward to Wyoming and Texas.

Hydrophilidae

Few published records exist for water scavenger beetles in the Black Hills or in South Dakota, mostly in Kirk and Bilsbaugh (1975). A few records have been found in revisions of various genera (e.g., Gundersen 1978) or isolated references to South Dakota or Wyoming in Smetana (1988); none are records specifically from the Black Hills.

A few hydrophilids are terrestrial, most of which are in the Sphaeridiinae, including some *Cercyon* species and *Sphaerium*.

Hydrophilids are not commonly collected in stream biomonitoring efforts. Except in *Tropisternus*, ecological knowledge of this group is primarily at the genus level, so species identification of hydrophilids other than *Tropisternus* is usually not necessary in biomonitoring programs. Genera that I have seen reported in biomonitoring efforts included *Enochrus*, *Hydrobius*, *Laccobius*, and *Paracymus*.

Cercyon Leach

Cercyon is primarily a terrestrial genus inhabiting decomposing organic matter; however, it can sometimes be found in wet situations, including waterfalls and beaches. I have not seen any records of *Cercyon* from aquatic biomonitoring samples, and given their natural history, I wouldn't expect to. However, the following two species are known from South Dakota – I do not know if the records were in the Black Hills.

Cercyon connivens Fall

Description/diagnosis: Smetana 1988 (adult). Distributed from southern Ontario through the eastern United States west to Minnesota, South Dakota, and eastern Oklahoma.

Smetana (1988) reported that this species occurs "mainly in wet habitats, such as swamps and bogs, edges of ponds, lakes, etc., in various debris, in *Sphagnum* and other mosses, under leaf litter, etc."

Cercyon matthewsi Smetana

Description/diagnosis: Smetana 1988 (adult). Smetana (1988) reported that the only known occurrences of this species at the time were from the type locality in Alaska, where it was found under a dead salmon on a river bank, and in South Dakota.

Crenitis Bedel

Crenitis digesta (LeConte)

Description/diagnosis: Smetana 1988 (adult). Distributed from Alberta east to the Maritime Provinces, south into the eastern United States.

Smetana (1988) reported that this species is found in both running and standing water, but otherwise no details are known of its habitat preferences.

Enochrus Thomson

Enochrus cinctus (Say)

Description/diagnosis: Smetana 1988 (adult). Eastern North America west to Saskatchewan, Wisconsin, South Dakota, Kansas, and Texas. Smetana (1988) states that this species "occurs in a wide range of aquatic habitats, particularly in shallow ponds with plenty of organic debris."

Enochrus diffusus (LeConte)

Description/diagnosis: Smetana 1988 (adult). Distributed from British Columbia south to southern California, east to the Northwest Territories, Manitoba, and Illinois. Isolated populations are known in eastern Ontario and New York, according to Smetana (1988). Habitat preferences for this species are unknown.

Enochrus hamiltoni (Horn)

Description/diagnosis: Smetana 1988 (adult). Distributed transcontinentally, from British Columbia and the Northwest Territories to the Maritime Provinces and south throughout the United States. This species occupies a wide range of aquatic habitats.

Enochrus pygmaeus (Fabricius)

Description/diagnosis: Epler 2010 (adult). Both Short (2004) and Epler (2010) reject the subspecies erected within E. pygmaeus, allowing the distribution to range from New England and the Atlantic seaboard south to Florida, the Bahamas, and the Greater Antilles, westward to Wyoming, Colorado, Texas, and central Mexico.

Hydrobius Leach

Hydrobius fuscipes (Linnaeus)

Description/diagnosis: Smetana 1988 (adult). Holarctic; in North America, "it occurs throughout Canada and the United States, including Alaska."

According to Smetana (1988), this species "frequents a wide range of aquatic habitats; however, it seems to be especially abundant in shallow stagnant pools with plenty of plant debris. . . . also abundant in swampy habitats and in *Sphagnum* bogs."

Hydrochara Berthold

Hydrochara occulta (d'Orchymont)

Description/diagnosis: Matta 1982 (larva), Smetana 1980 (adult). Generally a coastal species, with a distribution from Massachusetts south along the Atlantic seaboard to Florida and along the Gulf

Coast to Texas, with records from Oklahoma nd Tennessee (Epler 2010). Its presence in the Black Hills would be surprising, so it is likely that the reference from Kirk and Bilsbaugh (1975) that it occurs there is likely in error; however, I do not know what other species Kirk and Bilsbaugh (1975) could have been referring to.

Hydrophilus Müller

Hydrophilus triangularis Say

Description/diagnosis: Wilson 1923 (larva); Smetana 1988, Epler 2010 (adult). Distributed transcontinentally from British Columbia to Quebec in Canada and throughout the United States, south into Mexico.

This species prefers deep pools of standing water. I have frequently seen this species at lights, although not in the Black Hills – an observation echoing that of Epler (2010), in which he stated that *H. triangularis* is more commonly found in well-lit parking lots than in aquatic samples.

Laccobius Erichson

Laccobius was not listed among the species reported by Kirk and Bilsbaugh (1975) for the Black Hills; however, I have seen larvae of *Laccobius* from a biomonitoring sample. I do not know what species of *Laccobius* the specimen represented.

Paracymus Thomson

Paracymus subcupreus (Say)

Description/diagnosis: Smetana 1988 (adult), larva is apparently not yet described. Widely distributed over most of eastern North America from the Rocky Mountains to Ontario and New England, south to Florida and the Gulf Coaat.

Sphaeridium Fabricius

Sphaeridium bipustulatum Fabricius

Description/diagnosis: Smetana 1978 (larva, adult), 1988 (adult). Holarctic distribution, with Palaearctic distribution ranging across Europe and northern Asia, south into northern Africa, In the Nearctic, it is distributed transcontinentally from British Columbia to Nova Scotia, south through the United States. Smetana (1978, 1988) indicates that it is adventive in North America.

This is primarily a terrestrial species of Hydrophilidae, commonly being found in decaying organic matter such as carrion and, especially, fresh cow dung. It is unlikely to be found in streams but it is a possible hydrophilid where cows frequent. Smetana (1984) reported this species from Pennington County, nine miles north of Custer.

Tropisternus Solier

Tropisternus columbianus Brown

Description/diagnosis: Smetana 1988 (adult). Distributed in western North America from British Columbia south to Mexico, and east to Ontario and Ohio.

Smetana (1988) states that this species "occurs in a wide range of aquatic habitats in both standing and running water; however, it seems to prefer lentic habitats. Some speciemens were even collected in hot springs."

Tropisternus ellipticus (LeConte)

Description/diagnosis: Smetana 1988 (adult). Distributed from British Columbia south through the United States, Mexico, and Central America to Panama, east to Illinois, Missouri, and Arizona.

I have several adults of *Tropisternus ellipticus* in my personal collection, collected from the Belle Fourche River at a site five miles south of Newell on the Plains just north of the Black Hills. Smetana (1988) says that this species is "a very adaptable speices, occurring in a wide range of aquatic habitats in both standing an running water. Temporary and semipermenent habitats seem to be favored."

Tropisternus lateralis limbalis (LeConte)

Description/diagnosis: LeConte 1855 (adult). One of the five recognized subspecies of *T. lateralis* and described as a separate species by LeConte (1855) immediately after the description of the nominate subspecies (both within the genus *Hydrophilus*). Very widely distributed across southern Canada from British Columbia to Nova Scotia, south through the entire United States through Mexico to Panama and throughout the Caribbean.

Kirk and Bilsbaugh (1975) listed this species from the Black Hills as the subspecies *T. lateralis binotatus*, which is a synonym.

Tropisternus sublaevis (LeConte)

Description/diagnosis: LeConte 1855 (adult). Eastern North America from Nebraska and the Dakotas to the Atlantic seaboard south to Georgia.

Hydroscaphidae

The family Hydroscaphidae is represented in North America by a single genus, *Hydroscapha*, with only two described species in the Nearctic. Known distribution records for those two *Hydroscapha* species include Arizona, California, Idaho, and Nevada (De Jong et al. 2005; Meier et al. 2010); however, it appears likely that multiple species are actually involved, and many of those species may be restricted to isolated hot springs within the range of *Hydroscapha*.

Ecological information on lentic body size preference, current and substrate preferences for hydroscaphids are lacking.

Hydroscapha LeConte

Hydroscapha natans LeConte

Description/diagnosis: Meier et al. 2010 (adults, compares to a new species). Currently known to be distributed in Arizona, California, Idaho, and Nevada; this record would be a new state record and an eastern extension for the species, genus, and family.

I collected two hydroscaphid larvae from False Bottom Creek on 25 August 2009 for my personal collection. This record would serve as a **new state record** for South Dakota for this species.

Leiodidae

One species of leiodid beetle is an interesting ectoparasite that feeds on the dandruff of the beaver, *Castor canadensis* Kuhl, an aquatic mammal, so it is included here.

Platypsylla Ritsema

Platypsylla castoris Ritsema

Description/diagnosis: Peck 2006 (figures of larvae and adults), Moskowitz 2011 (photo of adult). Peck (2006) suggested that the actual distribution of this species is "probably the same as that of the host." Beavers are reported from all of North America except western and northern Alaska, northern Northwest Territories, including the Islands, northern Quebec, peninsular Florida, all of Mexico except scattered localities near the US border, and the Great Basin.

This species has not yet been reported from the Black Hills, although it has been reported from eastern South Dakota (Peck 2006). Beavers are common in the Black Hills (Turner 1974), so it is likely that this species occurs there, as well. Ecological data are lacking, but this is not likely to be a problem from an aquatic biomonitoring standpoint, since these beetles are not likely to be encountered during stream surveys.

Noteridae

I anticipate that this family would occur in the Black Hills, but I have no records of it being there.

Scirtidae

I anticipate that this family would occur in the Black Hills, but I have no records of it being there.

Staphylinidae

I don't know if aquatic representatives of this huge, primarily terrestrial family occur in the Black Hills. As with the Chrysomelidae and Curculionidae, other families that are primarily terrestrial with a few aquatic species, a good standard operating procedure is to consider staphylinids collected in aquatic biomonitoring samples to be terrestrial unless they exactly fit the descriptions of the aquatic species.

Shearer (2006) reported adults of Staphylinidae from the Black Hills.

Aquatic Coleoptera Traits Table

Table 10: Aquatic Coleoptera Traits Table

Species	TSN	Seen	Tolerance Value	FFG	Habit
Chrysomelidae	114509	+	--	SH	CN
Donacia	114510	/	--	SH	CN
Donacia hirticollis	114518	/	--	SH	CN
Donacia magnífica	114519	/	--	SH	CN
Donacia subtilis	114527	/	--	SH	CN
Donacia tuberculifrons	114528	/	--	SH	CN
Plateumaris	114564	/	--	SH	CN
Plateumaris germari	114572	/	--	SH	CN
Plateumaris nitida	719873	/	--	SH	CN
Plateumaris pusilla	114574	/	--	SH	CN
Plateumaris robusta	114575	/	--	SH	CN
Curculionidae	114666	-	--	SH	CN
Dryopidae	113999	+	--	SC	CN
Helichus	114006	+	5	SC	CN
Helichus striatus	114017	+	5	SC	CN
Dytiscidae	111963	+	5	PR	--
Acilius	112074	-	5	PR	--
Acilius semisulcatus	112075	-	5	PR	--
Agabus	111966	+	8	PR	SW
Agabus canadensis	112001	/	8	PR	SW
Agabus disintegratus	111978	/	8	PR	SW
Agabus griseipennis	112057	/	8	PR	SW
Agabus punctatus	112053	/	8	PR	SW
Boreonectes	--	+	5	PR	SW
Boreonectes griseostriatus	--	-	5	PR	SW
Boreonectes striatellus	--	+	5	PR	SW
Colymbetes	112379	-	5	PR	SW
Colymbetes exaratus	112380	-	5	PR	SW
Dytiscus	112118	-	5	PR	SW
Dytiscus marginicollis	112124	-	5	PR	SW
Hydroporus	112390	+	5	PR	SW
Hydroporus fuscipennis	112476	/	5	PR	SW
Hydroporus notabilis	112405	/	5	PR	SW
Hydroporus occidentalis	112396	/	5	PR	SW
Hygrotus	112200	+	5	PR	SW
Laccophilus	112278	-	5	PR	SW
Laccophilus fasciatus terminalis	112281	-	5	PR	SW
Laccophilus maculosus decipiens	193558	-	5	PR	SW
Liodessus	112580	+	5	PR	SW
Liodessus affinis	112581	/	5	PR	SW
Liodessus fuscatus	728507	/	5	PR	SW
Neoporus	728252	-	5	PR	SW
Neoporus dimidiatus	728416	-	5	PR	SW
Neoporus undulatus	728426	-	5	PR	SW
Platambus	728241	-	5	PR	SW
Platambus semivittatus	728316	-	5	PR	SW

Species	TSN	Seen	Tolerance Value	FFG	Habit
Rhantus	112086	+	5	PR	SW
Rhantus gutticollis	112091	/	5	PR	SW
Rhantus sericans	728331	/	5	PR	SW
Thermonectus	112109	-	5	PR	SW
Thermonectus nigrofasciatus ornaticollis	112115	-	5	PR	SW
Uvarus	112575	-	5	PR	SW
Uvarus granarius	112577	-	5	PR	SW
Elmidae	114093	+	4	GC	CN
Cleptelmis	114164	+	4	GC	CN
Cleptelmis addenda	114166	+	4	GC	CN
Dubiraphia	114126	+	4	GC	CN
Dubiraphia quadrinotata	114130	+	4	GC	CN
Heterlimnius	114167	+	4	GC	CN
Heterlimnius corpulentus	114169	+	4	GC	CN
Lara	114137	+	4	SH	CN
Lara avara	114139	+	4	SH	CN
Microcylloepus	114146	+	4	GC	CN
Microcylloepus pusillus	114147	+	2	GC	CN
Narpus	114142	+	4	GC	CN
Narpus concolor	114144	+	4	GC	CN
Optioservus	114177	+	4	SC	CN
Optioservus divergens	114178	+	4	SC	CN
Optioservus seriatus	114181	+	4	SC	CN
Stenelmis	114095	+	7	SC	CN
Zaitzevia	114205	+	4	GC	CN
Zaitzevia parvula	114206	+	4	GC	CN
Gyrinidae	112653	-	5	PR	SW
Gyrinus	112654	-	5	PR	SW
Gyrinus affinis	112658	-	5	PR	SW
Gyrinus aquiris	112691	-	5	PR	SW
Gyrinus maculiventris	112662	-	5	PR	SW
Gyrinus sayi	193589	-	5	PR	SW
Haliplidae	111857	-	7	SH	CB
Haliplus	111858	-	7	PI	CB
Haliplus immaculicollis	111883	-	7	PI	CB
Haliplus pantherinus	111874	-	7	PI	CB
Haliplus fulvus	--	-	7	PI	CB
Haliplus triopsis	111877	-	7	PI	CB
Peltodytes	111923	-	7	PI	CB
Peltodytes callosus	111935	-	7	PI	CB
Peltodytes edentulus	111936	-	7	PI	CB
Helophoridae	193642	-	5	SH	CB
Helophorus	113106	-	5	SH	CB
Helophorus linearoides	113118	-	5	SH	CB
Heteroceridae	114262	-	--	PR	--
Hydraenidae	112756	+	5	PR	CN
Hydraena	112757	+	5	PR	CN
Ochthebius	112777	+	5	PR	CN
Ochthebius lineatus	112789	/	5	PR	CN
Ochthebius tubus	193635	/	5	PR	CN

Species	TSN	Seen	Tolerance Value	FFG	Habit
Hydrochidae	722226	-	5	SH	SW
Hydrochus	113166	-	5	SH	SW
Hydrochus rufipes	113172	-	5	SH	SW
Hydrophilidae	112811	+	5	PR	SW
Cercyon	113039	-	5	PR	SW
Cercyon connivens	113051	-	5	PR	SW
Cercyon matthewsi	722343	-	5	PR	SW
Crenitis	113220	-	5	PR	BU
Crenitis digesta	722304	-	5	PR	BU
Enochrus	112973	+	5	GC	BU
Enochrus cinctus	112983	/	5	GC	BU
Enochrus diffusus	112980	/	5	GC	BU
Enochrus hamiltoni	112984	/	5	GC	BU
Enochrus pygmaeus	112993	/	5	GC	BU
Hydrobius	113196	+	8	PR	CB
Hydrobius fuscipes	113199	+	8	PR	CB
Hydrochara	113190	-	5	PR	SW
Hydrochara occulta	722331	-	5	PR	SW
Hydrophilus	113204	-	5	PR	SW
Hydrophilus triangularis	113207	-	5	PR	SW
Laccobius	112858	+	5	PI	SW
Paracymus	112909	+	5	PR	BU
Paracymus subcupreus	112910	/	5	PR	BU
Sphaeridium	112933	-	5	PR	BU
Sphaeridium bipustulatum	112935	-	5	PR	BU
Tropisternus	112938	-	5	PR	CB
Tropisternus columbianus	112950	-	5	PR	CB
Tropisternus ellipticus	112949	-	5	PR	CB
Tropisternus lateralis limbalis	722405	-	5	PR	CB
Tropisternus sublaevis	112961	-	5	PR	CB
Hydroscaphidae	112744	+	7	SC	CN
Hydroscapha	112745	+	7	SC	CN
Hydroscapha natans	112746	+	7	SC	CN
Leiodidae	113817	-	--	PA	--
Platypsylla	113815	-	--	PA	--
Platypsylla castoris	113816	-	--	PA	--
Noteridae	112606	-	--	PR	--
Sciritidae	113924	-	--	SC	CB
Staphylinidae	113265	-	8	PR	CN

Species	Temperature Range	Lotic	Lentic	Current	Substrate	Voltinism
Chrysomelidae	--	--	--	--	--	--
Donacia	--	4.0	--	--	--	Uni
Donacia hirticollis	--	4.0	--	--	--	Uni
Donacia magnifica	--	4.0	--	--	--	Uni
Donacia subtilis	--	4.0	--	--	--	Uni
Donacia tuberculifrons	--	4.0	--	--	--	Uni
Plateumaris	--	--	--	--	--	--
Plateumaris germari	--	--	--	--	--	--
Plateumaris nitida	--	--	--	--	--	--
Plateumaris pusilla	--	--	--	--	--	--
Plateumaris robusta	--	--	--	--	--	--
Curculionidae	--	--	--	--	--	--
Dryopidae	E: warm	--	--	--	--	--
Helichus	E: warm	3.4	4.0	--	6.8	Mero
Helichus striatus	E: warm	3.4	4.0	--	6.8	mero
Dytiscidae	E: warm	--	--	--	--	--
Acilius	E: warm	4.0	1.1	--	--	--
Acilius semisulcatus	E: warm	4.0	1.1	--	--	--
Agabus	E: warm	3.9	1.2	--	3.5	Uni
Agabus canadensis	E: warm	3.9	1.2	--	3.5	Uni
Agabus disintegratus	E: warm	3.9	1.2	--	3.5	Uni
Agabus griseipennis	E: warm	3.9	1.2	--	3.5	Uni
Agabus punctatus	E: warm	3.9	1.2	--	3.5	Uni
Boreonectes	E: warm	4.0	1.2	--	2.5	--
Boreonectes griseostriatus	E: warm	4.0	3.0	--	2.5	--
Boreonectes striatellus	E: warm	4.0	1.2	--	2.5	--
Colymbetes	E: warm	4.0	1.1	--	--	Uni
Colymbetes exaratus	E: warm	4.0	1.1	--	--	Uni
Dytiscus	E: warm	3.9	1.1	1.0	--	Uni
Dytiscus marginicollis	E: warm	3.9	1.1	1.0	--	Uni
Hydroporus	E: warm	4.0	1.1	1.0	7.0	Mero
Hydroporus fuscipennis	E: warm	4.0	1.1	1.0	7.0	Mero
Hydroporus notabilis	E: warm	4.0	1.1	1.0	7.0	Mero
Hydroporus occidentalis	E: warm	4.0	1.1	1.0	7.0	Mero
Hygrotus	E: warm	4.0	1.1	1.0	1.3	--
Laccophilus	E: warm	4.0	1.2	1.0	0.0	Multi
Laccophilus fasciatus terminalis	E: warm	4.0	3.0	1.0	0.0	Multi
Laccophilus maculosus decipiens	E: warm	4.0	3.7	1.0	0.0	Multi
Liodessus	E: warm	4.0	1.0	1.0	1.0	--
Liodessus affinis	E: warm	4.0	1.0	1.0	1.0	--
Liodessus fuscatus	E: warm	4.0	1.0	1.0	1.0	--
Neoporus	E: warm	4.0	1.3	1.0	1.7	--
Neoporus dimidiatus	E: warm	4.0	1.3	1.0	1.7	--
Neoporus undulatus	E: warm	4.0	4.0	1.0	1.7	--
Platambus	E: warm	--	--	--	--	--
Platambus semivittatus	E: warm	4.0	2.0	--	--	Uni
Rhantus	E: warm	4.0	1.3	--	--	Uni
Rhantus gutticollis	E: warm	4.0	1.3	--	--	Uni

Species	Temperature Range	Lotic	Lentic	Current	Substrate	Voltinism
Rhantus sericans	E: warm	4.0	1.3	--	--	Uni
Thermonectus	E: warm	--	1.1	--	--	Uni
Thermonectus nigrofasciatus ornaticollis	E: warm	--	1.1	--	--	Uni
Uvarus	E: warm	3.9	1.2	1.5	2.0	--
Uvarus granarius	E: warm	3.9	1.2	1.5	2.0	--
Elmidae	E: warm	--	--	--	5.5	Mero
Cleptelmis	E: cool	3.3	--	2.0	5.7	Mero
Cleptelmis addenda	E: cool	3.0	--	2.0	5.5	Mero
Dubiraphia	E; warm	3.8	1.1	1.3	6.0	Uni
Dubiraphia quadrinotata	E: warm	3.8	1.1	1.3	6.0	Uni
Heterlimnius	E: cool	3.1	---	2.3	5.3	Mero
Heterlimnius corpulentus	E: cool	3.0	--	2.3	5.2	Mero
Lara	E: cool	3.5	--	2.3	5.5	Mero
Lara avara	E: cool	3.6	--	2.3	5.5	Mero
Microcylloepus	E: warm	3.8	--	2.0	4.8	Mero
Microcylloepus pusillus	E: warm	3.9	--	2.0	4.9	Mero
Narpus	E: cool	3.3	--	2.2	5.3	Mero
Narpus concolor	E: cool	3.0	--	2.3	4.0	Mero
Optioservus	E: warm	3.5	--	2.3	5.2	Mero
Optioservus divergens	E: warm	3.7	--	2.3	5.5	Mero
Optioservus seriatus	E: warm	3.7	--	2.3	5.5	Mero
Stenelmis	E: warm	3.7	1.1	1.9	3.6	Mero
Zaitzevia	E: warm	3.2	--	2.0	5.3	Mero
Zaitzevia parvula	E: warm	3.0	--	2.0	5.2	Mero
Gyrinidae	E: warm	--	--	--	--	--
Gyrinus	E: warm	3.7	4.0	--	--	--
Gyrinus affinis	E: warm	3.7	4.0	--	--	--
Gyrinus aquiris	E: warm	3.7	4.0	--	--	--
Gyrinus maculiventris	E: warm	3.7	4.0	--	--	--
Gyrinus sayi	E: warm	3.7	4.0	--	--	--
Haliplidae	E: warm	--	--	--	--	
Haliplus	E: warm	3.0	1.2	0.5	--	Uni
Haliplus immaculicollis	E: warm	3.0	4.0	0.5	--	Uni
Haliplus pantherinus	E: warm	3.0	1.2	0.5	--	Uni
Haliplus fulvus	E: warm	3.0	1.2	0.5	--	Uni
Haliplus triopsis	E: warm	3.0	1.2	0.5	--	Uni
Peltodytes	E: warm	3.8	1.3	0.5	--	Multi
Peltodytes callosus	E: warm	3.8	1.3	0.5	--	Multi
Peltodytes edentulus	E: warm	3.8	1.3	0.5	--	Multi
Hclophoridae	E. warm	3.8	1.2	--	0.5	--
Helophorus	E: warm	3.8	1.2	--	0.5	--
Helophorus linearoides	E: warm	3.8	1.2	--	0.5	--
Heteroceridae	--	--	--	--	--	--
Hydraenidae	E: warm	--	--	--	--	--
Hydraena	E: cool	4.0	4.0	--	--	--
Ochthebius	E: warm	4.0	--	--	--	--
Ochthebius lineatus	E: warm	4.0	--	--	--	--
Ochthebius tubus	E: warm	4.0	--	--	--	--

Species	Temperature Range	Lotic	Lentic	Current	Substrate	Voltinism
Hydrochidae	E: warm	--	--	--	--	--
Hydrochus	E: warm	4.0	1.2	1.0	2.0	--
Hydrochus rufipes	E: warm	4.0	1.2	1.0	2.0	--
Hydrophilidae	E: warm	--	--	--	--	--
Cercyon	E: warm	--	--	--	--	--
Cercyon connivens	E: warm	--	--	--	--	--
Cercyon matthewsi	E: warm	--	--	--	--	--
Crenitis	E: warm	4.0	1.5	--	4.8	
Crenitis digesta	E: warm	4.0	1.5	--	4.8	--
Enochrus	E: warm	4.0	1.1	--	7.0	
Enochrus cinctus	E: warm	4.0	1.1	--	7.0	--
Enochrus diffusus	E: warm	4.0	1.1	--	7.0	--
Enochrus hamiltoni	E: warm	4.0	1.1	--	7.0	--
Enochrus pygmaeus	E: warm	4.0	1.1	--	7.0	--
Hydrobius	E: warm	4.0	1.1	0.5	7.0	--
Hydrobius fuscipes	E: warm	4.0	1.1	0.5	7.0	--
Hydrochara	E: warm	4.0	1.3	--	--	--
Hydrochara occulta	E: warm	4.0	1.3	--	--	--
Hydrophilus	E: warm	4.0	1.2	--	--	--
Hydrophilus triangularis	E: warm	4.0	1.2	--	--	--
Laccobius	E: warm	3.8	1.1	0.8	0.7	--
Paracymus	E: warm	4.2	1.1	0.5	1.5	--
Paracymus subcupreus	E: warm	4.2	1.1	0.5	1.5	--
Sphaeridium	--	--	--	--	--	--
Sphaeridium bipustulatum	--	--	--	--	--	--
Tropisternus	E: warm	4.0	1.1	1.0	1.0	Multi
Tropisternus columbianus	E: warm	4.0	1.1	1.0	1.0	Multi
Tropisternus ellipticus	E: warm	4.0	1.1	1.0	1.0	Multi
Tropisternus lateralis limbalis	E: warm	4.0	3.0	1.0	1.0	Multi
Tropisternus sublaevis	E: warm	4.0	4.0	1.0	1.0	Multi
Hydroscaphidae	--	--	--	--	--	--
Hydroscapha	E: warm	4.0	--	--	--	--
Hydroscapha natans	E: warm	4.0	--	--	--	Multi
Leiodidae	--	--	--	--	--	--
Platypsylla	--	--	--	--	--	--
Platypsylla castoris	--	--	--	--	--	--
Noteridae	E: warm	--	--	--	--	--
Sciritidae	--	--	--	--	--	--
Staphylinidae	--	--	--	--	--	--

Aquatic Lepidoptera

Introduction

Butterflies and moths are in the order Lepidoptera, a large order with over 12,700 species in North America (Pohl et al. (2018). Of these, the aquatic Lepidoptera are a very poorly studied group – not showy, mostly not of considerable economic interest, not in habitats frequented by most lepidopterists. In fact, whereas the showy butterflies and moths in North America almost invariably have common names assigned, only a handful of these little moths have been so graced. It is entirely feasible that some production taxonomists for biomonitoring programs may consider all non-gilled caterpillars to be terrestrial.

Most aquatic Lepidoptera are in the subfamilies Acentropinae and Schoenobiinae of the family Crambidae (which was moved from the Pyralidae), and a few genera are in the Arctiidae and Noctuidae. Some Lepidoptera do, however, feed on aquatic and semiaquatic plants, including pestiferous aquatic plants, so they could be of biocontrol interest. Most records of aquatic Lepidoptera in the literature are not based on checklists but rather life history and feeding studies. These records usually represent very scattered localities, so it would very tenuous to attempt to define distributions for the aquatic Lepidoptera, and I will resist the urge to attempt it but instead list representative localities.

The larvae of a few species live directly in the water, being adorned with filamentous gills. Most, however, are stem borers or leaf miners in aquatic and semiaquatic plants and can be found in the plant tissues below the water surface. The adults of some species are either obligatorily aquatic (e.g., *Acentropus niveus* (Olivier)) or at least enter the water using a plastron-like layer of air as a source of oxygen to oviposit. Lange (1956) provided an excellent overview of the taxonomy (at that time) and the known biology of the aquatic moths, and is a must-read for those studying this fascinating, but obscure group.

Overview of the Aquatic Lepidoptera in the Black Hills

There are no specific checklists, or for that matter, records, of aquatic Lepidoptera from the Black Hills, Wyoming, or South Dakota. Jurgens (1968) reported *Elophila* from the Fall River, Spring Creek, French Creek, and the South Fork of Rapid Creek, but it would be difficult to assign which current genus or species that record refers to. Searching through stands of aquatic plants is likely to result in many new records for aquatic Leidoptera for the Black Hills, South Dakota, and Wyoming. A literature review of insect herbivores of aquatic and wetland plants in the United States was prepared by Harms and Grodowitz (2009) and could be useful for future researchers looking for Lepidoptera on aquatic plants in the Black Hills. Aquatic plant genera that occur in or near the Black Hills and are known to serve as hosts for various species of aquatic Lepidoptera include the following:

Brasenia schreberi Gmel. – watershield
Carex spp. – sedges.
Ceratophyllum demersum L. – coon's tail, a species of hornwort
Elatine spp. – waterworts
Eleocharis spp. – spikerushes
Elodea spp. – waterweeds
Juncus spp. – rushes
Lemna spp. – duckweeds
Myriophyllum spp. – watermilfoils
Nuphar spp. – pond lilies
Nymphaea spp. – waterlilies
Potamogeton spp. – pondweeds
Sagittaria spp. – arrowheads
Scirpus spp. – bulrushes
Sparganium spp. – bur-reeds
Typha spp. – cattails
Utricularia spp. – bladderworts

Since there are hundreds of thousands of terrestrial Lepidoptera, in comparison to the few genera and species of aquatic Lepidoptera, it is prudent to consider that any Lepidoptera encountered in samples could be of terrestrial origin. Of course, if it is one of the species that has gills, then we know it is aquatic; but, otherwise, if it doesn't exactly match figures, descriptions, and diagnoses, then it should probably be considered to be terrestrial.

Distributions for the following species are based on locality maps on the website of the North American Moth Photographers Group at Mississippi State University (http://mothphotographersgroup.msstate.edu/MainMenu.shtml). Excellent species accounts and photographs of all species can be accessed online using the following url and replacing the #### with the "Hodges" number that I provide in the descriptions for each species: http://mothphotographersgroup.msstate.edu/species.php?hodges=####. The Hodges number is a standardized number based on data in Hodges et al. (1983); the numbers have been updated, expanded, and even subdivided with decimals with increasing taxonomic work on this group. Pohl et al. (2018) is compiling a working paper to be extensively reviewed by the lepidopterist community and includes a new numbering scheme, called "P3 numbers", with a two-digit number for the subfamily and a four-digit number for the species. Time will tell if the new scheme sufficiently supplants the Hodges numbers. I am including these P3 numbers as well in the species accounts.

Species Accounts

Crambidae

Although I could find no published records of aquatic Crambidae from the Black Hills in the peer-reviewed literature, I have collected larvae of *Petrophila* from the Fall River near Hot Springs, and I suspect that other genera are likely to occur there, as well.

Acentria Stephens

Acentria ephemerella (Denis & Schiffermüller)

Descriptions/Diagnosis: Passoa 1988 (larva, pupa, adult; as *Acentropus ephemerella*); Hodges = 5299.01; P3 = 800720. Native to Europe but introduced to northeastern North America in the 1920s and subsequently found in Quebec, New England, New York, and Wisconsin. Most North American literature refers to this species as the synonym *Acentropus nivea*.

This species was introduced to North America, likely on aquatic plants brought over from Europe, and have been investigated as to their potential for biological control of Eurasian water milfoil and other invasive aquatic plants. It has not yet been reported from the Dakotas, but the hosts for this invasive species occur there, so it may eventually be reported from the Black Hills. Larval hosts for this introduced species include members of the genera *Ceratophyllum*, *Elatine*, *Elodea*, *Lemna*, *Myriophyllum*, *Sparganium*, and *Pomatogeton*. Some females of this species have vestigial wings and are obligatorily aquatic; other females have fully-developed wings like the males (Treat 1954).

Chilo Zincken

About 40 species exist in this genus, of which four occur in North America, with most distributed along the Atlantic seaboard (Bleszynski 1970). Larvae are all stem borers and hosts include members of the Poaceae and *Eleocharis*, *Juncus*, and *Scirpus*.

Chilo plejadellus Zincken

Descriptions/Diagnosis: Bleszynski 1970; I have not yet found a description of the larvae; Hodges = 5470; P3 = 800831. Distributed from the Dakotas and Minnesota east to Quebec, south to Illinois and northern Florida; nearest known localities are in northeastern North Dakota. This species is commonly known as the rice stalk borer moth, and several of its hosts occur in the Black Hills.

Eoparargyractis Lange

Eoparargyractis irroratalis (Dyar)

Descriptions/Diagnosis: Dyar 1917 (adult, as *Elophila irroratalis*); larvae are apparently as yet undescribed, but larvae of conspecific *E. plevie* Dyar) are described in Fiance and Moeller 1977); Hodges = 4785; P3 = 800766. Primarily distributed in the Southeast, from Mississippi to Maryland and south into peninsular Florida, but there are records from Montana and British Columbia, so it could feasibly occur in the Black Hills.

Parapoynx Hübner

Numerous species of this genus occur in aquatic habitats. Larval host plants for species in this genus include *Brasemia*, *Eleocharis*, *Hydrochloa*, *Myriophyllum*, *Nuphar*, *Nymphaea*, *Potamogeton*, *Salvinia*, *Utricularia*, and *Vallisneria*.

Parapoynx maculalis (Clemens)

Descriptions/Diagnosis: Clemens 1860 (adult, as *Sironia maculalis*), Hart 1896 (adult, as *Nymphaeella maculalis*); Forbes 1910, Welch 1916 (larva, as *Nymphula maculalis*); Hodges = 4759; P3 = 800734. Eastern distribution from the Canadian Maritimes to the Florida keys, west to Alberta, the Mississippi River, and eastern Texas. This species is commonly known as the "polymorphic pondweed moth".

Parapoynx obscuralis (Grote)

Descriptions/Diagnosis: Hart 1896, Berg 1950 (larva); Grote 1881 (adult, as *Oligorstigma obscuralis*); Hodges = 4759; P3 = 800735. Eastern species with distribution from the Canadian Maritimes to the Florida keys, west to Minnesota and the Mississippi River. It has been accidentally introduced to Great Britain. This species is variously known as the "obscure pondweed moth", the "American china-mark", or the "Vallisneria leafcutter".

Parapoynx badiusalis (Walker)

Descriptions/Diagnosis: Berg 1950 (larva); Walker 1859 (adult, as *Cymoriza badiusalis*); Hodges = 4761; P3 = 800736; Distributed from Saskatchewan and Manitoba east to Maine, and south to Nebraska, Oklahoma, and pensinsular Florida. This species is commonly known as the "chestnut-marked pondweed moth".

Parapoynx allionealis Walker

Descriptions/Diagnosis: ; Walker 1859 (adult); Forbes 1923 (larva, as *Nymphula allionealis*, characters in keys); Hodges = 4764; P3 - 800739. Distributed from North Dakota, Wisconsin, and Illinois, south to Oklahoma and Texas, east to the Atlantic seaboard from Quebec and New Brunswick south to the Florida keys. This species is commonly known as the "watermilfoil leafcutter moth".

Petrophila Guilding

As mentioned above, I have collected larvae of *Petrophila* from the Fall River near Hot Springs.

Petrophila avernalis (Grote)

Descriptions/Diagnosis: Grote 1878 (adult, as *Chryseudeton avernalis*); Hodges = 4791; P3 = 800761. Distributed in a relatively narrow band from the Black Hills southwest to southern Arizona and New Mexico. The North American Moth Photographers Group includes a datapoint for Lawrence County in their records for this species. As of 2018, this species had not yet been entered into ITIS and has no TSN assigned to it.

Petrophila bifacialis (Robinson)

Descriptions/Diagnosis: Heppner 1976 (adult, as *Parargyractis bifacialis*), larvae have apparently not yet been described; Hodges = 4774; P3 = 800754. Distributed from western Texas to the Florida panhandle, north to Minnesota, Ontario, Quebec, and Maine. This species is commonly called the "two-banded petrophila moth".

Petrophila canadensis (Munroe)

Descriptions/Diagnosis: Munroe 1972 (adult, as *Parargyractis canadensis*); Hodges = 4779; P3 = 800759. Distributed primarily from Iowa, Kentucky, and West Virginia through the Great Lakes and the St. Lawrence River to New Brunswick, with disjunct populations in Alberta and Arizona. This species is commonly called the "Canadian petrophila moth".

Petrophila confusalis (Walker)

Descriptions/Diagnosis: Walker 1866 (adult, as *Cataclysta confusalis*); Hodges = 4780; P3 = 800760. This species has a transcontinental distribution from British Columbia to California, east to Minnesota, Maryland, and Texas. This species is commonly called the "confusing petrophila moth".

With *P. avernalis*, above, this is probably one of the more likely species of *Petrophila* to be found in the Black Hills with a record from southern Wyoming. This species is not in ITIS and has no TSN assigned to it.

Petrophila kearfottalis (Barnes & McDunnough)

Descriptions/Diagnosis: Barnes and McDunnough 1917 (adult, as *Cataclysta kearfottalis*); Hodges = 4773; P3 = 800753. Distributed from British Columbia and Alberta south to California, Arizona, New Mexico, and Texas. This species is not in ITIS and has no TSN assigned to it.

Synclita Lederer

Synclita has numerous aquatic species. *Synclita obliteralis* is known as the "waterlily leafcutter moth", and its larvae are known to feed on *Eichhornia*, *Hydrilla*, *Lemna*, *Pistia*, *Potamogeton*, and various Nymphaeaceae.

Noctuidae

Within the Noctuidae, several genera are aquatic and could be found in the Black Hills. These genera include *Archanara* Walker, which feed on *Juncus*, *Scirpus*, *Sparganium*, and *Typha*. One of these is the "oblong sedge-borer moth", *A. oblonga*. Numerous species of *Bellura* Walker are aquatic: larvae of the "pickerelweed borer", *B. densa*, feed on *Eichhornia*, *Pontederia*, and *Typha*; larvae of the "white-tailed diver", *B. gortynoides*, feed on *Nuphar*, *Pontederia*, and *Typha*; and larvae of the "cattail borer", *B. obliqua*, feed on *Nelumbo*, *Pontederia*, *Sagittaria*, *Sparganium*, *Symplocarpus*, and *Typha*. In *Neoerastria* McDunnough, the larvae of *N. caduca* (Grote) feed on *Nuphar*. Larvae of *Oligia* Hübner are known to feed on *Scirpus*.

I have not seen any of these genera in biomonitoring samples from the Black Hills, but I think it would be likely that they could be found in aquatic plants in the Black Hills.

Aquatic Lepidoptera Traits Table

Table 11: Aquatic Lepidoptera Traits Table

Species	TSN	Seen	Tolerance Value	FFG	Habit
Crambidae	693963	-	6	SH	--
Acentria	117697	-	5	SH	CB
Acentropus ephemeerella	117698	-	5	SH	CB
Chilo	117704	-	5	SH	CB
Chilo plejadellus	117705	-	5	SH	CB
Eoparargyractis	117695	-	5	SH	CB
Eoparargyractis irroratalis	117696	-	5	SH	CB
Parapoynx	117714	-	5	SH	CB
Parapoynx allionealis	117719	-	5	SH	CB
Parapoynx badiusalis	117717	-	5	SH	CB
Parapoynx maculalis	117715	-	5	SH	CB
Parapoynx obscuralis	117716	-	5	SH	CB
Petrophila	117682	-	5	SC	CN
Petrophila avernalis	--	-	5	SC	CN
Petrophila bifascialis	117985	-	5	SC	CN
Petrophila canadensis	117689	-	5	SC	CN
Petrophila confusalis	--	-	5	SC	CN
Petrophila kearfottalis	--	-	5	SC	CN
Synclita	117654	-	5	SH	CB
Noctuidae	117318	-	6	SH	BU
Archanara	117331	-	6	SH	BU
Archanara oblonga	117332	-	6	SH	BU
Bellura	117321	-	6	SH	BU
Bellura densa	117328	-	6	SH	BU
Bellura gortynoides	117329	-	6	SH	BU
Bellura obliqua	117323	-	6	SH	BU
Neoerastria	117319	-	6	SH	BU
Neoerastria caduca	117320	-	6	SH	BU
Oligia	117424	-	6	SH	BU

Species	Temperature Range	Lotic	Lentic	Current	Substrate	Voltinism
Crambidae	--	--	--	--	--	--
Acentria	--	--	--	--	7.0	--
Acentropus ephemeerella	--	--	--	--	7.0	--
Chilo	--	--	--	--	7.0	--
Chilo plejadellus	--	--	--	--	7.0	--
Eoparargyractis	--	--	--	--	7.0	--
Eoparargyractis irroratalis						
Parapoynx	--	--	4.0	--	7.0	--
Parapoynx allionealis	--	--	4.0	--	7.0	--
Parapoynx badiusalis	--	--	4.0	--	7.0	--
Parapoynx maculalis	--	--	4.0	--	7.0	--
Parapoynx obscuralis	--	--	4.0	--	7.0	--
Petrophila	--	4.0	1.4	2.0	7.0	Multi
Petrophila avernalis	--	4.0	1.4	2.0	7.0	Multi
Petrophila bifascialis	--	4.0	1.4	2.0	7.0	Multi
Petrophila canadensis	--	4.0	1.4	2.0	7.0	Multi
Petrophila confusalis	--	4.0	1.4	2.0	7.0	Multi
Petrophila kearfottalis	--	4.0	1.4	2.0	7.0	Multi
Synclita	--	--	--	--	7.0	--
Noctuidae	--	--	--	--	--	--
Archanara	--	--	--	--	--	--
Archanara oblonga	--	--	--	--	--	--
Bellura	--	--	--	--	--	--
Bellura densa	--	--	--	--	--	--
Bellura gortynoides	--	--	--	--	--	--
Bellura obliqua	--	--	--	--	--	--
Neoerastria	--	--	--	--	--	--
Neoerastria caduca	--	--	--	--	--	--
Oligia	--	--	--	--	--	--

Trichoptera

Introduction

Finally, the "T" in the EPT orders: the Trichoptera, or caddisflies. This group, with the mayflies, dragonflies, damselflies, and stoneflies, most clearly represent the aquatic lifestyle at the ordinal level. Of course, there are exceptions, such as the larvae of *Ironoquia plattensis*, which have been found in mud and cattle-trampled soil on the banks of the Platte River in Nebraska. But, for the most part, caddisflies live as larvae and pupae in the water and emerge as winged adults.

The larvae have a variety of appearances, and the most obvious structure on most is not even a part of the animal itself, but is its case. Caddisfly larvae construct an astounding variety of architectural wonders from plant pieces, tiny stones, and silk – from simple sand tubes to fluted tubes made from leaves to square "log cabins" of wood bits to coiled cases that look like snail shells, to two-valved silk purses. For some species, or at least genera, the form of the case is just as diagnostic as the form of the animal itself. Wiggins (2005) is the most comprehensive account of the ecology, morphology, life history, and distribution of the caddisflies of North America.

Overview of the Trichoptera in the Black Hills

No comprehensive checklists have previously been prepared covering the caddisflies of South Dakota or the Black Hills; Ruiter and Lavigne (1985) assembled a distributional checklist of the Trichoptera of Wyoming, with about 45 species with Black Hills records (Crook and Weston counties). With 181 Trichoptera species reported from Wyoming, this represents a relatively small portion of what probably actually exists. Harris et al. (1980) listed the caddisflies of North Dakota, breaking the state into six regions; nineteen taxa were listed from the southwestern region of North Dakota, the closest geographically to the Black Hills.

Most other records I could find are those that exist in revisions of various caddisfly groups (e.g., Flint 1985; Weaver 1988) and the taxa voucher list in Shearer (2006). Ruiter (1996) provided records of many caddisfly species from the Black Hills from a collecting trip undertaken as a function of the 8th International Symposium on Trichoptera that had been held in Minneapolis, Minnesota. Caddisfly knowledge in the Black Hills is remarkably sparse and an open field for investigation.

Species Accounts

Apataniidae

I could not find any records of Apataniidae from the Black Hills in the published literature. This, of course, does not mean they do not occur there. In fact, Shearer (2006) encountered apataniid caddisflies in biomonitoring samples, identified to the genus level. Anglers commonly refer to this family as the "early smoky wing sedges".

Allomyia Banks

Descriptions/Diagnoses: Wiggins 1996 (larva), Ross 1950 (adult, as *Imania*). Twelve species in this genus occur in western North America from Alaska south to Nevada and Colorado, but larvae of only a couple western species have been described. No records of *Allomyia* species have been reported from the Black Hills in the published literature.

Apatania Kolenati

Descriptions/Diagnoses: Wiggins 1960, 1996 (larva), Schmid 1953 (adult). Holarctic and Oriental; 17 species in North America with distributions extending from the northern arctic latitudes south through the major mountain ranges to Arizona and Georgia.

Shearer (2006) reported larvae of unidentified species of *Apatania* from Black Hills streams.

Brachycentridae

Both *Brachycentrus americanus* and *B. occidentalis*, as well as the genus *Micrasema* (likely represented by *M. bactro*) occur in the Black Hills. *Amiocentrus aspilus* may also occur in the Black Hills, based on its distribution elsewhere in the Rocky Mountain cordillera and in Wyoming (western three quarters of the state), but it has not yet been reported from the Black Hills either in biomonitoring samples or in the published literature, so it is not included here.

Brachycentrus Curtis

Both Jergens (1968) and Shearer (2006) reported *Brachycentrus* from Black Hills streams at the genus level. At the genus level, anglers refer to these species as "grannoms", but they have different names at the species level.

Brachycentrus americanus (Banks)

Descriptions/Diagnoses: Flint (1985). Flint (1985) reported *B. americanus* from regions both east (Wisconsin and Minnesota) and west (Rocky Mountain cordillera), although he apparently did not have records from the Black Hills. This species is common in Wyoming (Ruiter and Lavigne 1985). Harris et al. (1980) reported this species from North Dakota. This species is called the "American grannom" in angler parlance.

Brachycentrus occidentalis Banks

Descriptions/Diagnoses: Flint (1985). Flint (1985) reported *B. occidentalis* from regions both east (Wisconsin and Minnesota) and west (Rocky Mountain cordillera), although he apparently did not have records from the Black Hills. Ruiter and Lavigne (1985) also had records of this species from across the state, but not from the Black Hills. Anglers commonly call this species the "Mother's Day caddis" or just a "grannom".

Micrasema **McLachlan**

Shearer (2006) reported *Micrasema* from Black Hills streams at the genus level. Based on Ruiter and Lavigne (1985), *Micrasema* is represented by two species in Wyoming: the widespread *M. bactro* and *M. alexanderi* Denning, the latter of which is known in Wyoming only from the type series from Yellowstone National Park.

Micrasema bactro **Ross**

Descriptions/Diagnoses: Chapin 1978 (larva, adult), Ross 1938 (adult). Distributed from the Yukon south to California, and east to Alberta, Montana, South Dakota, Colorado, and New Mexico. Known to anglers as the "little grannom".

This species was reported from Spearfish Creek and Little Spearfish Creek by Ruiter (1996).

Micrasema bactro is far more abundant and widespread in the Black Hills than either *Brachycentrus* species, and I have seen it in many biomonitoring samples from the northern Black Hills.

Glossosomatidae

In addition to the genus *Glossosoma*, which has previously been reported from the Black Hills, I believe *Agapetus*, *Anagapetus*, and *Culoptila* may also occur in or near the Black Hills. Shearer (2006) reported larvae of an unidentified species of *Glossosoma* from Black Hills streams, as well as pupae identified to the family level. The family is known as the "saddle-case makers" with other common anglers' names for some genera.

Agapetus **Curtis**

Description/Diagnosis: Wiggins 1996 (larva). Distributed in all faunal regions except the neotropics. In North America, about 30 species are known. Anglers sometimes call this the "tiny black short-horned caddis".

This genus has not yet been reported from the Black Hills. Most *Agapetus* in North America are eastern, with ranges that do not overlap into the Black Hills (Etnier et al. 2010), but a few western species (e.g., *A. boulderensis*, which is known from Albany and Carbon counties in Wyoming; Ruiter and Lavigne 1985) may venture into the Black Hills. The temperature preference for *Agapetus* given in Grafe et al. (2002) is eurythermal warm; based on data in Vieira et al. (2006), the temperature preference for this genus would be eurythermal hot.

Anagapetus **Ross**

Description/Diagnosis: Wiggins 1996 (larva). Distributed in western North America, with six species. This genus ranges locally from California north to British Columbia and west to Colorado and Wyoming. Based on its presence throughout the central Rocky Mountain cordillera, I suspect that this genus, and, in particular, *Anagapetus debilis* (Ross), may eventually be found in the Black Hills. Ruiter and Lavigne (1985) reported *A. debilis* from Albany, Park, and Teton counties in Wyoming and predicted that it "should be found in small, cold streams throughout the State."

The temperature preference for *Anagapetus* is stenothermal cold in Grafe et al. (2002); based on data in Vieira et al. (2006), the temperature preference for this genus would be eurythermal cool.

Culoptila **Mosely**

Based on the generally Neotropical distribution of this genus, *C. cantha* and *C. thoracica* are the only species likely to occur in or near the Black Hills.

Culoptila cantha Ross

Description/Diagnosis: Houghton and Stewart 1998 (larva), Blahnik and Holzenthal 2006 (adult). Distributed from Saskatchewan south through the Rocky Mountains to Arizona, New Mexico, and Texas.

Although Huryn et al. (2008) reported that species of *Culoptila* are univoltine (and it is the designation used here), Houghton and Stewart 1998 found that *C. cantha* was multivoltine in their studies in Texas; it is possible that *C. cantha* may be either univoltine or multivoltine in the higher latitudes of its range.

Culoptila thoracica (Ross)

Descriptions/Diagnoses: Blahnik and Holzenthal 2006 (adult), larva is undescribed. This species is distributed primarily from Wyoming and Utah south to Chihuahua and Michoacán. Described from Boulder, Wyoming, the type series is the only Wyoming record known (Ruiter and Lavigne 1985), but its distribution indicates that it could occur in the Black Hills.

Glossosoma Curtis

Jurgens (1968) reported *Glossosoma* at the genus level from Spring Creek, Battle Creek, and Castle Creek. Shearer (2006) reported larvae of an unidentified species of *Glossosoma* from Black Hills streams. *Glossosoma ventrale* is the only species of this genus to have been reported from the Black Hills, although six other species are known from elsewhere in Wyoming (Ruiter and Lavigne 1985). These are the "little brown short-horned sedges" of the angling world.

The temperature preference for both *Glossosoma* and *G. ventrale* is eurythermal cool in Grafe et al. (2002); based on data in Vieira et al. (2006), the temperature preference for this genus would be eurythermal warm.

Glossosoma ventrale Banks

Descriptions/Diagnoses: Banks 1904 (adult), Ross 1956 (adult, as description of subgenus *Ripaeglossa*); larva is not yet described. Distribution is from South Dakota and Wyoming, south to Arizona.

Ruiter and Lavigne (1985) reported this species from the Black Hills National Forest in the Wyoming portion of the Black Hills. Ruiter (1996) reported this species from Little Spearfish Creek.

Helicopsychidae

This family, with its distinctive larvae bearing comb-shaped anal claws and coiled cases that can pass for snail shells, is probably represented by only a single species in the Black Hills: *Helicopsyche borealis*.

Helicopsyche von Siebold

Helicopsyche borealis (Hagen)

Descriptions/Diagnoses: Vorhies 1908-09 (larva), Ross 1944 (adult). This species is widespread and distributed over much of North America; Wiggins (1996) reported specimens as far north as 55° latitude in Saskatchewan. Anglers call this species the "speckled peter".

Ruiter and Lavigne (1985) reported this species from the Wyoming portion of the Black Hills National Forest in Crook County. Jergens (1968) reported *Helicopsyche* at the genus level from the Fall River, French Creek, and Battle Creek; Shearer (2006) reported larvae of unidentified species of

Helicopsyche from Black Hills streams. Given the distribution of *Helicopsyche* species in North America, it is likely that these were *H. borealis*.

Hydropsychidae

This family is probably the most widespread caddisfly family in the Nearctic. Specimens are collected in nearly every benthic sample in mountainous regions, and with moderate tolerance to perturbations, they even occur in moderately polluted stream sites. They appear to prefer nutrient-enriched streams and can often be found in greater numbers downstream of lakes and reservoirs. The larvae are usually found in crevices on rocks, often underneath, where they build silk retreats that have an "awning"-type net that they situate in the stream's current to catch food. They then graze on whatever is captured on the net. Harris et al. (1980) reported the following hydropsychids from southwestern North Dakota: *Cheumatopsyche campyla, C. smithi,* and *Symphitopsyche bifida (= Hydropsyche morosa)*.

Shearer (2006) reported larvae of unidentified species of *Ceratopsyche (= Hydropsyche bronta* group) and *Cheumatopsyche* from Black Hills streams. I have larvae of *Cheumatopsyche* and *Hydropsyche*, s.l., in my collection; they are from the Fall River near Hot Springs, Battle Creek near Hermosa, and Whitewood Creek.

Cheumatopsyche larvae are, as yet, unseparable to species, but the adults are. While larvae of some species of *Hydropsyche* are separable (particularly in the *Hydropsyche bronta* group), I found no records of which non-*bronta* group *Hydropsyche* species actually occur in the Black Hills. The *Hydropsyche bronta* group is considered as presented under the latest revision of *Hydropsyche* (Geraci et al. 2010), such that the species previously in the *Hydropsyche morosa* group and the genus *Ceratopsyche* are considered to be in the *Hydropsyche bronta* group.

Cheumatopsyche Wallengren

A generic description of the larva is given in Wiggins (1996), based on just a handful of species – most larvae remain unknown. As such, they are identified only to the genus level. These are known as the "little sister sedges" by anglers.

Jergens (1968) reported *Cheumatopsyche* at the genus level from the Fall River, French Creek, Spring Creek, Rapid Creek, Spearfish Creek, Battle Creek, and Castle Creek. Shearer (2006) reported larvae of unidentified species of *Cheumatopsyche* from Black Hills streams. I have seen larvae of *Cheumatopsyche* in 12 aquatic biomonitoring samples collected using EMAP protocols at abundances of 10 to 211 organisms per sample, representing up to 12 percent of total invertebrate abundance. In six sets of quantitative, replicated samples from riffle habitat, *Cheumatopsyche* larvae were present from 11 to 640 organisms/m². Streams included Bobtail Gulch, Blacktail Gulch, Cleopatra Creek, Rubicon Gulch, Whitetail Creek, and Whitewood Creek.

I also have larvae of *Cheumatopsyche* in my collection; they are from the Fall River near Hot Springs, Battle Creek near Hermosa, and Whitewood Creek.

Cheumatopsyche campyla Ross

Descriptions/Diagnoses: Gordon 1974 (adult); the larva has not yet been described. Distributed from British Columbia, Washington, and Oregon, east to Ontario, New York, and the Atlantic Coast as far south as Georgia.

Ruiter and Lavigne (1985) provided three records of this species from the Wyoming portion of the Black Hills: "0.5 miles north of Buelah", edge of the town of Sundance, and at Devil's Tower.

Cheumatopsyche lasia Ross

Descriptions/Diagnoses: Gordon 1974 (adult); the larva has not yet been described. Distributed from Minnesota, Nebraska, and Wyoming south to Arizona, New Mexico, and Texas.

Ruiter and Lavigne (1985) reported this species from the Driskill Ranch on the Belle Fourche River, at Devil's Tower.

Cheumatopsyche oxa Ross

Descriptions/Diagnoses: Gordon 1974 (adult); the larva has not yet been described. Distributed from the Black Hills in South Dakota east and south to New York, Ohio, Tennessee, and Arkansas.

Gordon (1974) reported this species from Spearfish Creek, and Ruiter and Lavigne (1985) reported it from the Wyoming portion of the Black Hills National Forest.

Cheumatopsyche pettiti (Banks)

Descriptions/Diagnoses: Gordon 1974 (adult); the larva has not yet been described. Distributed from Washington and Idaho east to Saskatchewan to Ontario, south to Oregon, Colorado, Texas, Arkansas, Indiana, and Florida.

Gordon (1974) reported this species from Newcastle, Wyoming, and Hot Springs, South Dakota. Ruiter and Lavigne (1985) reported additional specimens collected at Beaver Creek Reservoir, just east of Newcastle.

Hydropsyche Pictet

Geraci et al. (2010) recently redefined the genus *Hydropsyche*, established or confirmed numerous synonymies, and divided the genus into subgenera and species groups. One of those groups, the *bronta* group, roughly corresponds to the former *Hydropsyche morosa* group and the genera *Symphitopsyche* and *Ceratopsyche*. Two species in the *bronta* group are known from the Black Hills, and Shearer (2006) reported larvae of unidentified species of *Ceratopsyche* from Black Hills streams. Jergens (1968) reported *Hydropsyche* at the genus level from all streams he looked at: the Fall River, French Creek, Spring Creek, Rapid Creek, Spearfish Creek, Battle Creek, Castle Creek, and the South Fork of Rapid Creek. Additionally, other species of *Hydropsyche* may occur in the Black Hills, but I have not yet seen any published records, and this is a situation that may require extensive revision of the key should when additional species are identified. In my personal collection, I have non-*bronta* group larvae of *Hydropsyche* from the Fall River near Hot Springs, Battle Creek near Hermosa, and Whitewood Creek. Anglers call all *Hydropsyche* species the "spotted sedges".

Hydropsyche bronta Ross

Descriptions/Diagnoses: Schuster and Etnier 1978 (larva, as *Symphitopsyche bronta*), Schefter and Wiggins 1986 (larva), Nimmo 1987 (adult). Distributed from Alberta to Quebec and the Maritimes, south to Wyoming, South Dakota, Illinois, Tennessee, and South Carolina, with a disjunct population in the Ozark Plateau in Arkansas and Oklahoma. There are two color forms, a "central" form with a distinct checkerboard pattern, and an "Appalachian" form with a transverse striped pattern; the geographic names assigned to these forms reflect the greatest concentrations of those forms and each form can actually occur throughout the range of the species.

Schefter and Wiggins (1986) reported that a population of striped larvae of this species has been recorded from the Black Hills. Ruiter and Lavigne (1985) reported this species from two locations in the Wyoming portion of the Black Hills ("0.5 mile north of Buelah" and the Driskill Ranch, Belle Fourche River, Devil's Tower); these specimens were adults and it is unknown if they were the striped morph of *H. bronta* or not. This species was previously in the genus *Ceratopsyche*.

Hydropsyche occidentalis Banks

Descriptions/Diagnoses: Alstad 1980 (larva); Ross 1938 (adult). Distributed in western North America from British Columbia to California, east to Saskatchewan, Wyoming, Colorado, and New Mexico, south into Mexico.

Ruiter and Lavigne (1985) reported this species from two locations in Crook County (Redwater Creek at Beulah and 0.5 miles north of Beulah) and two locations in Weston County ("Beaver Creek at Stockade Creek, Route 16, southwest of Newcastle" and Beaver Creek Reservoir [= LAK Reservoir?] east of Newcastle).

Hydropsyche oslari (Banks)

Descriptions/Diagnoses: Schuster and Etnier 1978 (larva, as *Symphitopsyche oslari*), Schefter and Wiggins 1986 (larva), Nimmo 1987 (adult). Distributed from the Yukon south into Mexico through the Pacific Coast provinces and states, east to Alberta, Montana, South Dakota, Colorado, and New Mexico.

Ruiter and Lavigne (1985) reported this species from numerous locations in Crook County, including the Rod & Gun Club 4.7 miles northwest of Sundance, the Reuter Campground in the Black Hills National Forest 4 miles north of Sundance, and another location in the Black Hills National Forest. Ruiter (1996) reported this species from Little Spearfish Creek and Spearfish Creek.

Hydropsyche slossonae Banks

Descriptions/Diagnoses: Ross 1944 (larva, adult). Transcontinentally distributed in Canada from British Columbia and Northwest Territories east to the Maritimes and Labrador, south to Montana, the Dakotas, Minnesota, Arkansas, Tennessee, and North Carolina

Ruiter and Lavigne (1985) reported this species from a location "0.5 miles north of Buelah" and from the Black Hills National Forest in the Wyoming portion of the Black Hills. This species was previously in the genus *Ceratopsyche*.

Hydroptilidae

Undoubtedly this common family is widespread in the Black Hills, but, as postulated by Ruiter and Lavigne (1985), there do not appear to be many records in the published literature because the small organism size tends to being overlooked by collectors. It is likely that many more species than are listed here actually occur in the Black Hills and could yield numerous new records. Shearer (2006) reported larvae of unidentified species of *Hydroptila*, *Leucotrichia*, and *Ochrotrichia* from Black Hills streams.

Hydroptila Dalman

Species of *Hydroptila* are known to anglers as the "varicolored microcaddis".

Hydroptila ajax Ross

Descriptions/Diagnoses: Ross 1938 (adult); Ross 1944 (larva, adult). Distributed broadly from Washington, Oregon, Nevada, and Arizona east to Minnesota, West Virginia, and Texas.

Ruiter and Lavigne (1985) reported this species from the Driskill Ranch, Devil's Tower.

Hydroptila consimilis Morton

Descriptions/Diagnoses: Ross 1944 (larva, adult). This common species is distributed from British Columbia, Washington, and Oregon east to South Carolina, and from Texas north to the Hudson Bay.

Ruiter and Lavigne (1985) reported this species from Red Water Creek at Beulah and from a location "0.5 miles north of Buelah" in Crook County.

Hydroptila jackmanni Blickle

Descriptions/Diagnoses: Blickle 1963; Huryn 1983 (adult); larva is undescribed. Known from Alabama north to Wyoming, Minnesota, and Ohio.

Ruiter and Lavigne (1985) reported this species from the Belle Fourche River at Devil's Tower in Crook County – this was the only record of the species from the state of Wyoming.

Ithytrichia Eaton

Ithytrichia clavata Morton

Descriptions/Diagnoses: Wiggins 1996 (larva); Moulton et al. 1999 (adult). Distributed from California, Texas, and Florida north to Manitoba and Quebec.

Ruiter and Lavigne (1985) reported this species from the Driskill Ranch on the Belle Fourche River at Devil's Tower in Crook County – this was the only record of the species from the states of South Dakota and Wyoming.

Leucotrichia Mosely

Leucotrichia is a primarily Neotropical genus of very distinctive caddisflies, with one species (*L. pictipes*) whose distribution extends far enough northward to include the Black Hills, although there appear to be no published records of the genus in the Black Hills.

Leucotrichia pictipes (Banks)

Descriptions/Diagnoses: Flint 1970 (larva, adult), Thomson & Holzenthal 2015 (adult). This species is the most northerly distributed species of Leucotrichi, being reported in an apparently disjunt distribution (Flint 1970) including, in the west, Oregon and California east to Montana, Colorado, New Mexico, and Chihuahua, and in the east, Minnesota east to New York and Virginia. The nearest published records to the Black Hills are in the Wind River in Teton County, Hot Springs County, and Yellowstone National Park, all in Wyoming (Flint 1970, Ruiter and Lavigne 1985, Thomson & Holzenthal 2015). Anglers know this species as the "ring-horn microcaddis".

Mayatrichia Mosely

The three species of *Mayatrichia* reported here were all collected in the same light trap sample, and Ruiter and Lavigne (1985) mentioned the possibility that further study may result in synonymy of several of the six known *Mayatrichia* species; however, to my knowledge, this has not yet happened.

Mayatrichia acuna Ross

Descriptions/Diagnoses: Ross 1944 (adult); larva is undescribed. Texas and Wyoming.

Ruiter and Lavigne (1985) reported this species from the Driskill Ranch at Devil's Tower in Crook County, and this was the only record of the species from the state.

Mayatrichia ayama Mosely

Descriptions/Diagnoses: Ross 1944 (larva, adult). Distributed from Alberta east to New Brunswick, south through Mexico and Central America to Costa Rica.

Ruiter and Lavigne (1985) reported this species from the Driskill Ranch at Devil's Tower in Crook County.

Mayatrichia ponta Ross

Descriptions/Diagnoses: Wiggins 1996 (larva); Ross 1944 (adult). Described from Oklahoma and known otherwise only from Texas and Wyoming (Wang and Kennedy 2004)

Ruiter and Lavigne (1985) reported this species from the Driskill Ranch at Devil's Tower in Crook County, and this was the only record of the species from the state.

Ochrotrichia Mosely

In addition to *O. stylata*, which is reported from both South Dakota and Wyoming, several other species are known to occur in Wyoming and and may occur in the Black Hills: *O. logana* (Ross), *O. lometa* (Ross), *O. oregona* (Ross), and *O. potomus* Denning.

Ochrotrichia stylata (Ross)

Descriptions/Diagnoses: Ross 1938 (adult); larva is undescribed. Described from Sweetwater County, Wyoming, but ranging from Idaho and North Dakota south to Arizona.

Denning and Blickle (1972) reported *Ochrotrichia stylata* from South Dakota, mentioning that this species has the widest distribution in the Rocky Mountain states; Ruiter and Lavigne (1985) reported it from four counties in southern and central Wyoming.

Lepidostomatidae

Two species of lepidostomatid caddisflies are known to occur in South Dakota, with an additional six species in Wyoming; two of these are also known to occur within the Black Hills. There are probably additional species to be found.

Lepidostoma Rambur

As a production taxonomist, I generally identified *Lepidostoma* only to the genus level. Occasionally, morphospecies could be identified using case morphology. *Lepidostoma* was quite common in biomonitoring samples that I looked at, being represented in most streams. Shearer (2006) reported larvae of unidentified species of *Lepidostoma* from Black Hills streams. All species are known as the "little brown sedges" by anglers.

Lepidostoma pluviale (Milne)

Descriptions/Diagnoses: Weaver 1988 (adult), larva is apparently not yet described, but the case is illustrated in Weaver 1988. Distributed from British Columbia and Alberta, south to Baja California, and east to Montana, South Dakota, Colorado, and New Mexico.

Weaver's (1988) South Dakota record for *L. pluviale* is "Hardy", but I could not locate that place name in South Dakota, and the Wyoming material reported in Weaver (1988) is from in and near Yellowstone National Park.

Ruiter and Lavigne (1985) reported this species, as its synonym *Lepidostoma veleda* Denning, from the Black Hills National Forest in Crook County, Wyoming (Weaver 1988). Ruiter (1996) reported this species from Iron Creek.

Lepidostoma unicolor (Banks)

Descriptions/Diagnoses: Weaver 1988 (larva, adult). Distribution is from British Columbia, south to California, east to Quebec, Minnesota, South Dakota, Colorado, and New Mexico.

Ruiter and Lavigne (1985) reported collection of this species from the Rod & Gun Club 4.7 miles northwest of Sundance and from the Black Hills National Forest, with both localities in Crook County, Wyoming. Weaver (1988) reported *L. unicolor* from Spearfish Creek in South Dakota. Ruiter (1996) reported this species from Iron Creek, Little Spearfish Creek, and Spearfish Creek.

Leptoceridae

Three species of Leptoceridae have been reported from South Dakota, all in revisions of various genera (Floyd 1995, Glover 1996). *Triaenodes marginatus* Sibley was reported from South Dakota as a state record only (Ross 1944), while the other two appear to have records specifically from the Black Hills. Conversely, 15 species of leptocerids have been reported from Wyoming, with six records from the Black Hills (Ruiter and Lavigne 1985). Harris et al. (1980) reported the following leptocerids from southwestern North Dakota: *Ceraclea tarsipunctata*, *Mystacides longicornis (=interjecta)*, *Nectopsyche diarina*, *Oecetis immobilis*, *O. inconspicua*, and *O. ochracea*. Shearer (2006) reported larvae of unidentified species of *Mystacides*, *Nectopsyche*, and *Oecetis* from Black Hills streams.

Ceraclea Stephens

Descriptions/Diagnoses: Wiggins 1996 (larva), Morse 1975 (adult). Approximately 36 species are known from all over North America, but, unfortunately, the distribution information in Morse (1975) is too vague to hypothesize which species might even occur in the Black Hills. *Ceraclea tarsipunctata* (Vorhies) was reported from South Dakota by Ross (1944, as *Athripsodes tarsi-punctata*), but only at the state level, while *C. arielles* (Denning), *C. copha* (Ross), and *C. tarsipunctata* were reported from various localities (but not the Black Hills) in Wyoming (Ruiter and Lavigne 1985). Anglers call *Ceraclea* the "scaly-wing sedges".

Mystacides Berthold

Shearer (2006) reported larvae of unidentified species of *Mystacides* from Black Hills streams.

Mystacides sepulchralis (Walker)

Descriptions/Diagnoses: Yamamoto and Wiggins 1964 (larva, adult). Distributed transcontinentally from Alaska, British Columbia, Montana, and California, east to Labrador and the Atlantic Coast south to Georgia. This species is known as the "black dancer" in angler parlance.

Yamamoto and Wiggins (1964) reported collection of this species in the Black Hills National Forest and Custer State Park; using this information, Ruiter and Lavigne (1985) anticipated that the species should also occur in Wyoming. These records are somewhat isolated from the rest of the distribution, with the next nearest localities being northwest Minnesota, Wisconsin, southern Missouri, and northwest Arkansas.

Nectopsyche Müller

Jergens (1968) reported *Nectopsyche* (as its now synonymized genus *Leptocella*) from the Fall River and French Creek; Shearer (2006) reported larvae of unidentified species of *Nectopsyche* from Black Hills streams.

Nectopsyche lahontanensis Haddock

Descriptions/Diagnoses: Haddock 1977 (larva, adult). Distributed primarily in the Great Basin from eastern California and Arizona to British Columbia, Utah, and Wyoming. Anglers call it the "white miller".

Ruiter and Lavigne (1985) reported this species from the Belle Fourche River at Devil's Tower.

Oecetis McLachlan

The genus *Oecetis* appears to be represented in the Black Hills by at least four species, although it is certainly possible that more species may eventually be found there. Jergens (1968) reported *Oecetis* at the genus level from French Creek, while Shearer (2006) reported larvae of unidentified species of *Oecetis* from Black Hills streams. Larvae can fairly easily be separated into species groups using Floyd (1995), but the species within the species groups are very difficult to separate. Even adults (e.g., *O. avara* and *O. houghtoni*) can sometimes be difficult to assign to species, depending on locality (Blahnik & Holzenthal 2014).

Biomonitoring records have usually left *Oecetis* at the genus level or as a slashed taxon (e.g., *O. avara/disjuncta*). With the revision of the *O. avara* group by Blahnik and Holzenthal (2014), splitting many records of *O. avara* into *O. houghtoni* and relegating most records of *O. disjuncta* to *O. sordida*, it is probably now preferable and simpler to refer to the species group level for larvae.

Oecetis avara (Banks)

Descriptions/Diagnoses: Floyd 1995 (larva); Smith and Lehmkuhl 1980, Blahnik & Holzenthal 2014 (adult). Floyd (1995) shows the distribution to be from Alberta and Saskatchewan east to Maine and Nova Scotia, south to northern California, Arizona, Texas, and the Gulf Coast states except Florida; however, Blahnik & Holzenthal (2014) reported examining material of this species primarily from Minnesota, with a few specimens from Alabama, Florida, Kansas, Maryland, New York, Quebec, and Wisconsin, remarking that "Some North American material [previously known as *O. avara*] has also been transferred to . . . *O. houghtoni*." This species is given the common name "tan spotted-wing long-horned sedge" by anglers.

Ruiter and Lavigne (1985) reported *O. avara* from a location "0.5 miles north of Beulah" and from the Driskill Ranch on the Belle Fourche River at Devil's Tower. It is unknown whether or not this material would have been transferred to *O. houghtoni* by Blahnik and Holzenthal (2014).

Oecetis houghtoni Blahnik & Holzenthal

Descriptions/Diagnoses: Blahnik & Holzenthal 2014 (adult), larva unknown, but probably similar to *O. avara* in Floyd 1995. As discussed above for *O. avara*, Blahnik & Holzenthal (2014) transferred some material previously known as *O. avara* to their new species, *O. houghtoni*. Most of their specimens of *O. houghtoni* were from Minnesota, where the species are easily separated, but included specimens from Arkansas, Michigan, Mississippi, and Ontario. As of 2018, this change had not yet been reflected in ITIS, so a TSN has not been assigned for this species.

Since it appears that *O. avara* and *O. houghtoni* are sympatric over a good portion of their ranges, it is unknown whether or not Ruiter and Lavigne's (1985) reports of *O. avara* from a location "0.5 miles north of Beulah" and from the Driskill Ranch on the Belle Fourche River at Devil's Tower represent *O. avara* or *O. houghtoni*.

Oecetis sordida Blahnik & Holzenthal

Descriptions/Diagnoses: Floyd 1995 (larva, as *Oecetis disjuncta*), Blahnik and Holzenthal 2014 (adult). Blahnik and Holzenthal (2014) state that "This is the species that has been generally referred to in the literature under the name *Oecetis disjuncta* . . . which is known to us by only a few specimens from California and Oregon." Reconciling the maps in Floyd (1995) for *O. disjuncta* with the new information in Blahnik and Holzenthal, the distribution appears to be from Montana south to Arizona and northwestern Mexico, east to Michigan.

The type locality for *O. sordida* is Boxelder Creek, 1.8 miles west of Nemo in Lawrence County, South Dakota, in the Black Hills (Blahnik and Holzenthal 2014). Ruiter and Lavigne (1985) reported this species (as *O. disjuncta*) from a gully in Sundance and from the Wyoming portion of the Black Hills National Forest in Crook County, Wyoming, while Floyd (1995) included a dot for *O. disjuncta* in the Black Hills in South Dakota.

Oecetis inconspicua (Walker)

Descriptions/Diagnoses: Ross 1944 (adults). Floyd (1995) discussed this species as a probable species complex, presenting seven species for which he had associated morphologically distinct larvae with typical *O. inconspicua* adults. Taking into consideration the possibility that this is a species complex, the range for the members is extensive, from British Columbia to New Brunswick, south to California, Texas, Alabama, and Florida, into Mexico and through the Greater Antilles. Anglers commonly refer to this species as the "plain red-brown long-horned sedge".

Ruiter and Lavigne (1985) reported this species from Devil's Tower.

Oecetis ochracea (Curtis)

Descriptions/Diagnoses: Floyd 1995 (larva), Milne 1934 (adult, as *O. ochracea carri*). Holarctic; in North America, this species is distributed from Alaska and the Yukon south to northern California, Colorado, South Dakota, Tennessee, and North Carolina. In comparison to other species of *Oecetis*, with drawn out common names, anglers simply call this one the "long horn sedge".

Ruiter and Lavigne (1985) reported this species from Devil's Tower and from the Red Water Creek at Beulah. The South Dakota record for *O. ochracea* in Floyd (1995) is a state-level record only. Harris et al. (1980) reported *O. ochracea* from the southwestern region of North Dakota.

Triaenodes McLachlan

Triaenodes tardus Milne

Descriptions/Diagnoses: Ross 1944 (larva, adult, as *T. tarda*). Distributed primarily in the northeast and central states from New Brunswick, Ontario, and New York southwest to Oklahoma and Arkansas, with populations in Arizona and British Columbia.

Jergens (1968) reported *Triaenodes* at the genus level from the Fall River. Ruiter and Lavigne (1985) reported this species from the Driskill Ranch on the Belle Fourche River at Devil's Tower.

Ylodes Milne

Ylodes frontalis (Banks)

Descriptions/Diagnoses: Ross 1944 (adult, as *Triaenodes frontalis*); larva is undescribed. Distributed from Colorado to Saskatchewan, east to Ohio.

Ruiter and Lavigne (1985) reported this species, as *Triaenodes frontalis*, from a site 0.5 miles north of Beulah and from the Driskill Ranch on the Belle Fourche River at Devil's Tower.

Ylodes reuteri (McLachlan)

Descriptions/Diagnoses: Ross 1944 (adult, as *Triaenodes grisea*); larva is undescribed. Distributed from Saskatchewan and Manitoba south to Colorado.

Ruiter and Lavigne (1985) reported this species, as *Triaenodes grisea*, from the Belle Fourche River at Devil's Tower.

Limnephilidae

The Limnephilidae is the largest family of Trichoptera in North America, so it seems likely that there should be a large contingent of species in the Black Hills. The seven species-level records of limnephilid caddisflies that I have found for the Black Hills have been in generic revisions (Parker and Wiggins 1985, Ruiter 1995). Harris et al. (1980) reported *Anabolia bimaculata*, *Limnephilus hyalinus* Hagen, *L. janus* Ross, *L. labus* Ross, and *L. secludens* Banks from southwestern North Dakota, so it is possible that they could be found in the Black Hills.

Anabolia Stephens

Anabolia bimaculata (Walker)

Descriptions/Diagnoses: Flint 1960 and Wiggins 1996 (larva), Nimmo 1971 (adult). Distributed from British Columbia to New England, north into the Northwest Territories, and south to Idaho, New Mexico, and Michigan.

Ruiter and Lavigne (1985) reported this species from the Nussbaum Shelterbelt 15 miles south of Sundance, Wyoming.

Anabolia consocia (Walker)

Description/Diagnoses: Flint 1960 (larva), Nimmo 1971 (adult). Distributed from Alberta to Quebec, south to Nebraska, Illinois, Michigan, West Virginia, and New Jersey. Nimmo (1971) includes a dot in South Dakota, indicating its presence in the state, although it is unknown if this species occurs in the South Dakota portion of the Black Hills.

Ruiter and Lavigne (1985) reported this species from the Belle Fourche River at Devil's Tower and from a location in the Black Hills National Forest in Crook County, Wyoming; this was the only record of this species in the state at the time.

Chyranda Ross

Chyranda is monospecific (*C. centralis*). The temperature classification for *Chyranda* and *C. centralis* is stenothermal cold in Grafe et al. (2002) but would be considered to be eurythermal cool if based on data from Vieira et al. (2006).

Chyranda centralis (Banks)

Descriptions/Diagnoses: Wiggins 1963, 1996 (larva), Nimmo 1971 (adult). The distribution extends from Alaska to Quebec, south to California, Utah, and Colorado. There have been no records of this species from the Black Hills in the published literature. Anglers know this species as the "pale western stream sedge".

I collected one larva of *Chyranda centralis* from a spring seep within Ross Valley.

Glyphopsyche **Banks**

Given the distribution of the genus, *Glyphopsyche*, if found in the Black Hills, would likely be monospecific (*G. irrorata*) even though additional species are known from other Nearctic locations (e.g., Missouri, Tennessee). No temperature preference was given for *Glyphopsyche* in Grafe et al. (2002); based on data in Vieira et al. (2006), the temperature preference at the genus level is eurythermal warm but for *G. irrorata* is eurythermal hot; therefore, I consider the genus level classification to be eurythermal hot in the Black Hills, reflecting the likely monospecific status.

Shearer (2006) reported larvae of unidentified species of *Glyphopsyche* from Black Hills streams.

Glyphopsyche irrorata **(Fabricius)**

Descriptions/Diagnoses: Flint 1960, Wiggins 1996 (larva), Nimmo 1971 (adult). Distributed in a broad transcontinental band from Alaska south to California, east to Newfoundland and New Hampshire. It is likely that this is the species of *Glyphopsyche* reported from the Black Hills by Shearer (2006).

Hesperophylax **Banks**

The temperature designation for *Hesperophylax* in Grafe et al. (2002) is eurythermal warm; this classification would be extrapolated to all its species. Based on data in Vieira et al. (2006), the genus-level classification would be eurythermal hot, but both *H. designatus* and *H. occidentalis* would be eurythermal cool. *Hesperophylax* are commonly called the "silver striped sedges" by anglers.

Jurgens (1968) reported *Hesperophylax* from Battle Creek, French Creek, and Spring Creek. Shearer (2006) reported larvae of unidentified species of *Hesperophylax* from Black Hills streams, and I have collected larvae of *Hesperophylax* in a spring seep within Ross Valley.

Hesperophylax consimilis **(Banks)**

Descriptions/Diagnoses: Ross 1944, Nimmo 1971 (adult); larva is undescribed. Distributed from Alberta south to Nevada and Colorado.

Ruiter and Lavigne (1985) reported this species from a gully at the eastern edge of the town of Sundance, Wyoming.

Hesperophylax designatus **(Walker)**

Descriptions/Diagnoses: Parker and Wiggins 1985 (larva, adult). Distributed from the Yukon Territory east to Nova Scotia, and south to California, New Mexico, South Dakota, Illinois, and Pennsylvania.

Parker and Wiggins (1985) reported *Hesperophylax designatus* from Lawrence, Meade, and Pennington counties in South Dakota. Ruiter (1996) reported this species from Spearfish Creek.

Hesperophylax magnus **Banks**

Descriptions/Diagnoses: Parker and Wiggins 1985 (larva, adult). Distributed from Idaho and California east to Wyoming, Colorado, and New Mexico.

Ruiter and Lavigne (1985) reported this species from a "3 foot wide clear mountain stream" in the Black Hills National Forest in Crook County, Wyoming.

Hesperophylax occidentalis (Banks)

Descriptions/Diagnoses: Nimmo 1971 (adult); the larva remains undescribed, but Parker and Wiggins (1985) state that it is indistinguishable from the larva of *H. designatus*, so that description should apply. Distributed in the western United States from British Columbia and Alberta south to California, Arizona, New Mexico, and South Dakota.

Ruiter and Lavigne (1985) reported this species from a couple sites in Crook County: the Rod & Gun Club 4.7 miles northwest of Sundance, at the edge of the town of Sundance, and 4 miles northwest of Sundance near the Reuter Campground. Parker and Wiggins (1985) reported *H. occidentalis* from Lawrence and Pennington counties in South Dakota. Ruiter (1996) reported *H. occidentalis* from Iron Creek, Little Spearfish Creek, and Spearfish Creek.

Limnephilus Leach

I found no published records of *Limnephilus* species from the South Dakota portion of the Black Hills, but records of six species in the Wyoming portion of the Black Hills. Jurgens (1968) reported *Limnephilus* at the genus level from Spring Creek. Additionally, based on their widespread distributions (Nimmo 1971), I would anticipate that the following species may also eventually be collected in the Black Hills: *Limnephilus hyalinus* Hagen, *Limnephilus janus* Ross, *Limnephilus lithus* (Milne), *Limnephilus perpusillus* Walker, and *Limnephilus secludens* Banks.

The temperature designation for *Limnephilus* in Grafe et al. (2002) is eurythermal warm; this classification would be extrapolated to all its species. Based on data in Vieira et al. (2006), the genus-level classification (and extrapolated to all its species except *L. moestus*) would be eurythermal hot. *Limnephilus moestus*, which is not expected to occur in the Black Hills anyway, is specifically characterized in Vieira et al. (2006) as having no temperature preference. Otherwise, there are not known ecological differences (at the scale I am assigning traits) among the species of *Limnephilus*, so identifications could reasonably stop at the genus level. Wiggins (1996) stated that he had associated larval and adult material for 40 species of *Limnephilus*, but that the published literature only gave larval descriptions for five species. Anglers call *Limnephilus* the "summer flier sedges".

Limnephilus castor Ross & Merkley

Descriptions/Diagnoses: Ross & Merkley 1952, Ruiter 1995 (adult); larva is undescribed. Distributed in the central Rocky Mountains from Colorado, New Mexico, Utah, and Wyoming.

Ruiter and Lavigne (1985) reported this species from a site in the Black Hills National Forest in Crook County.

Limnephilus externus (Hagen)

Descriptions/Diagnoses: Nimmo 1971 (adult); the larva is apparently undescribed. Distributed from Alaska and British Columbia south through the Rocky Mountain Cordillera to California, Utah, and Colorado. Ruiter (1995) suggested that previously reported distributional ranges given for this species should be re-examined, since the other members of the *externus* group are so similar to this species.

Ruiter and Lavigne (1985) reported this species from the Driskill Ranch on the Belle Fourche River in Devil's Tower, a site 10 miles northwest of Sundance, and a site in the Black Hills National Forest, all in Crook County.

Limnephilus rhombicus (Linnaeus)

Descriptions/Diagnoses: Nimmo 1971, 1991 (as *L. chilcotinensis*) (adult); larva is undescribed. Distributed from the Yukon Territory to Greenland, the Canadian Maritimes, and New England, south to Colorado, Minnesota, Illinois, and Pennsylvania. Ruiter (1995) synonymized *Limnephilus chilcotinensis* Nimmo with this species.

Ruiter and Lavigne (1985) reported this species from the Nussbaum Shelterbelt 15 miles south of Sundance.

Limnephilus sericeus (Say)

Descriptions/Diagnoses: Nimmo 1971 (adult); larva is undescribed. Distributed from Alaska to Quebec and New England, south to Oregon, Wyoming, Michigan, and West Virginia.

Ruiter and Lavigne (1985) reported this species from a site in the Black Hills National Forest in Crook County.

Limnephilus spinatus Banks

Descriptions/Diagnoses: Nimmo 1971 (adult); larva is undescribed. Distributed from Alberta south to California, Arizona, and Colorado.

Ruiter and Lavigne (1985) reported this species from a site in the Black Hills National Forest in Crook County and from the Field City site 3.1 miles southeast of Newcastle.

Limnephilus taloga Ross

Descriptions/Diagnoses: Ross 1938, Nimmo 1991 (as *L. chavas*), Ruiter 1995 (adult); larva is undescribed. Distributed from Utah and Wyoming, through Colorado, to New Mexico and Texas.

Ruiter and Lavigne (1985) reported this species from the Driskill Ranch on the Belle Fourche River at Devil's Tower and from the Field City site 3.1 miles southeast of Newcastle.

Onocosmoecus Banks

Onocosmoecus unicolor Banks

Descriptions/Diagnoses: Flint 1960 (larva, as *Onocosmoecus quadrinotata*), Wiggins and Richardson 1986 (adult). Distributed transcontinentally from Alaska to Labrador and Nova Scotia, south to California, Utah, New Mexico, the Black Hills of South Dakota, Manitoba, Ontario, Michigan, New York, and the New England states. It is found across the state of Wyoming (Ruiter and Lavigne 1985).

Wiggins and Richardson (1986) reported larvae of this species from Lawrence County, South Dakota.

Psychoglypha Ross

Out of the 14 North American species of *Psychoglypha*, only one, *P. subborealis*, appears likely to occur in the Black Hills. Shearer (2006) reported larvae of an unidentified species of *Psychoglypha* from Black Hills streams.

Psychoglypha subborealis (Banks)

Descriptions/Diagnoses: Flint 1960, 1996 (larva), Ross 1944 (adult). Distributed from Alaska to Maine, south to California, Colorado, and Michigan. It has not yet actually been reported from the

Black Hills or South Dakota but has been collected in southern Wyoming (Ruiter and Lavigne 1985). Anglers call this species the "snow sedge".

Pycnopsyche Banks

Most of the 15 species of *Pycnopsyche* occur in eastern North America, but two species venture westward to the Rocky Mountains and are possibly found in the Black Hills. Both of these species (*P. guttifer* and *P. subfasciata*) have been reported from South Dakota and Wyoming, but it is unknown if the published records for South Dakota included the Black Hills. Ruiter and Lavigne (1985) reported *P. subfasciata* from Carbon County, Wyoming, but had not seen any actual specimens of *P. guttifer* from Wyoming and had only one published citation referencing Wyoming, but with no locality data; they did report capturing the latter species three miles south of the state border in Colorado. Shearer (2006) reported larvae of an unidentified species of *Pycnopsyche* from Black Hills streams. *Pycnopsyche* are the "great autumn brown sedges" in angler circles.

Pycnopsyche guttifer (Walker)

Descriptions/Diagnoses: Flint 1960 (larva), Ross 1944 (adult). Distributed from Alberta and Washington, east to Nova Scotia, Massachusetts, New York, North Carolina, and Florida.

Pycnopsyche subfasciata (Say)

Descriptions/Diagnoses: Ross 1944 (adult), it does not appear that the larva is yet described. Distributed from South Dakota east to Minnesota, New York, and Pennsylvania.

Molannidae

I have found no published records specifically reporting this family from the Black Hills. The family is included here because one species, *Molanna flavicornis*, has been reported in the vicinity and may eventually be found there. The temperature preference of eurythermal cool reported for *Molanna* and *M. flavicornis* is based on Vieira et al. (2006); there was no temperature preference for these taxa in Grafe et al. (2002).

Molanna Curtis

Molanna flavicornis Banks

Description/Diagnosis: Wiggins 1996 (larva, as genus), Banks 1914 (adult). Distributed from Saskatchewan, South Dakota, and Colorado east to Quebec and New York. This species, as well as others in the genus, is known as the "gray checkered sedges" by anglers.

Harris et al. (1980) reported *Molanna flavicornis* from the southwestern region of North Dakota. Ross (1944) and Ruiter and Lavigne (1985) reported it from South Dakota and Wyoming , respectively, so it is included here as possibly occurring in the Black Hills.

Philopotamidae

There are two genera and species of philopotamid caddisflies known from the Black Hills (Armitage 1991, Shearer 2006): *Chimarra utahensis* and *Wormaldia gabriella*. The temperature preference of stenothermal cold for the family level is based on Vieira et al. (2006); there was no temperature preference for this taxon in Grafe et al. (2002). Given that all three North American genera (including *Dolophilodes*) are otherwise classified as eurythermal warm in the database, it is not reasonable that the family should be classified as stenothermal cold, but this is the result of Vieira et al.'s (2006) meta-analysis of the literature.

Chimarra Stephens

Chimarra utahensis Ross

Description/Diagnosis: Ross 1938, Armitage 1991 (adult), it appears that the larva has not yet been described. Distributed from Oregon to Baja California, east to Montana, South Dakota, Colorado, New Mexico, and central Mexico. Anglers call this species the "little brown sedge" (along with a few other species).

Jurgens (1968) reported *Chimarra* at the genus level from French Creek, Battle Creek, Castle Creek, and the Fall River. Shearer (2006) reported larvae of *Chimarra* from South Dakota state data. In my personal collection, I have larvae of *C. utahensis* that I collected in the Fall River near Hot Springs.

Dolophilodes Ulmer

Dolophilodes aequalis (Banks)

Descriptions/Diagnosis: Wiggins 1996 (larva, as genus), Banks 1924 (adult, as *Philopotamus aequalis*). Distributed in western North America from southern British Columbia along the Pacific Coast to California, then east to Wyoming and Colorado. Not yet reported from the Black Hills, but widely distributed in Wyoming (Ruiter and Lavigne 1985). This is the "medium evening sedge" of angler parlance.

Wormaldia McLachlan

Wormaldia gabriella (Banks)

Description/Diagnosis: Muñoz-Quesada and Holzenthal 2008 (adult), it appears that the larva has not yet been described for this species although the genus has been described. Distributed primarily in western North America from the Yukon and the Northwest Territories south to California, Utah, and Colorado, with a disjunct record from Quebec. Ruiter and Lavigne (1985) anticipated that *W. gabriella* "should be found in small, spring-fed streams at lower elevations" in Wyoming, but had no records. Anglers have called *Wormaldia* the "little autumn sedges".

Shearer (2006) reported larvae of *Wormaldia* from Black Hills stream data. Muñoz-Quesada and Holzenthal (2008) provided records of this species from Spearfish and Boxelder Creek.

Phryganeidae

Three species in two genera of Phryganeidae are known to occur or may occur in the Black Hills, given their geographic distributions. Additionally, four species of *Agrypnia* are known from Wyoming and may eventually be found in the Black Hills but are not included here (Ruiter and Lavigne 1985). Harris et al. (1980) reported *Phryganea cinerea* and *Ptilostomis semifasciata* from southwestern North Dakota.

Phryganea Linnaeus

Phryganea cinerea Walker

Description/Diagnosis: Wiggins 1960 (larva), Ross 1944 (larva, adult). Distributed from Alberta, Montana, Wyoming, and Colorado, east to Ontario, Maine, Massachusetts, and Illinois. Ruiter and Lavigne (1985) reported this species from numerous locations across Wyoming. Anglers call species in the genus *Phryganea* the "rush sedges".

Ptilostomis **Kolenati**

The temperature preference of eurythermal cool reported for *Ptilostomis* and its species is based on Vieira et al. (2006); there was no temperature preference for these taxa in Grafe et al. (2002). Wiggins (1960) stated that the larvae of *P. ocellifera* and *P. semifasciata* were as yet unseparable and his description applied to both species. *Ptilostomis* are the "giant rusty sedges" of anglers.

Ptilostomis ocellifera **(Walker)**

Description/Diagnosis: Wiggins 1960 (larva), Ross 1944 (larva, adult). Distributed from Ontario to Nova Scotia, south to South Dakota, Illinois, Indiana, Ohio, and Pennsylvania.
Ruiter (1996) reported *Ptilostomis ocellifera* from Spearfish Creek.

Ptilostomis semifasciata **(Say)**

Description/Diagnosis: Wiggins 1960 (larva), Ross 1944 (larva, adult). Distributed from South Dakota east to Quebec, Ohio, and Kentucky.
Ruiter and Lavigne (1985) reported this species form the Belle Fourche River at Devil's Tower.

Polycentropodidae

Four species of polycentropodids are reported in the literature to occur in South Dakota (Nimmo 1986), including *Neureclipsis bimaculata*, *Nyctiophylax affinis*, *Plectrocnemia cinera*, and *P. crassicornis*. It is unknown if these records include specimens from the Black Hills, since they were all reported at the state level. Four species are also known from Wyoming, *N. affinis*, *Polycentropus halidus* Milne, *Polycentropus pentus* Ross, and *Polycentropus variegatus* Banks, including some records in the Black Hills. Harris et al. (1980) also reported *N. bimaculata* and *P. cinerea* (as *Polycentropus cinereus*) from the southwestern region of North Dakota, and those species may thus appear in the Black Hills.
Polycentropodid caddisflies, including those that are now known to be monospecific within the region, have mostly been identified only to the genus level.

Neureclipsis **McLachlan**

Neureclipsis bimaculata **(Linnaeus)**

Distribution/diagnosis: Wiggins 1996 (larva, as genus), Nimmo 1986 (adult). Distribution is transcontinental from Alaska to Nova Scotia, south to Montana, South Dakota, Illinois, and New England. *Neureclipsis* caddisflies are known commonly by anglers as the "little red twilight sedges".
It is unknown if the records in Nimmo (1986) are from the Black Hills.

Nyctiophylax **Brauer**

Shearer (2006) reported larvae of an unidentified species of *Paranyctiophylax* (now considered to be a subgenus of *Nyctiophylax*) from Black Hills streams.

Nyctiophylax affinis **(Banks)**

Distribution/diagnosis: Wiggins 1996 (larva, as genus), Nimmo 1986 (adult). This species is primarily distributed from Ontario east to Quebec and Nova Scotia, south to South Dakota, Texas, Mississippi, and South Carolina, with disjunct populations in British Columbia. This is the only species of *Nyctiophylax* that is reported in the published literature as being from (or anywhere near) South Dakota; moreover, the dot on the map in Morse (1972) representing that record is in the far eastern portion of the state. Species of *Nyctiophylax*, including this one, are known by anglers as the "dinky light summer sedges".

Nevertheless, in additional to Shearer's (2006) genus-level record from the Black Hills, Ruiter and Lavigne (1985) reported this species from the Driskill Ranch at Devil's Tower in the Wyoming portion of the Black Hills.

Plectrocnemia Stephens

Based on distribution maps, three species of *Plectrocnemia* seem likely to occur in the Black Hills, although I do not have literature records of their presence there.

Plectrocnemia cinerea (Hagen)

Description/diagnosis: Nimmo 1986 (adult, as *Polycentropus cinereus*). Distributed transcontinentally from Alaska to Newfoundland, south to Oregon, Utah, South Dakota, Texas, Arkansas, Tennessee, and North Carolina.

Plectrocnemia crassicornis (Walker)

Description/diagnosis: Nimmo 1986 (adult, as *Polycentropus crassicornis*). Distributed from Idaho and Saskatchewan, east to Quebec and Nova Scotia, south to Kansas, Arkansas, Tennessee, Georgia, and Florida.

Plectrocnemia variegata (Banks)

Description/diagnosis: Nimmo 1986 (adult, as *Polycentropus variegatus*). Distributed from southeast Alaska south to California, and east to Alberta, Wyoming, and Arizona.

Polycentropus Curtis

Several species of *Polycentropus* has been identified from South Dakota and Wyoming at the state level; however, the proximity of some of these collection records to the Black Hills is entirely unknown. These are the "brown checkered summer sedges", as commonly called by anglers.

In biomonitoring efforts, Shearer (2006) reported the genus *Polycentropus* from Black Hills streams.

Polycentropus flavus Banks

Description/diagnosis: Nimmo 1986 (adult). Distributed from Alaska, across all of mainland Canada (south of treeline) to the Maritimes, and south into the United States to Montan, Texas, Iowa, Tennessee, and New York, with an isolated record from California.

Polycentropus halidus Milne

Description/diagnosis: Nimmo 1986 (adult). Distributed from southeast Alaska and Vancouver Island south to California and Mexico, east to Montana, Wyoming, and New Mexico.

Ruiter and Lavigne (1985) reported this species from a site 0.5 miles north of Beulah, Wyoming.

Polycentropus pentus Ross

Descriptions/Diagnoses: Ross 1944 (adult); larva is undescribed. Distributed from Wyoming east to Ontario and New Hampshire.

Ruiter and Lavigne (1985) reported this species from Sundance and from a site in the Black Hills National Forest in Crook County as the only records of this species in Wyoming.

Psychomyiidae

It is likely that this family occurs in some of the warmer streams emanating from the Black Hills. *Psychomyia flavida* Hagen occurs statewide in Wyoming (Ruiter and Lavigne 1985), but without any records from the Black Hills. Shearer (2006) reported larvae of unidentified species of *Psychomyia* from Black Hills streams.

Psychomyia Latreille

Psychomyia flavida Hagen

Descriptions/Diagnoses: Flint 1964 (larva); Ross 1944 (adult). Distributed transcontintntally from British Columbia and the Pacific coast states to the Canadian Maritime Provinces south to Florida; less common in the deep south from New Mexico to Mississippi. *Psychomyia* are commonly called the "dinky purple breasted sedges" by anglers.

It is likely that the species of *Psychomyia* reported by Shearer (2006) from the Black Hills is *P. flavida*.

Rhyacophilidae

Rhyacophila Pictet

It appears that the only two species of *Rhyacophila* previously reported from the Black Hills are *R. brunnea* and *R. vao*, both members of the distinctive *R. brunnea* species group as defined by Wold (1974, as the *R. acropedes* species group). I have seen a few *Rhyacophila* larvae from the *R. rotunda* species group, but I was unable to identify them to species. No characters are known to separate the larvae of species within the *R. brunnea* species group (Giersch and Wisseman 2013), so larvae of the *R. brunnea* species group from the Black Hills have often been referred to as the slashed taxon *R. brunnea/vao*. Adults can reliably be separated and are the basis of the following records.

Shearer (2006) reported larvae of unidentified species of *Rhyacophila* from Black Hills streams. Anglers refer to all species of *Rhyacophila* as the "green sedges".

Rhyacophila brunnea Banks

Descriptions/diagnosis: Smith 1968 (larva, as *R. acropedes*), Nimmo 1971 (adult). Distributed transcontinentally from California north to British Columbia and east to New York, New Hampshire, and Labrador.

Smith (1968) indicated that he had records of this species from South Dakota and Wyoming, but did not give specific localities in those states. Ruiter (1996) and Prather and Morse (2001) reported *Rhyacophila brunnea* from Iron Creek, Little Spearfish Creek, and Spearfish Creek, both from the same collecting trip that was associated with the 8th International Symposium on Trichoptera.

Rhyacophila rotunda species group

Species within this species group include *Rhyacophila ebria* Denning, *Rhyacophila latitergum* Davis, *Rhyacophila norcuta* Ross, *Rhyacophila rotunda* Banks, and *Rhyacophila tralala* Schmid. This species group is known from the Pacific coast of British Columbia south to California, eastward to Montana and Colorado.

Rhyacophila ebria is a regional endemic in high alpine snowmelt and springfed streams of the Rocky Mountain cordillera in Glacier National Park in the USA and Waterton, Banff, and Jasper National Parks in Canada. *Rhyacophila latitergum* is known only from its type locality in Whatcom County, Washington. *Rhyacophila norcuta* occurs only in the western portions of British Columbia, Washington, Oregon, and California. *Rhyacophila tralala* occurs only in the high Cascades Mountains of Oregon and Washington. Therefore, *R. rotunda* is the most likely member of the species group to be found in the Black Hills of South Dakota, with a distribution from the coast inland to Colorado. Because the "species-group" is not a Linnaean taxonomic level, a TSN does not exist for this taxon in ITIS.

Specimens representing the *R. rotunda* species group were found in Labrador Gulch in 2013 and 2014

Rhyacophila vao Milne

Descriptions/diagnosis: Smith 1968 indicated that the larva is indistinguishable from *R. brunnea*, Nimmo 1971 (adult). Distributed from Alaska and British Columbia south and east to California, Utah, New Mexico, Colorado, South Dakota, and Montana.

Ruiter (1996) reported this species from Iron Creek and the Spearfish Lodge.

Uenoidae

I can find only one published reference to uenoid presence in the Black Hills in the published literature; in Ruiter's (1996) report of a collecting trip in the Black Hills, he reported *Oligophlebodes minutus*.

Neophylax McLachlan

Characters have not yet been published to differentiate between the species of larval *Neophylax* (even though some manuscript keys have been made available in Trichoptera identification workshops). Based on the known distributions of *Neophylax* species, *N. rickeri* and *N. splendens* could reasonably occur in the Black Hills. Shearer (2006) reported larvae of unidentified species of *Neophylax* from Black Hills streams. *Neophylax* are known to anlgers as the "autumn mottled sedges".

Neophylax rickeri Milne

Description/Diagnoses: Flint 1996 (larva), Vineyard et al. 2005 (adult). Distributed from Alaska, south through the Rocky Mountain cordillera to California, Idaho, and Colorado.

Neophylax splendens Denning

Descriptions/Diagnoses: Vineyard et al. 2005 (adult). Distributed from British Columbia and Montana south to California, Utah, and Colorado.

Oligophlebodes Ulmer

Oligophlebodes minutus (Banks)

Descriptions/Diagnoses: Flint 1960, 1996 (larva); Nimmo 1971 (adult). Distributed from the Yukno Territory south through the Rocky Mountains to Mexico, west into Oregon, east into South Dakota, with a disjunct distribution in Michigan. *Oligophlebodes* are called the "little western dark sedges" by anglers.

Ruiter (1996) reported this species from Iron Creek, Little Spearfish Creek, and the Spearfish Lodge.

Trichoptera Traits Table

Table 12:   Trichoptera Traits Table

Species	TSN	Seen	Tolerance Value	FFG	Habit
Apataniidae	598182	+	--	--	SP
Allomyia	116438	+	0	SH	CN
Apatania	115935	-	1	SC	CN
Brachycentridae	116905	+	1	FC	CN
Brachycentrus	116906	+	1	FC	CN
Brachycentrus americanus	116912	+	1	FC	CN
Brachycentrus occidentalis	116918	+	1	FC	CN
Micrasema	116958	+	1	SH	CN
Micrasema bactro	116967	+	1	SH	CN
Glossosomatidae	117120	-	0	SC	CN
Agapetus	117121	-	0	SC	CN
Anagapetus	117154	-	0	SC	CN
Culoptila	115236	-	0	SC	CN
Culoptila cantha	115238	-	0	SC	CN
Culoptila thoracica	606207	-	0	SC	CN
Glossosoma	117159	-	0	SC	CN
Glossosoma ventrale	117217	-	0	SC	CN
Helicopsychidae	117015	+	3	SC	CN
Helicopsyche	117016	+	3	SC	CN
Helicopsyche borealis	117020	+	3	SC	CN
Hydropsychidae	115398	+	4	GC	cn
Cheumatopsyche	115408	+	5	FC	Cn
Cheumatopsyche campyla	115409	/	6	FC	CN
Cheumatopsyche lasia	115442	/	5	FC	CN
Cheumatopsyche oxa	115424	/	5	FC	CN
Cheumatopsyche pettiti	115426	/	6	FC	CN
Hydropsyche	115453	+	4	FC	Cn
Hydropsyche bronta	115578	/	4	FC	CN
Hydropsyche occidentalis	115513	/	4	FC	CN
Hydropsyche oslari	115490	/	4	FC	CN
Hydropsyche slossonae	115587	/	4	FC	CN
Hydroptilidae	115629	-	4	PI	CB
Hydroptila	115641	-	6	PI	CB
Hydroptila ajax	115643	-	6	SC	CN
Hydroptila consimilis	115647	-	6	PI	CN
Hydroptila jackmanni	606531	-	6	PI	CN
Ithytrichia	115823	-	4	PI	CB
Ithytrichia clavata	115824	-	4	PI	CB
Leucotrichia	115630	-	6	SC	CB
Leucotrichia pictipes	115631	-	6	SC	CB
Mayatrichia	115811	-	6	SC	CN
Mayatrichia acuna	115813	-	6	SC	CN
Mayatrichia ayama	115812	-	6	SC	CN
Mayatrichia ponta	115814	-	6	SC	CN
Ochrotrichia	115714	-	4	GC	CN
Ochrotrichia stylata	115765	-	4	GC	CN
Lepidostomatidae	116793	+	3	SH	CB
Lepidostoma	116794	+	1	SH	CB
Lepidostoma pluviale	116875	/	1	SH	CB
Lepidostoma unicolor	116821	/	1	SH	CB
Leptoceridae	116547	+	4	GC	CB
Ceraclea	116684	-	5	GC	SP
Mystacides	116598	+	4	GC	SP
Mystacides sepulchralis	116604	/	4	GC	SP

Species	TSN	Seen	Tolerance Value	FFG	Habit
Nectopsyche	116651	+	3	SH	CB
Nectopsyche lahontanensis	568813	/	3	SC	SP
Oecetis	116607	+	8	PR	CN
Oecetis avara	116608	/	8	PR	CN
Oecetis houghtoni	--	/	8	PR	CN
Oecetis sordida	116641	/	8	PR	CN
Oecetis inconspicua	116613	/	8	PR	CN
Oecetis ochracea	116632	/	8	PR	CN
Triaenodes	116565	-	6	SH	SW
Triaenodes tardus	116580	-	6	SH	SW
Ylodes	598372	-	4	GC	CB
Ylodes frontalis	603762	-	4	GC	CB
Ylodes reuteri	603767	-	4	GC	CB
Limnephilidae	115933	+	4	SH	CB
Anabolia	115956	-	4	SH	CB
Anabolia bimaculata	115957	-	4	SH	CB
Anabolia consocia	115959	-	4	SH	CB
Chyrandra	116017	-	1	SH	SP
Chyrandra centralis	116018	-	1	SH	SP
Glyphopsyche	116030	-	1	SH	SP
Glyphopsyche irrorata	116031	-	1	SH	SP
Hesperophylax	116001	+	5	SH	SP
Hesperophylax consimilis	116011	/	5	SH	SP
Hesperophylax designatus	116008	/	5	SH	SP
Hesperophylax magnus	116004	/	5	SH	SP
Hesperophylax occidentalis	116006	/	5	SH	SP
Limnephilus	116069	+	5	SH	CB
Limnephilus castor	116128	/	5	SH	CB
Limnephilus externus	116146	/	5	SH	CB
Limnephilus rhombicus	116077	/	5	SH	CB
Limnephilus sericeus	116205	/	5	SH	CB
Limnephilus spinatus	116213	/	5	SH	CB
Limnephilus taloga	116076	/	5	SH	CB
Onocosmoecus	116315	-	1	SH	SP
Onocosmoecus unicolor	116318	-	2	SH	CB
Psychoglypha	115974	+	1	GC	SP
Psychoglypha subborealis	115981	/	2	GC	SP
Pycnopsyche	116409	-	4	SH	SP
Pycnopsyche guttifer	116414	-	4	SH	SP
Pycnopsyche subfasciata	116419	-	4	SH	SP
Molannidae	116473	-	--	SC	SP
Molanna	116474	-	--	SC	SP
Molanna flavicornis	116475	-	--	SC	SP
Philopotamidae	115257	+	3	GC	CN
Chimarra	115273	+	3	FC	CN
Chimarra utahensis	115302	+	3	FC	CN
Dolophilodes	115319	+	1	FC	CN
Dolophilodes aequalis	183769	+	1	FC	CN
Wormaldia	115258	+	3	FC	CN
Wormaldia gabriella	115261	+	3	FC	CN
Phryganeidae	115867	-	--	SH	CB
Phryganea	115892	-	4	SH	CB
Phryganea cinerea	115894	-	4	SH	CB
Ptilostomis	115868	-	--	SH	CB
Ptilostomis ocellifera	115873	-	--	SH	CB
Ptilostomis semifasciata	115869	-	--	SH	CB
Polycentropodidae	117043	+	--	FC	CN
Neureclipsis	117095	+	--	FC	CN
Neureclipsis bimaculata	117096	/	--	FC	CN

Species	TSN	Seen	Tolerance Value	FFG	Habit
Nyctiophylax	117104	-	5	FC	CN
Nyctiophylax affinis	117105	-	5	FC	CN
Plectrocnemia	568776	-	--	FC	CN
Plectrocnemia cinera	117050	-	6	PR	CN
Plectrocnemia crassicornis	117053	-	6	PR	CN
Plectrocnemia variegata	183770	-	6	PR	CN
Polycentropus	117044	+	6	PR	CN
Polycentropus flavus	117062	/	6	PR	CN
Polycentropus halidus	117072	/	6	PR	CN
Polycentropus pentus	117057	/	6	PR	CN
Psychomyiidae	115334	-	--	GC	CN
Psychomyia	115335	-	2	GC	CN
Psychomyia flavida	115341	-	2	GC	CN
Rhyacophilidae	115096	+	0	PR	CN
Rhyacophila	115097	+	0	PR	CN
Rhyacophila brunnea	115151	/	0	PR	CN
Rhyacophila rotunda species-group	--	+	0	PR	CN
Rhyacophila vao	115192	/	0	PR	CN
Uenoidae	568757	+	0	SC	CN
Neophylax	116046	+	3	SC	CN
Neophylax rickeri	116054	/	3	SC	CN
Neophylax splendens	116063	/	3	SC	CN
Oligophlebodes	116039	+	1	SC	CN
Oligophlebodes minutus	116041	+	1	SC	CN

Species	Temperature Range	Lotic	Lentic	Current	Substrate	Voltinism
Apataniidae	--	--	--	--	--	--
Allomyia	S: cold	3.3	5.0	--	7.0	Uni
Apatania	S: cold	3.8	1.1	--	5.5	Uni
Brachycentridae	E: cool*	--	--	--	--	Uni
Brachycentrus	E: warm	3.7	5.0	1.8	6.6	Uni
Brachycentrus americanus	E: warm	3.5	5.0	2.0	7.9	Uni
Brachycentrus occidentalis	E: warm	3.8	5.0	1.5	6.6	Uni
Micrasema	E: warm	3.8	1.1	2.0	7.2	Uni
Micrasema bactro	E: warm	4.1	1.1	2.0	8.5	Uni
Glossosomatidae	--	--	--	--	7.0	Multi
Agapetus	E: warm	3.8	5.0	2.0	7.6	Uni
Anagapetus	S: cold	3.6	--	2.0	7.0	Uni
Culoptila	E: warm	4.1	5.0	--	7.0	Uni
Culoptila cantha	E: warm	5.0	5.0	--	7.0	Uni
Culoptila thoracica	E: warm	5.0	5.0	--	7.0	Uni
Glossosoma	E: cool	3.8	1.3	2.2	6.5	Multi
Glossosoma ventrale	E: cool	3.8	1.3	2.2	6.5	Multi
Helicopsychidae	E: warm	--	--	--	7.0	--
Helicopsyche	E: warm	3.8	1.1	1.5	5.3	Multi
Helicopsyche borealis	E: warm	3.7	4.8	1.5	5.3	Uni
Hydropsychidae	--	--	--	--	5.9	--
Cheumatopsyche	E: warm	3.8	1.1	2.0	5.9	Uni
Cheumatopsyche campyla	E: warm	3.8	5.0	1.2	5.0	Uni
Cheumatopsyche lasia	E: warm	4.4	1.1	1.0	7.0	Uni
Cheumatopsyche oxa	E: warm	3.3	1.1	2.5	7.0	Uni
Cheumatopsyche pettiti	E: warm	3.8	4.0	2.3	3.7	Uni
Hydropsyche	E: warm	4.0	1.1	2.0	5.8	Multi
Hydropsyche bronta	E: warm	4.0	1.1	2.0	5.8	Multi
Hydropsyche occidentalis	E: warm	3.5	5.0	1.8	7.0	Multi
Hydropsyche oslari	E: warm	4.3	1.1	2.5	5.5	Multi
Hydropsyche slossonae	E: warm	4.0	1.1	2.0	5.8	Multi
Hydroptilidae	E: warm	--	--	--	4.3	--
Hydroptila	E: warm	3.6	1.0	2.1	4.0	Uni
Hydroptila ajax	E: warm	4.3	3.0	2.1	4.0	Uni
Hydroptila consimilis	E: warm	4.2	5.0	3.0	3.0	Uni
Hydroptila jackmanni	E: warm	3.6	1.0	2.1	4.0	Uni
Ithytrichia	E: warm*	3.4	1.3	1.7	5.7	Uni
Ithytrichia clavata	E: warm*	4.5	5.0	1.7	4.8	Uni
Leucotrichia	E: warm	3.8	--	2.5	7.6	Uni
Leucotrichia pictipes	E: warm	4.1	--	3.0	8.0	Uni
Mayatrichia	E: warm	3.7	--	2.3	7.0	Uni
Mayatrichia acuna	E: warm	3.7	--	2.3	7.0	Uni
Mayatrichia ayama	E: warm	3.8	--	2.3	7.0	Uni
Mayatrichia ponta	E: warm	3.7	--	2.3	7.0	Uni
Ochrotrichia	E: warm	3.9	5.0	1.7	6.0	Uni
Ochrotrichia stylata	E: warm	4.3	5.0	1.7	6.0	Uni
Lepidostomatidae	E: cool*	--	--	--	--	--
Lepidostoma	E: cool	3.4	1.0	1.5	6.3	Uni
Lepidostoma pluviale	E: cool	4.0	1.0	2.5	8.5	Uni
Lepidostoma unicolor	E: cool	3.7	3.0	1.0	6.3	uni
Leptoceridae	E: warm	--	--	--	4.0	--
Ceraclea	E: warm	3.8	1.1	1.7	5.2	Uni
Mystacides	E: warm	4.0	1.0	0.7	3.9	Uni
Mystacides sepulchralis	E: warm	4.0	1.0	0.7	3.9	Uni
Nectopsyche	E: warm	4.0	1.0	1.1	3.8	Uni
Nectopsyche lahontanensis	E: warm	4.3	1.0	1.1	3.8	Uni
Oecetis	E: warm	3.8	1.0	1.4	4.0	Uni
Oecetis avara	E: warm	3.7	5.0	3.0	4.0	Uni

Species	Temperature Range	Lotic	Lentic	Current	Substrate	Voltinism
Oecetis houghtoni	E: warm	3.8	1.0	1.4	4.0	Uni
Oecetis sordida	E: warm	3.0	5.0	2.0	7.0	Uni
Oecetis inconspicua	E: warm	3.9	4.2	1.4	3.4	Uni
Oecetis ochracea	E: warm	4.3	4.3	1.4	0.5	Uni
Triaenodes	E: warm	3.7	1.1	1.5	3.0	Uni
Triaenodes tardus	E: warm	3.7	1.1	1.5	3.0	Uni
Ylodes	E: warm	3.3	1.6	--	4.0	--
Ylodes frontalis	E: warm	4.0	1.6	--	4.0	--
Ylodes reuteri	E: warm	3.3	1.6	--	4.0	--
Limnephilidae	--	--	--	--	7.0	--
Anabolia	E: cool	3.9	1.1	0.9	3.2	--
Anabolia bimaculata	E: cool	4.0	3.4	1.0	1.0	--
Anabolia consocia	E: cool	3.8	1.1	1.0	7.0	--
Chyrandra	S: cold	3.6	3.0	1.0	--	Uni
Chyrandra centralis	S: cold	3.5	3.0	1.0	--	Uni
Glyphopsyche	E: hot	3.7	1.0	0.7	2.0	--
Glyphopsyche irrorata	E: hot	3.6	3.0	0.5	0.5	--
Hesperophylax	E: warm	3.6	1.1	2.0	6.1	Uni
Hesperophylax consimilis	E: warm	3.0	5.0	2.0	6.1	Uni
Hesperophylax designatus	E: warm	3.8	4.6	2.0	7.0	Uni
Hesperophylax magnus	E: warm	3.0	1.1	2.0	6.1	Uni
Hesperophylax occidentalis	E: warm	3.8	5.0	2.0	5.0	Uni
Limnephilus	E: warm	3.9	1.0	0.9	5.2	Uni
Limnephilus castor	E: warm	3.9	1.0	0.9	5.2	Uni
Limnephilus externus	E: warm	3.9	1.0	0.9	5.2	Uni
Limnephilus rhombicus	E: warm	3.9	1.0	0.9	5.2	Uni
Limnephilus sericeus	E: warm	3.9	1.0	0.9	5.2	Uni
Limnephilus spinatus	E: warm	3.9	1.0	0.9	5.2	Uni
Limnephilus taloga	E: warm	3.9	1.0	0.9	5.2	Uni
Onocosmoecus	E: cool	4.0	1.1	0.6	3.0	Uni
Onocosmoecus unicolor	E: cool	4.1	4.0	0.6	3.0	Uni
Psychoglypha	E: cool	3.7	1.1	0.8	2.9	Uni
Psychoglypha subborealis	E: cool	3.8	4.6	0.8	4.0	Uni
Pycnopsyche	E: cool	3.7	1.2	1.6	3.7	Uni
Pycnopsyche guttifer	E: cool	4.3	5.0	1.6	3.7	Uni
Pycnopsyche subfasciata	E: cool	3.7	1.2	1.6	3.7	Uni
Molannidae	--	--	--	--	--	--
Molanna	E: cool	3.7	1.1	0.6	2.1	Uni
Molanna flavicornis	E: cool	5.0	5.0	0.6	0.7	Uni
Philopotamidae	S: cold	--	--	--	7.0	--
Chimarra	E: warm	3.8	2.0	1.8	7.9	Multi
Chimarra utahensis	E: warm	3.7	3.0	1.8	7.9	Multi
Dolophilodes	E: warm*	3.4	3.0	2.0	7.4	Multi
Dolophilodes aequalis	None*	3.0	3.0	2.0	7.4	Multi
Wormaldia	E: cool	3.6	--	2.0	7.0	Uni
Wormaldia gabriella	E: warm	4.0	--	2.0	7.0	Uni
Phryganeidae	--	--	--	--	--	--
Phryganea	E: warm	4.2	1.1	0.5	0.5	--
Phryganea cinerea	E: warm	4.0	5.0	0.5	0.5	--
Ptilostomis	E: cool	3.7	1.2	1.0	--	--
Ptilostomis ocellifera	E: cool	3.7	1.2	1.0	--	--
Ptilostomis semifasciata	E: cool	3.7	1.2	1.0	--	--
Polycentropodidae	--	--	--	--	7.0	--
Neureclipsis	E: warm	4.1	1.4	1.4	5.2	Multi
Neureclipsis bimaculata	E: warm	4.1	1.4	1.4	5.2	Uni
Nyctiophylax	E: warm	4.0	1.1	1.0	5.5	Uni
Nyctiophylax affinis	E: warm	4.0	1.4	1.4	5.2	Uni
Plectrocnemia	--	--	--	--	7.0	--

Species	Temperature Range	Lotic	Lentic	Current	Substrate	Voltinism
Plectrocnemia cinera	E: warm	--	--	--	7.0	Uni
Plectrocnemia crassicornis	E: warm	--	--	--	7.0	Uni
Plectrocnemia variegate	E: warm	--	--	--	7.0	Uni
Polycentropus	E: warm	3.5	1.0	1.5	3.6	Uni
Polycentropus flavus	E: warm	3.5	1.0	1.5	3.6	Uni
Polycentropus halidus	E: warm	3.0	1.0	1.5	3.6	Uni
Polycentropus pentus	E: warm	3.5	1.0	1.5	3.6	Uni
Psychomyiidae	E: cool*	--	--	--	7.0	--
Psychomyia	E: warm	3.9	1.3	2.7	6.4	Uni
Psychomyia flavida	E: warm	3.8	5.0	3.0	7.0	Uni
Rhyacophilidae	E: cool	--	--	--	7.0	--
Rhyacophila	E: cool	3.3	1.0	2.3	5.8	Uni
Rhyacophila brunnea	E: cool	4.3	1.0	2.3	5.8	Uni
Rhyacophila rotunda species-group	E: cool	3.3	1.0	2.3	5.8	Uni
Rhyacophila vao	E: cool	3.5	5.0	2.5	5.5	Uni
Uenoidae	E: cool	--	--	--	7.0	--
Neophylax	E: cool	3.6	1.7	2.5	6.3	Uni
Neophylax rickeri	E: cool	3.9	1.7	2.0	6.3	Uni
Neophylax splendens	E: cool	3.6	1.7	2.5	6.3	Uni
Oligophlebodes	S. cold	3.7	--	3.0	7.0	Uni
Oligophlebodes minutus	S: cold	4.0	--	3.0	7.0	Uni

* Temperature tolerances derived from Vieira et al. (2006) because they were unavailable in Grafe et al. (2002).

Aquatic Diptera

Introduction

The Diptera are the true flies, a very diverse group of insects with over 125,000 species described. Most families have a distinctive morphology, with much less variation within families. For most Diptera, individuals cannot be identified to the species level, and identifications must be left at the genus or subfamily level.

Morphological similarity often begets ecological similarity, such that there is only subtle variability of habitat requirements among species within a genus or among genera within a family. The Chironomidae are an exception, with broad variability in habitat requirements; in fact, some researchers have suggested that the Chironomidae can function as a microcosm of the entire benthic invertebrate community.

Overview of the Aquatic Diptera in the Black Hills

As with most of the other groups, the Diptera have not had regional taxonomic treatments from the Black Hills. Most records came from biomonitoring studies and scattered data points in broad taxonomic revisions. The Diptera are also particularly susceptible to having scattered distribution records in the literature, since the museum specimens researchers relied on for distribution records are often fewer than in other, more charismatic and easily identifiable groups. Many taxonomists simply do not want to put in the time and effort to pin or point all of the tiny little gnats and flies that they caught in the field.

Species Accounts

Athericidae

One aquatic genus and species of Athericidae occurs in the Black Hills.

Atherix Meigen

Atherix variegata Walker

Description/diagnosis: Webb 1977 (larva, adult). Webb (1977) indicated that he could not find characters to separate the three species of *Atherix* in the larval state, but the distribution of the species was fairly distinct within North America. Of the North American species of *Atherix*, the Black Hills falls within the distributional range of only *A. variegata*.

Jergens (1968) found *Atherix* in only one stream, Spring Creek, reporting at the genus level. Shearer (2006) reported *Atherix* larvae from the Black Hills, and I have *Atherix* larvae that I collected from Annie Creek in my personal collection.

Ceratopogonidae

The ceratopogonid midges are known by several colloquial names, including no-see-ums, punkies, biting gnats, predaceous midges, etc. The extremely tiny adults are not always deterred by mosquito netting and can deliver a ferocious bite; the most abundant and widespread biters are in the genus *Culicoides*. In addition to the pain associated with the bite, a few species are also known to vector medical and veterinary pathogens. In the Black Hills region, the most prevalent disease vectored by a ceratopogonid (*Culicoides sonorensis*) is bluetongue virus (*Orbivirus* spp.) in livestock (Schmidtmann et al. 2011).

Most other ceratopogonids are of little economic interest, but they can sometimes be found in great densities in biomonitoring samples. Usually, these samples are representative of highly organic and high nutrient waters.

Some production taxonomists feel that larval Ceratopogoninae can reliably be identified to the genus level ince keys to genus are included in Merrit et al. (2006), but the standard operating procedure at GEI was to identify ceratopogonine individuals to the subfamily level. Larvae of the other three subfamilies (Forcipomyiinae, Dasyheleinae, and Leptocopinae) are easily identified to the genus level. Shearer (2006) had data identified to the genus level, reporting larvae of *Bezzia*, *Culicoides*, and *Probezzia*.

Atrichopogon Kieffer

I have found no records of *Atrichopogon* from South Dakota or Wyoming; Boesel (1973), in reporting on the *Atrichopogon* of Ohio, mentioned the widespread distribution of the following three species which likely occur in the Black Hills: *A. minutus* (Meigen) ("California to Maryland and Virginia"), *A. levis* (Coquillett) ("nearly all parts of the United States . . . records from 22 widely distributed states"), and *A. fusculus* (Coquillett) ("from Canada, to Brazil, and from coast to coast in the United States"). Records from Wyoming for *A. fusculus* in Wirth (1994) are in Yellowstone and Grand Teton National Parks. I have seen this genus in biomonitoring samples. *Atrichopogon* has hundreds of species worldwide, with about 20 species known in North America. There are likely dozens of undescribed species in North America, with hundreds more elsewhere, evidenced by the abundance of new species described monthly.

Bezzia Kieffer

Jergens (1968) reported *Bezzia* at the genus level from the Fall River, Spring Creek, and Castle Creek. As mentioned above, Shearer (2006) reported *Bezzia* in samples collected in the Black Hills.

Culicoides Latreille

This genus includes many nasty biters and the vector of bluetongue virus in livestock, *C. sonorensis*. Normal habitat for *Culicoides* usually is lentic waters in stock ponds, small lakes, soil saturated from overflowing stock tanks, and sometimes the backwaters of small streams. These are not habitats normally sampled for stream biomonitoring purposes, so I have not seen themin and around the Black Hills.

The species listed below have been reported in the literature specifically from the Black Hills. Additionally, Schmidtmann et al. (2011) also reported *Culicoides crepuscularis* Malloch, *Culicoides stellifer* (Coquillett), *Culicoides haematopotus* Malloch, *Culicoides (Selfia)* sp., and *Culicoides gigas* Root & Hoffman as sporadic and in small numbers from their studies in Nebraska and the Dakotas; there was no indication if these records were from their sites in or around the Black Hills.

Culicoides coquerellii (Coquillett)

This species was reported from Spearfish Creek (Wirth and Blanton 1969).

Culicoides sonorensis (Wirth & Jones)

This species was reported in Butte, Fall River, Lawrence, Meade, and Pennington counties, and absent from Custer County in intensive trapping for *Culicoides* (Schmidtmann et al. 2011). As of 2018, this species was not in ITIS and does not yet have a TSN assigned.

Culicoides variipennis (Coquillett)

This species was reported in Lawrence, Meade, and Pennington counties, and absent from Custer County in intensive trapping for *Culicoides* (Schmidtmann et al. 2011).

Parabezzia Malloch

Parabezzia williamsi Wirth

This species was reported from Oral, South Dakota, in Fall River County, among other specimens examined from Florida, Michigan, New York, and Virginia when the species was described in 1965 (Wirth 1965).

Probezzia Kieffer

I have no records of this genus from the Black Hills, but *P. albiventris* (Loew) is commonly reported from the Atlantic seaboard from New Jersey to Georgia west to Michigan, Oklahoma, and Louisiana (Wirth 1951) and may eventually be found in the vicinity of the Black Hills.

Chaoboridae

I found only one record of Chaoboridae from the Black Hills, a family-level record from Bear Butte Creek, reported in James (2013). The distribution of Chaoboridae in North America indicates that it is possible that multiple species may occur there. According to distribution maps in Borkent (1981), three species of *Chaoborus* are known to occur in South Dakota (*C. americanus*, *C. flavicans*, and *C. punctipennis*), although the records are from the eastern half of the state, and three species of chaborids are known to occur in Wyoming (*C. flavicans*, *Eucorethra underwoodi*, and *Mochlonyx velutinus*), again from parts of the state other than in the Black Hills. All five of these species are included here.

Although they can be found in larger streams and rivers, the normal habitat for *Chaoborus* is ponds and lakes (occasionally in profundal areas).

Chaoborus Lichtenstein

Chaoborus americanus (Johannsen)

Descriptions/diagnoses: Cook (1956); Saether (1970). Distributed transcontinentally from Alaska, British Columbia, and Washington to the Canadian Maritimes and New England, south to Utah, Colorado, Kansas, Missouri, the Great Lakes, and New Jersey (Borkent 1981). In South Dakota, the known distribution is limited to the far northeastern corner of the state.

Chaoborus flavicans (Meigen)

Descriptions/diagnoses: Cook (1956); Saether (1970). Distributed broadly across North America from Alaska, British Columbia, Washington, Oregon and northern California to Quebec and southern New England, south to Colorado, Nebraska, Arkansas, the Great Lakes, and New Jersey; isolated occurrences in southeastern Texas and northern Georgia (Borkent 1981). Broadly distributed in Europe, as well. The known distribution near the Black Hills is limited to the far northeastern corner of South Dakota and the North Platte River basin in Wyoming.

Chaoborus punctipennis (Say)

Descriptions/diagnoses: Cook (1956); Saether (1970). Eastern North America from Quebec and Ontario south to Florida and the Gulf Coast to southern Texas, west to Alberta, South Dakota, Missouri, Oklahoma, and central Texas; isolated occurrences in Colorado and west-central Mexico (Borkent 1981). In South Dakota, known distribution is limited to the extreme northeastern and southeastern corners of the state.

Eucorethra Underwood

Eucorethra underwoodi Underwood

Descriptions/diagnoses: Cook (1956). Distributed transcontinentally from the Pacific Coast (Alaska to central California) to Quebec, Labrador, the Canadian Maritimes, and New England, south to Utah, New Mexico, Minnesota, Michigan, and New York (Borkent 1981). Known occurrences near the Black Hills include a collection in Wyoming just west of Devil's Tower National Monument.

Mochlonyx Loew

Mochlonyx velutinus (Ruthe)

Descriptions/diagnoses: Cook (1956). Northern transcontinental in North America, from Alaska to Quebec and Labrador, south to Utah, New Mexico, Minnesota, Michigan, New York, and Connecticut (Borkent 1981). In Palearctic region from Lapland south to Tunisia. This species appears to follow the Rocky Mountain cordillera south to New Mexico; the closest known occurrences of this species are in Yellowstone National Park and northcentral Colorado.

Chironomidae

Species-level Black Hills, South Dakota, and Wyoming records for Chironomidae are extremely scattered and rare. Hudson (1971) provided a list of 147 Chironomidae species from South Dakota, primarily based on extensive collecting at Lake Francis Case, Lewis and Clark Lake, and Gavins Point National Fish Hatchery, as well as less intensive collecting at ten other locations in eastern South Dakota – the westernmost collecting point reported as Lake Francis Case, over 300 km (>200 mi) east of the Black Hills on the mainstem Missouri River between Pierre and Yankton. Jergens (1968) collected Chironomidae in all streams that he investigated but differentiated only those with or without anal gills. Hudson (1971) also included records uncovered during a literature search, differentiating between records for which the State of South Dakota is actually mentioned in the distribution (e.g., Sublette and Sublette 1965, Roback 1971) and those "recorded . . . as occurring in any of the six states surrounding South Dakota or whose broad distributional record included South Dakota." I have to follow suit, such that the following list is based either on the rare explicit literature report or my own observation of the genus/species or the fact that the Black Hills fall generally within its distribution. Unfortunately, this is not a group that I relish working on, so I have very few actual specimens from the Black Hills in my personal collection. This would be a good group for faunistic studies!

Genera are sorted alphabetically within each subfamily in the ecological discussions, rather than for the whole family. The reason for splitting the chironomid taxa into subfamilial and tribal levels is because some biomonitoring programs have incorporated taxonomic metrics based on the subfamily or tribe level (e.g., percent Tanypodinae). For many genera, there is no comprehensive taxonomic treatment, and, as stated above, distributional records are highly scattered. Given the few actual collection records from widely disparate localities, it is not implausible for a species currently known only from a very distance place – Greenland, for example – to also be collected in the Black Hills. Therefore, while it may be feasible to guess which species might occur, based on widespread Rocky Mountain or Nearctic distributions, I will generally not give in to that temptation. This is hard enough on the genus level.

There are a few taxa for which even genus-level identification of larvae is not possible, despite the best efforts to prepare specimens. For example, some larvae of *Cricotopus* and *Orthocladius* are often very similar and are sometimes rendered as a slashed taxon, *Cricotopus/Orthocladius*. The same is true for larvae of *Conchepelopia* and *Thienemanimyia*, which are reported as a slashed taxon, too. On the other hand, a couple genera have some well-characterized species. For example, three Black Hills species of *Cricotopus* are readily identifiable: *C. bicinctus*, *C. nostocicola*, and *C. trifascia*. These three species are identified, while other species of *Cricotopus* are left at the genus level, and yet others have to be slashed with *Orthocladius* as described above because they are indistinguishable. Some researchers have been using pupal exuviae for biomonitoring and the method looks promising, although procedurally obscure.

The most recent taxonomic treatment of the Chironomidae is Andersen et al. (2013), which provides keys and diagnoses for larvae of all of the genera of Chironomidae in the Holarctic region. As such, this will be the source for descriptions/diagnoses and distributions for all genera encountered in the Black Hills, and I won't repeat the citation for each genus. Further editions of this work may include citations for the adults, which do not have a similar treatment and the literature is more scattered.

During almost all stream biomonitoring efforts, only larvae and a few pupae are collected. Most larvae can only be identified to the genus level, at best, with some to species or species-group level. In most cases, material was mounted on microscope slides and examined under a compound microscope even for that level of identification. Shearer (2006) had data reported only at the subfamily level. GEI mostly subcontracted the identification of chironomid larvae to Dr. Len Ferrington, Jr. of the University of Minnesota. Dr. Ferrington is a recognized world expert on the Chironomidae and identified to the lowest practical taxonomic level, usually the genus.

Chironominae

Chironomus Meigen

Distribution is cosmopolitan, with several hundred species. Hudson (1971) reported 13 named species and five new, but as yet undescribed, species from South Dakota.

Cladotanytarsus Kieffer

Worldwide distribution, with about 28 Holarctic species. Hudson (1971) reported *C. crusculus* (Saether) (as *Lenziella crusculus*) and a new, but as yet undescribed, species from eastern South Dakota.

Cryptochironomus Kieffer

Worldwide distribution, with about 30 species in the Holarctic and 13 in the Nearctic. Hudson (1971) reported *C. digitatus* (Malloch), *C. ponderosus* Sublette, *C. psittacinus* Meigen, a species in the *C. fulvus* species-group, and a new, but as yet undescribed, species from eastern South Dakota.

Cladopelma Kieffer

Holarctic and Afrotropical in distribution, with 7 species in the Nearctic. Hudson (1971) reported *C. amachaerus* (Townes), *C. collator* (Townes), *C. galeator* (Townes), *C. viridulus* (L.), and an unidentified species from eastern South Dakota, all in the genus *Cryptocladopelma*, which is a synonym of *Cladopelma*.

Cryptotendipes Lenz

Mostly Holarctic, with one species from the Neotropics and two from the Oriental region; six or seven species in North America. Hudson (1971) reported *C. ariel* (Sublette), *C. darbyi* (Sublette), *C. emorsus* (Townes), *C. pseudotener* (Goetghebuer), and a new, but as yet undescribed, species from eastern South Dakota.

Demicryptochironomus Lenz

Istributed in Holarctic, Paleotropical, and Oriental regions. Hudson (1971) reported *D. cuneatus* (Townes) (as *Leptochironomus cuneatus*) from eastern South Dakota.

Dicrotendipes Kieffer

Cosmopolitan distribution, with about 30 species in the Holarctic. Hudson (1971) reported eight species, including one new, but as yet undescribed, species from eastern South Dakota.

Einfeldia Kieffer

Four or five species, generally with a Holarctic distribution. Hudson (1971) reported *E. brunneipennis* (Johannsen), "*Einfeldia dorsalis* (Meigen)", and *E. pagana* Meigen from eastern South Dakota.

Endochironomus Kieffer

Distributed in temperate Holarctic (15 species), Aftrotropical (4 species), and Oriental (1 species) regions. Hudson (1971) reported *E. nigricans* Johannsen and *E. subtendens* (Townes) from eastern South Dakota.

Glyptotendipes Kieffer

Holarctic species, of which there are about 25 species, are located primarily in temperate regions; some Oriental and Afrotropical species. Hudson (1971) reported eight species, including one new, but as yet undescribed, species from eastern South Dakota.

Harnischia Kieffer

Worldwide distribution, with about 8 Holarctic species. Hudson (1971) reported "*Harnischia alboviridis* (Malloch)" [= *Gillotia alboviridis* (Malloch)?], *H. curtilamellata* (Malloch), "*Harnischia grisea* (Malloch)", and *H. incidata* (Townes) from eastern South Dakota.

Kiefferulus Goetghebuer

Worldwide distribution, except Neotropics. Hudson (1971) reported *K. dux* Johannsen from eastern South Dakota.

Lauterborniella Thienemann & Bause

Supposedly monospecific (*L. agrayloides* (Kieffer)), with a Holarctic distribution (and disjunct localities in South America). Hudson (1971) reported "*Lauterborniella varipennis* (Coquillett)", which may actually be in *Zavreliella* or *Kribiodorum*, according to Andersen et al. (2013), from eastern South Dakota.

Microchironomus Kieffer

Holarctic, Afrotropical, and Oriental distribution of the genus. Hudson (1971) reported *M. nigrovittatus* (Malloch) (as *Leptochironomus nigrovittata*) from eastern South Dakota.

Micropsectra Kieffer

Distributed throughout the Holarctic, with approximately 100 species. Hudson (1971) reported *M. dives* (Johannsen), *M. nigripila* (Johannsen), *M. polita* (Malloch), and *M. xantha* (Roback) (as *Calopsectra xanthus)* from eastern South Dakota.

Microtendipes Kieffer

There are about 17 species in the Holarctic. Hudson (1971) reported *M. caducus* Townes, *M. pedellus* (DeGeer), and an unidentified species from eastern South Dakota.

Parachironomus Lenz

Worldwide distribution with about 30 species in the Holarctic. Hudson (1971) reported seven species from eastern South Dakota.

Paracladopelma Harnisch

Distribution is Holarctic, with about 20 species. Hudson (1971) reported *P. amphitrite* (Townes), *P. doris* (Townes), *P. nais* (Townes), *P. tethys* (Townes), and two new, but as yet undescribed, species from eastern South Dakota.

Paralauterborniella Lenz

Nearctic and Neotropical, possibly Holarctic. Hudson (1971) reported the only species in this genus, *P. nigrohalterale* (Malloch), from eastern South Dakota.

Paratanytarsus Thienemann & Bause

Worldwide distribution with more than 60 species. Hudson (1971) reported *P. dissimilis* (Johannsen) and three new, but as yet undescribed, species from eastern South Dakota.

Paratendipes Kieffer

About 14 species occur in the Holarctic and 6 species each in the Afrotropical and Oriental Regions. Hudson (1971) reported *P. albimanus* (Meigen), *P. basidens* Townes, *P. duplicatus* (Johannsen), and a new, as yet undescribed, species from eastern South Dakota.

Phaenopsectra Kieffer

Holarctic, with 3 species in the Palaearctic and 9 species in the Nearctic Region. Hudson (1971) reported six species from eastern South Dakota.

Polypedilum Kieffer

Cosmopolitan, with several hundred species, of which >100 species occur in the Holarctic. Hudson (1971) reported 17 species from eastern South Dakota. I have larval specimens of this genus, collected from the Black Hills, in my personal collection.

Pseudochironomus Malloch

Holarctic and Neotropical in distribution. There are about 11 species in the Nearctic, but only 1 or 2 species in the Palaearctic and an additional 5 in the Neotropics. Hudson (1971) reported *P. fulviventris* (Johannsen), *P. pseudoviridis* (Malloch), *P. rex* Hauber, and *P. richardsoni* Malloch from eastern South Dakota.

Rheotanytarsus Thienemann & Bause

Worldwide distribution, with nearly 100 species; Andersen et al. (2013) cite James Sublette that there are probably 30 species in North America, even though only 4 have been formally recorded. Hudson (1971) reported *R. exiguus* (Johannsen) and a new, but as yet undescribed, species from eastern South Dakota.

Stenochironomus Kieffer

Worldwide distribution, with about 11 species in the Nearctic. Hudson (1971) reported *S. macateei* (Malloch), *S. poecilopterus* (Mitchell), *S. hilaris* (Walker) (as its synonym *S. taeniapennis*), and an unidentified species from eastern South Dakota.

Stictochironomus Kieffer

Thirteen species are known in the Nearctic, with about fifteen species known from other regions. Hudson (1971) reported *S. annulicrus* (Townes), *S. devinctus* (Say), *S. marmoreus* (Townes), *S. palliatus* (Coquillett), *S. varius* (Townes), and an unidentified species from eastern South Dakota.

Tanytarsus van der Wulp

Cosmopolitan, with more thatn 100 species in the Holarctic. Hudson (1971) reported *T. neoflavellus* (Malloch) (as *Calopsectra neoflavella*), and six new, but as yet undescribed species (four in "*Calopsectra*"), from eastern South Dakota.

Tribelos Townes

There are six Nearctic species and one Palaearctic species. Hudson (1971) reported *T. ater* (Townes), *T. fuscicorne* (Malloch), and *T. jucundus* (Walker), all as the subgenus *Phaenopsectra (Tribelos)*, from eastern South Dakota.

Xenochironomus Kieffer

The genus occurs in several regions, but with only one species, *X. xenolabis* Kieffer, purported to be Holarctic. Hudson (1971) reported *X. festivus* Say, *X. scopula* (Townes), *X. taenionotus* Say, and *X. xenolabis* from eastern South Dakota.

Zavrelia Kieffer, Thienemann, & Bause

Holarctic distribution with 10 species. Hudson (1971) reported the genus from eastern South Dakota.

Diamesinae

Boreoheptagyia Brundin

In the Nearctic region, three species are known, but only one has been named; otherwise 22 species are known in the genus. Hudson (1971) reported *B. lurida* (Garrett) from eastern South Dakota.

Diamesa Meigen

More than 100 species, primarily Holarctic. Hudson (1971) reported four species of this genus from eastern South Dakota, including *D. ancysta* Roback, *D. leona* Roback, *D. mendotae* Muttkowski, and *D. nivoriunda* (Fitch).

Pagastia Oliver

Holarctic with about 9 species. I have larval specimens of the genus collected from Annie Creek in my personal collection.

Pseudodiamesa Goetghebuer

Twelve species are known from the Holarctic Region. Hudson (1971) reported *P. pertinax* (Garrett) from eastern South Dakota.

Orthocladiinae

Acricotopus Kieffer

Holarctic and Oriental distribution, with 5 species in the Holarctic. Hudson (1971) reported *A. nitidellus* (Malloch) (as it synonym *A. senex*) from eastern South Dakota.

Brillia Kieffer

This genus is widely distributed in the Holarctic, Neotropical, and Oriental realms. Hudson (1971) reported this genus from eastern South Dakota.

Bryophaenocladius **Thienemann**

Worldwide distribution, with 16 species known from the Nearctic. Hudson (1971) reported this genus from eastern South Dakota.

Camptocladius **van der Wulp**

Monospecific (*C. stercorarius* (de Geer)) genus found throughout the Europe, Greenland, and much of the Nearctic and Australasian regions. Hudson (1971) reported *C. stercorarius* from eastern South Dakota.

Cardiocladius **Kieffer**

Worldwide except Antarctica, with 9 described species in the Holarctic.

Chaetocladius **Kieffer**

Holarctic and Aftrotropical, with about 45 species in the Holarctic. Hudson (1971) reported *C. stamfordi* (Johannsen) (as its synonym *C. oliveri)* from eastern South Dakota.

Corynoneura **Winnertz**

Cosmopolitan, with over 15 species in North America. Hudson (1971) reported *C. scutellata* Winnertz from eastern South Dakota.

Cricotopus **van der Wulp**

Worldwide distribution, except Antarctica; in the Nearctic, there are about 45 named species, but many remain undescribed in this genus.

As mentioned above, *Cricotopus* is one of the few genera of chironomids in which subgenera and a few species are readily separable, and Oliver and Roussel (1983) provide a key to the subgenera and species groups of *Cricotopus* in Canada. Three species that can be identified (*C. bicinctus*, *C. nostocicola*, and *C. trifascia*) are known from the Black Hills, while Hudson (1971) reported *C. flavibasis* (Malloch), *C. infuscatus* (Malloch), *C. remus* Sublette, and six new, but as yet undescribed, species from eastern South Dakota.. Some *Cricotopus* species are not separable to subgenus, so those organisms are usually identified as *Cricotopus* sp. Some larvae of *Cricotopus* are indistinguishable from those of *Orthocladius*, so those are usually reported as a slashed taxon.

Diplocladius **Kieffer**

Purportedly monospecific (*D. cultriger* Kieffer, in the Holarctic, but possibly several undescribed species, especially in the Nearctic. Hudson (1971) reported the genus from eastern South Dakota.

Epoicocladius **Šulk and Zavřel**

Two species known (one, *E. ephemerae* (Kieffer), in Europe and Near East and the other, *E. flavens* (Malloch), from the Nearctic), with possibly others. Hudson (1971) reported *E. flavens* from eastern South Dakota.

Eukiefferiella **Thienemann**

Cosmopolitan except southern Netropics and Antarctica; more than 50 species known in the Holarctic. I have specimens of this genus collected from the Black Hills in my personal collection.

Heleniella **Gowin**

Nine or ten species are known from the Holarctic into the Oriental Regions.

Heterotrissocladius **Spärck**

Mostly Holarctic with up to 15 species, but also reported from the Ecuadorean Andes.

Hydrobaenus **Fries**

Holarctic, with 24 species. Hudson (1971) reported *H. pilipes* (Malloch), in the Palaearctic genus *Trissocladius*, from eastern South Dakota.

Krenosmittia **Thienemann & Krüger**

Holarctic, Afrotropical, and Oriental, of which 3 species are Holarctic.

Limnophyes **Eaton**

This genus is cosmopolitan but is most speciose in the Holarctic (34 species in Holarctic and Afrotropics). Hudson (1971) reported *L. immucronatus* Saether and a new, but as yet undescribed, species from eastern South Dakota.

Metriocnemus **van der Wulp**

Worldwide with about 50 speices from the Holarctic. Hudson (1971) reported the genus from eastern South Dakota.

Nanocladius **Kieffer**

Worldwide, with 13 species known from the Holarctic. Hudson (1971) reported the synonymized "*Microcricotopus distinctus* (Malloch)" from eastern South Dakota.

Orthocladius **van der Wulp**

Worldwide, except Antarctica, with over 100 species in the Holarctic.

In one of the few species-level reports of chironomids that I could find from the Black Hills, Soponis (1977) reported *O. obumbratus* Johannsen from Spring Creek near Hill City. Elsewhere, Hudson (1971) reported *O. anteilis* (Roback), *O. barbicornis* (L.), *O. obumbratus*, *O. wiensi* Saether, and two new, but as yet undescribed, species from eastern South Dakota.

Larvae of *Orthocladius* are not always separable from *Cricotopus*, so they are often reported as a slashed taxon. Oliver and Roussel (1983) provided a larval key to the subgenera of Orthocladius in Canada.

Parakiefferiella Thienemann

Distributed in all biogeographic regions, although there are only sparse records in the Neotropical, and 29 species in the Holarctic. Hudson (1971) reported the genus from eastern South Dakota.

Parametriocnemus Goetghebuer

Worldwide distribution, except Antarctica, with 26 species described for the Holarctic region. Hudson (1971) reported *Parametriocnemus lundbecki* (Johannsen) from eastern South Dakota.

Paraphaenocladius Thienemann

Holarctic and Afrotropical, with 27 species total. Hudson (1971) reported *P. exagitans* (Johannsen) and *P. nasthecus* (Saether) from eastern South Dakota.

Parorthocladius Thienemann

Holarctic and Afrotropical, with many undescribed Nearctic species.

Psectrocladius Kieffer

This genus is found worldwide except Australasia and Antarctica, with about 70 species in the Holarctic. Hudson (1971) reported *P. barbimanus* (Edwards), *P. flavus* (Johannsen), *P. limbellatus* (Holmgren), *P. pilosus* Roback, and a new, but as yet undescribed, species from eastern South Dakota.

Pseudosmittia Edwards

Worldwide distribution, with about 50 of the known 94 species being present in the Holarctic. Hudson (1971) reported the genus from eastern South Dakota.

Psilometriocnemus Sæther

Two species: one (*P. triannulatus* Saether) Nearctic and the other (*P. europaeus* Tuiskunen) from Scandinavia.

Rheocricotopus Thienemann & Harnisch

Worldwide, except Neotropical Region, with about 35 species in the Nearctic.

Smittia Holmgren

A worldwide genus, and the number of Holarctic species is unknown because the larvae of many species are terrestrial. Hudson (1971) reported *S. aterrima* (Meigen), as well as at least two other unidentified species, from eastern South Dakota.

Synorthocladius Thienemann

Essentially worldwide distribution for the genus, with one widely distributed Holarctic species, *S. semivirens* (Kieffer), and at least one undescribed Nearctic species.

Thienemanniella Kieffer

Probably a worldwide distribution, with about 50 known species and likely many more yet to be described. Hudson (1971) reported the genus from eastern South Dakota.

Tvetenia Kieffer

Holarctic, Afrotropical, and Oriental, with about 20 species in the Holarctic.

Podonominae

Lasiodiamesa Kieffer

Northern Holarctic in distribution, with five species known from North America. Hudson (1971) reported this genus from eastern South Dakota.

Paraboreochlus Thienemann

Monospecific (*P. stahli* Coffman) and widespread in North America, with a second species in the genus from Europe and a third from the Orient.

Prodiamesinae

Odontomesa Pagast

Three species, with one (*O. fulva* (Kieffer)) Holarctic, one (*O. ferringtoni* Saether) Nearctic, and one (*O. lutosopra* (Garrett)) from British Columbia. Hudson (1971) reported *O. fulva* from eastern South Dakota.

Prodiamesa Kieffer

Holarctic genus with six described species, one of which, *P. olivacea* (Meigen), is the only species extending into North America.

Tanypodinae

Ablabesmyia Johannsen

Cosmopolitan; Poole and Gentili (1996-1997) reported 19 valid species from North America. Hudson (1971) reported 11 species from this genus in eastern South Dakota – *A. illinoensis* (Malloch), *A. peleensis* (Walley) *A. pulchripennis* (Lundbeck), *A. annulata* (Say); *A. aspera* (Roback), *A. basilis* (Walley), *A. mallochi* (Walley), *A. monilis* (L.), *A. rhamphe* Sublette, *A. tarella* Roback, and one undescribed species near *peleensis*.

Alotanypus Roback

Three species are known from North America: *A. venustus* (Coquillett) from Ohio to western North America and south to Costa Rica, *A. aris* Roback from West Virginia to Florida, and and undescribed species from New Hampshire. Hudson (1971) reported *A. venustus* from eastern South Dakota.

Clinotanypus Kieffer

Genus is worldwide except Antarctica; in North America, *C. pinguis* (Loew) is in the eastern United States, and three undescribed species are known from Florida. Hudson (1971) reported *C. pinguis* from eastern South Dakota.

Coelotanypus Kieffer

A New World genus, with 5 species known from North America. Hudson (1971) reported two species, *C. concinnus* (Coquillet) and *C. scapularis* (Loew), from eastern South Dakota.

Conchepelopia Fittkau/*Thienemannimyia* Fittkau

Holarctic, Neotropical, Afrotropical, Oriental. Hudson (1971) reported three species of *Conchepelopia*: *C. goniodes* (Sublette), *C. rurika* (Roback), and *C. currani* (Walley), as well as *Thienemannimyia senata* (Walley), from eastern South Dakota.

Larvae of *Conchepelopia* and *Thienemanimyia* are incompletely separable and are often reported as a slashed taxon or as the *Thienemannimyia* complex.

Guttipelopia Fittkau

Two species: the Nearctic *G. rosenbergi* Bilyj from Ontario and Manitoba and the Holarctic *G. guttipennis* (van der Wulp) from Canada south to Florida in North America. Hudson (1971) reported *G. guttipennis* from eastern South Dakota.

Labrundinia Fittkau

Mostly New World distribution (6 named Nearctic species, but numerous undescribed species), with one Holarctic species and one Indonesian species. Hudson (1971) reported *L. pilosella* (Loew) from eastern South Dakota.

Larsia Fittkau

Worldwide genus, with 21 species and many undescribed forms; 7 named species in Nearctic with several undescribed species known. Hudson (1971) reported *L. decolorata* (Malloch) and an undescribed species from eastern South Dakota.

Macropelopia Thienemann

Holarctic and possibly Neotropical, with 3 species in Nearctic.

Natarsia Fittkau

Holarctic, with 2-3 species from the Nearctic. Hudson (1971) reported *N. baltimoreus* (Macquart) from eastern South Dakota.

Nilotanypus Kieffer

Holarctic, Oriental-Australasian, and possibly Neotropical, with four species in the Nearctic: *N. americanus* Beck & Beck, *N. fimbriatus* (Walker), *N. kansensis* Roback, and an unnamed species from Texas (although ITIS recognizes only *N. fimbriatus* and *N. kansensis*).

178 AQUATIC INSECTS OF THE BLACK HILLS

Paramerina Fittkau

Holarctic, Neotropical, and Oriental-Australasian; 5 species in the Nearctic. Hudson (1971) reported *P. smithae* (Sublette) and an undescribed species from eastern South Dakota.

Pentaneura Philippi

Western Hemisphere in distribution, with 2 named species in North America (*P. inconspicua* (Malloch) and *P. inyoensis* (Sublette), both common throughout the United States); however, Len Ferrington (pers. comm.) frequently finds unnamed species in biomonitoring samples, not just in the Black Hills. Hudson (1971) reported *P. inconspicua* from eastern South Dakota.

Procladius Skuse

Mostly Holarctic with some speices in Oriental, Neotropical, and Afrotropical Regions; 25 species in North America. Hudson (1971) reported seven species from this genus in eastern South Dakota.

Psectrotanypus Kieffer

Holarctic genus, with 2 Palaearctic species and 3 Nearctic species extending into the Neotropics. Hudson (1971) reported *P. aclines* (Sublette) (as *Derotanypus aclines*), *P. dyari* (Coquillette) and *P. florens* (Johanssen) from eastern South Dakota.

Radotanypus Fittkau & Murray

Nearctic (California, Colorado, Ohio, Oregon, and Wyoming). *Radotanypus* is not recognized in ITIS, and there is no TSN assigned for this genus.

Tanypus Meigen

Worldwide in distribution, with 11 species in North America. Hudson (1971) reported six species of this genus from eastern South Dakota.

Telopelopia Roback

The genus is entirely Holarctic in distribution, and monospecific in North America: T. Okoboji (Walley) is found from Manitoba and Minnesota south to New Mexico. Hudson (1971) reported *T. okoboji* from eastern South Dakota.

Zavrelimyia Fittkau

Holarctic, Oriental, Neotropical; with three named species and possibly several unnamed species in Nearctic. Hudson (1971) reported two of the three, *Z. sinuosa* (Coquillette) and *Z. thryptica* (Sublette), from eastern South Dakota.

Culicidae

The most familiar fact that everyone seems to know about mosquitoes is that they bite and spread disease (Berenbaum 2016). The most prominent recent outbreak of mosquito-borne illness in the United States has been West Nile Virus, first found (in this outbreak) in New York City in 1999 and in all contiguous states by 2006. West Nile Virus was identified in South Dakota in 2002, and 1,759 human cases of West Nile Virus were reported in South Dakota from 2002 to 2011, including 26 deaths

(Kightlinger 2012). The Black Hills counties generally had low incidence of human West Nile virus cases: from 2002 to 2011, Custer County reported 6 cases (average annual disease rate of 7 cases/100,000 population), Lawrence County reported 34 cases (14 cases/100,000 population), Pennington County had 177 cases (18 cases/100,000 population); however, Fall River County had a relatively high incidence rate with 31 cases (44 cases/100,000 population). Kightlinger (2012) considers it prudent to anticipate persistence of West Nile virus as a public health threat in the foreseeable future.

Nine of the 43 species of mosquitoes reported from South Dakota are known to be able to transmit West Nile Virus (http://doh.sd.gov/DiseaseFacts/WNV.aspx), but the most common vector in South Dakota is *Culex tarsalis*. Other diseases potentially carried by mosquitoes in the Black Hills include malaria, dog heartworm, Trivittatus virus, eastern equine encephalitis, Venezuelan encephalitis, western equine encephalitis, California encephalitis virus, St. Louis encephalitis, other encephalatides, and yellow fever (U.S. Army Walter Reed Biosystematics Unit [WRBU], http://wrbu.si.edu and http://www.mosquitocatalog.org/); however, some of these diseases (e.g., malaria, California encephalitis virus, yellow fever) have never been reported in the Black Hills.

The new emerging disease, Zika virus, is so far understudied. Mosquitoes that can vector this virus include *Aedes aegypti* and *Ae. albopictus*, the same species that can spread dengue and chikungunya viruses; the good news is that neither of these species occurs or is anticipated to become endemic in the Black Hills. However, people travelling to and from places where Zika virus is actively transmitted are at risk of acquiring the disease from mosquito bites and potentially spreading it (the same can be said of any exotic diseases!) The special things about Zika virus is that 1) it is not only mosquito-borne, but it is also sexually transmitted, so you can acquire the virus from a mosquito bite and transmit it through sexual activity, and 2) it can be transmitted transplacentally from mother to fetus, whereupon babies are often born with microcephaly – abnormally small heads – and other birth defects.

According to the U.S. Centers for Disease Control, prior to 2015, Zika virus outbreaks occurred in tropical Africa, Southeast Asia, and a few Pacific Islands, where it is endemic. In May 2015, it was found in Brazil, and it is currently (2018) epidemic throughout Central and South America, Mexico, and several Pacific island nations. In the continental United States, as of June 29, 2016, there were 934 travel-associated cases of Zika virus reported, none of which were locally acquired mosquito-borne cases. It has since been reported as endemic in far southern counties of Texas and Florida. Idaho, Wyoming, North Dakota, South Dakota, and Alaska were the only states that had not reported Zika virus. For what it's worth, the risk of transmission by mosquito bite to travelers in endemic countries is likely much lower than it is in countries with Zika virus epidemics (and even lower if the mosquito vector doesn't occur, such as in the Black Hills.)

Owen and Gerhardt (1957) provided records of a few species from the Black Hills counties in Wyoming. Gerhardt (1965, 1966) reported 40 species of mosquitoes from South Dakota, and the number hasn't changed much since then. In 2012, the South Dakota Department of Health stated that 43 species of mosquitoes are reported from the state. Of these, 27 species were reported from counties in or near the Black Hills or were considered to widespread throughout South Dakota (and thus likely to occur in the Black Hills). Many species have only rarely been collected and are considered to be of not much importance.

In addition to the requisite, "must-have" volume, *Identification and Geographical Distribution of the Mosquitoes of North America, North of Mexico* (Darsie and Ward 2005), I also extensively cite Gerhardt's (1965) publication, "South Dakota Mosquitoes and Their Control," for the descriptions/diagnoses for all species listed here, since that paper is specifically limited to those species occurring in South Dakota. Gerhardt (1965) is available for free on the internet from South Dakota State University's website (http://pubstorage.sdstate.edu/AgBio_Publications/articles/B531.pdf). I also cite Wood et al. (1979) rather frequently, as well, since it was made available rather cheaply a few years ago by Agriculture Canada. Because mosquitoes are of such great human and veterinary health importance, there are many good studies on their taxonomy and distributions, such that identification keys, descriptions, and diagnoses are not difficult to find. Also, species traits (for the traits that most biomonitoring programs are interested in) are not differentiated below the genus level, so a good STE for aquatic biomonitoring purposes could be the genus; for epidemiological monitoring, species identification is mandatory.

Collection records during biomonitoring activities are sparse, and I have seen no mosquitoes in biomonitoring samples from the Black Hills. Most mosquito larvae prefer lentic habitats and are not frequently encountered during stream biomonitoring. And adults – well, they are sometimes encountered as they bite, but then they are usually slapped, squashed, ground into the dirt, and otherwise made unavailable for collection. Besides mosquito specialists, only the most intrepid entomologists seem to manage to put an adult mosquito in a vial when they are being attacked – and then it is likely out of revenge!

Aedes **Meigen**

Aedes has a worldwide distribution, with more than 900 species. Many species are known vectors of diseases, the best known of which are yellow fever and dengue fever. The subgenus *Ochlerotatus* was recently elevated to generic level (Reinert et al. 2004), taking many of the previous North American *Aedes* species with it. Therefore, while Owen and Gerhardt (1957) and Gerhardt (1965) may indicate that, between them, eleven species of *Aedes* occur in the Black Hills, this reclassification has resulted in four species of *Aedes* in the Black Hills.

Aedes cinereus **(Meigen)**

Descriptions/diagnoses: Gerhardt 1965, Wood et al. 1979, Darsie & Wood 2004 (larva, adult). A Holarctic species distributed from the Yukon to the Canadian Maritimes, south throughout the United States except the arid Southwest (Arizona and New Mexico) in the Nearctic; distributed from Europe across northern Asia to Korea and Japan in the Palaearctic.

Gerhard (1965) reported this species from Butte County, just north of the Black Hills, but noted that it occurs in small numbers, is not particularly aggressive, and is likely of little importance. Vector status of this species is unknown.

Aedes intrudens **(Dyar)**

Descriptions/diagnoses: Gerhardt 1965, Wood et al. 1979, Darsie & Wood 2004 (larva, adult). Holarctic distribution includes much of northern and eastern Europe, Asia, and Japan, and a transcontinental band from Alaska and to Maritime Canada south to Oregon, Utah, Colorado, South Dakota, Wisconsin, and Georgia.

Gerhardt (1965) reported that the only record of this species in South Dakota was from Butte County, just north of the Black Hills. Females can be fairly annoying, particularly because they will attack both outdoors and indoors. Vector status of this species is unknown.

Aedes triseriatus **(Say)**

Descriptions/diagnoses: Gerhardt 1965, Darsie & Wood 2004 (larva, adult). British Columbia to Quebec, south to Utah, Texas, and the Gulf Coast, through Florida to the keys; also occurs south into Mexico and north into Greenland.

Gerhardt (1965) reported this aggressively biting species from Butte, Fall River, and Pennington counties in the Black Hills.

Larvae are treehole specialists, but can be found in artificial containers, as well. According to the WRBU, this species is a vector of California (LaCrosse) encephalitis, yellow fever, eastern equine encephalitis, Venezuelan encephalitis, and Western equine encephalitis, and a potential vector of the filarioid dog heartworm, *Dirofilaria immitis* (Leidy).

Aedes vexans (Meigen)

Descriptions/diagnoses: Gerhardt 1965, Darsie & Wood 2004 (larva, adult). A very widespread species, reported from the entire Palaearctic and Nearctic Realms (Alaska and Canada, through the entire United States and Mexico), and a large portion of the Oriental Realm.

Gerhardt (1965) reported that this species is unquestionably the most important pest mosquito (particularly vectoring the encephalitus viruses) occurring in South Dakota, reporting that it occurs throughout the state. A summertime species, *Ae. vexans* is known to be able to complete a life cycle in less than a week under optimal conditions. This species may be of moderate importance as a bridge vector from birds to humans and horses for West Nile virus (Turell et al. 2005).

Anopheles Meigen

Worldwide, this genus has 459 recognized species and more than 50 unnamed members of species complexes that are probably distinct species. Four species are known to occur in the Black Hills. *Anopheles* is the only genus in which species can vector malaria (*Plasmodium* spp.), which is not known to occur in the Black Hills.

Anopheles earlei Vargas

Descriptions/diagnoses: Gerhardt 1965, Wood et al. 1979, Darsie & Wood 2004 (larva, adult). Distributed transcontinentally from Alaska to Labrador, south to Washington, Idaho, Utah, Colorado, South Dakota, Wisconsin, Pennsylvania, and New Jersey.

Gerhardt (1965) reported this species from Custer and Fall River counties in South Dakota. Owen and Gerhardt (1957) reported this species from Alva, Crook County, Wyoming. This species is an aggressive biter, but the vector status of this species is unknown.

Anopheles punctipennis Say

Descriptions/diagnoses: Gerhardt 1965, Wood et al. 1979, Darsie & Wood 2004 (larva, adult). Distributed from British Columbia to the Maritimes, south through the contiguous 48 States into central Mexico. Although it occurs throughout the country, it is rare in much of the Rocky Mountain Region, including the Black Hills.

Gerhardt (1965) reported this species from Fall River and Meade counties in South Dakota, but considered it to be of little importance.

Larvae occur in permanent and temporary ponds and pools, seemingly preferring cool, clear water in hill streams. Females feed mostly after dusk but also during daytime in dense woodlands. According to the WRBU, this species can vector *Plasmodium* spp. (malaria) in parts of the world where that disease occurs (not the Black Hills).

Anopheles quadrimaculatus Say

Descriptions/diagnoses: Gerhardt 1965, Wood et al. 1979, Darsie & Wood 2004 (larva, adult). Ontario and Quebec south through the entire eastern United States (as far west as the Dakotas, Nebraska, Kansas, Oklahoma, and Texas) into Mexico.

Gerhardt (1965) reported this species from Meade County, stating that it is probably distributed statewide but uncommon enough to be of little importance.

Larvae occur in permanent lentic habitats, especially where there is emergent vegetation. Adults actively feed on man and wild and domestic animals, principally at night. According to the WRBU, this species is an important vector of *Plasmodium* spp. (malaria) in parts of the world where that disease occurs; more pertinent in the Black Hills, it is considered to be a possible vector of dog heartworm, *Dirofilaria immitis* (Leidy).

Anopheles walkeri **Theobald**

Descriptions/diagnoses: Gerhardt 1965, Wood et al. 1979, Darsie & Wood 2004 (larva, adult). With a distribution very similar to An. quadrimaculatus – Manitoba east to the Maritimes, south through the eastern United States (west to the Dakotas, Nebraska, Kansas, Oklahoma, and Texas) into Mexico. It has also been found in British Columbia, although Wood et al. (1979) questions the record.

Gerhardt (1965) reported this "apparently rare" species from Custer and Fall River counties.

Larvae occur in freshwater marshes containing emergent/floating vegetation and debris. Adults are known to enter dwellings at night to feed on humans. According to the WRBU, this species could potentially vector *Plasmodium* sp. (malaria) where that disease occurs.

Culex **Linnaeus**

Culex includes over 750 species and is distributed worldwide. Several species are medically important, particularly in the tropical and subtropical regions of the world, as vectors of encephalatides, West Nile Virus, Rift Valley Fever, and filarial worms. Five species of *Culex* occur in the Black Hills.

Culex pipiens **(Linnaeus)**

Descriptions/diagnoses: Gerhardt 1965, Wood et al. 1979, Darsie & Wood 2004 (larva, adult). A nearly cosmopolitan species, it might be easier to list where it isn't found than where it is! It is found across North America; in South America only in Argentina; all of Europe; Asia north of China and Mongolia, scattered countries throughout Africa; and Australia.

Gerhardt (1965) reported this species from across the state, including Butte County, just north of the Black Hills.

The "northern house mosquito" is a common container breeder and can become a pest if such breeding places are available; however, it is also easier to control this mosquito if the breeding containers are removed. Water can be extremely stagnant and the larvae can still exist there. It prefers to feed on birds but will not hesitate to feed on humans. The diseases this species is known to vector are generally not found in the Black Hills; for example, the WRBU lists exotic Sindbis virus, Rift Valley fever, and the elephantiasis-producing filarioid roundworm *Wucheraria bancrofti* (Cobbold) as potentially vectored by *Cx. pipiens*. However, of diseases that do occur in the Black Hills, *Cx. pipiens* can vector St. Louis encephalitis, West Nile virus (Terrell et al. 2005), bird malaria (*Hemoproteus* spp.), fowl pox, and dog heartworm (*Dirofilaria immitis* (Leidy)).

Culex restuans **(Theobald)**

Descriptions/diagnoses: Gerhardt 1965, Wood et al. 1979, Darsie & Wood 2004 (larva, adult). Distributed from British Columbia east to New Brunswick, south through the United States and Mexico to Guatemala and Honduras.

Gerhardt (1965) reported this species across the state, including Butte and Fall River counties in and near the Black Hills.

Larvae occur in a wide variety of permanent and temporary, lentic and lotic habitats, but particularly artificial and small, natural containers. This species prefers to feed on birds but will attack reptiles, mammals, and humans and can become annoying when abundant. According to the WRBU and Terrell et al. (2005), it is considered to be a vector of St. Louis encephalitis and West Nile virus.

Culex salinarius **(Coquillett)**

Descriptions/diagnoses: Gerhardt 1965, Darsie & Wood 2004 (larva, adult). Distributed from the Canadian Maritimes south to Florida, and west to Idaho, Utah, and New Mexico.

Gerhardt (1965) reported this species from Butte and Fall River counties in and near the Black Hills, as well as across the state, but noted that it seldom occurs (in South Dakota, at least) in sufficient abundance to be a serious pest. This species is likely to vector eastern equine encephalitis, St. Louis encephalitis, and West Nile virus (Terrell et al. 2005).

Culex tarsalis (Coquillett)

Descriptions/diagnoses: Gerhardt 1965, Wood et al. 1979, Darsie & Wood 2004 (larva, adult). Distributed from British Columbia south to California, east to the Northwest Territories, Manitoba, Michigan, Indiana, Virginia, the southern Atlantic Coast, and the Gulf Coast. The distribution extends south into Mexico.

Gerhardt (1965) reported this species as occurring throughout the State of South Dakota.

Larvae occur in a wide variety of temporary lentic habitats, particularly ones in which the water is highly enriched organically. *Culex tarsalis* is a well-studied mosquito, since, according to the WRBU, it is the most important vector of Western equine encephalitis in the United States. Contrary to most *Culex* species, which prefer to feed on birds but will feed on mammals opportunistically, this species appears to prefer mammals and feeds on birds more opportunistically. It can also vector St. Louis encephalitis, California encephalitis, and West Nile virus (Terrell et al. 2005).

Culex territans (Walker)

Descriptions/diagnoses: Gerhardt 1965, Darsie & Wood 2004 (larva, adult).

Gerhardt (1965) reported that this species has been collected in Butte and Custer counties, as well as other counties across the state.

Adults are not known to feed on warm-blooded organisms, so humans appear to be safe from this species.

Culiseta Felt

Culiseta contains 37 species, four of which occur in the Black Hills. The others are generally found worldwide, except in most of Latin America, southeast Asia (including the Phillipine and Indonesian archipelagoes), and scattered countries in Africa. Most species feed on birds and mammals, and three North American species are vectors of encephalitis viruses.

Culiseta impatiens (Walker)

Descriptions/diagnoses: Gerhardt 1965, Darsie & Wood 2004 (larva, adult). Distributed from Alaska, across Canada to the Maritimes, and south to California, Utah, Colorado, Nebraska, Missouri, Wisconsin, New York, and New England.

Gerhardt (1965) reported that this species' distribution within South Dakota was, at that time, known only from Pennington County. This species can occasionally be a vicious biter, but vector status is unknown.

Culiseta incidens (Thomson)

Descriptions/diagnoses: Gerhardt 1965, Darsie & Wood 2004 (larva, adult). Distributed from Alaska south to California, east to the Northwest Territories, the Dakotas, Nebraska, Oklahoma, and Texas, with a report of this species in Nova Scotia.

Gerhardt (1965) reported that this species' distribution within South Dakota appeared to be restricted to the Black Hills, being reported from Butte, Custer, and Fall River counties. In the laboratory, this species has successfully vectored St. Louis encephalitis, Western equine encephalitis, and Japanese B encephalitis, but vector status of wild populations of this species is unknown.

Culiseta inornata (Williston)

Descriptions/diagnoses: Gerhardt 1965, Darsie & Wood 2004 (larva, adult). Distributed throughout North America from the Yukon and Northwest Territories south through the entire United States into Mexico.

Gerhardt (1965) reported that this species is "found throughout South Dakota", noting that it may be of some nuisance importance; however, it is usually reluctant to bite humans (often preferring livestock and larger mammals), so it is not as annoying as more aggressive species.

This species prefers temporary lentic habitats, including artificial water containers. Water can be grossly polluted. According to the WRBU and Terrell et al. (2005), this species is considered to be a potential vector of Western equine encephalitis and West Nile Virus.

Culiseta morsitans (Theobald)

Descriptions/diagnoses: Gerhardt 1965, Darsie & Wood 2004 (larva, adult). A Holarctic species also known from Europe, in North America this species distributed from Alaska, throughout Canada, and south to Oregon, Idaho, Colorado, South Dakota, Kentucky, Pennsylvania, and New England.

Gerhardt (1965) reported this species from Lawrence and Fall River counties and stated that it is apparently of little importance in the state. Vector status of this species is unknown, but females rarely, if ever, bite humans.

Mansonia Blanchard

Mansonia contains 25 species, distributed primarily in the tropics and subtropics, but extending north into the United States and China, and south to Argentina, South Africa, and Australia. One species, *M. perturbans*, occurs in the Black Hills. A few species (not *M. perturbans*, as far as is known) are vectors of various arboviruses and filarial worms.

Mansonia perturbans (Walker)

Descriptions/diagnoses: Gerhardt 1965, Darsie & Wood 2004 (larva, adult). Distributed from southern Canada (British Columbia to the Maritimes) south through the entire United States into Mexico.

Gerhardt (1965) reported that this aggressive biter has been collected from Butte County just north of the Black Hills, and otherwise across the state. Vector status of this species is unknown, but it could potentially vector eastern equine encephalitis and West Nile virus (Terrell et al. 2005).

Ochlerotatus Lynch Arribálzaga

Based on both morphological and molecular data supporting its monophyly and status as a distinct clade, the *Aedes* subgenus *Ochlerotatus* was recently elevated to generic level (Reinert et al. 2004, Shepard et al. 2006), taking many of the previous North American *Aedes* species with it. Although this reclassification is widely used, there remains sufficient opposition to it such that ITIS has not yet (2018) been updated to reflect all of these changes. A TSN has been generated for *Ochlerotatus* at the genus level, but there are no species assigned to the genus.

Seven of the eleven *Aedes* species listed by Owen and Gerhardt (1957) and Gerhardt (1965) for the Black Hills are now in *Ochlerotatus*.

Ochlerotatus dorsalis (Meigen)

Descriptions/diagnoses: Gerhardt 1965 (as *Aedes dorsalis*), Wood et al. 1979 (as *Aedes dorsalis*), Darsie & Wood 2004 (larva, adult). A Holarctic species, *Ae. dorsalis* is found throughout North America south into Mexico and in Europe and Asia from Germany south to Turkey, east through Russia to Tajikstan, Korea, and Japan. The TSN for this species as *Aedes dorsalis* is 126278.

Gerhardt (1965) stated that this species is reported from across the state of South Dakota, being the most abundant mosquito species in many localities in the western portions of the state. Larvae are found in all varieties of lentic habitats, from freshwater to brackish. According to the WRBU, this species is a major pest to both man and animals and is capable of transmitting several encephalitides, including western equine encephalitis, California encephalitis virus, and West Nile Virus.

Ochlerotatus excrucians (Walker)

Descriptions/diagnoses: Gerhardt 1965 (as *Aedes excrucians*), Wood et al. 1979 (as *Aedes excrucians*), Darsie & Wood 2004 (larva, adult). Distributed from Alaska through most of mainland Canada to the Maritimes, south to Oregon, Utah, Colorado, Illinois, Ohio, and New Jersey. Also from Europe across northern Asia to Mongolia and Japan. The TSN for this species as *Aedes excrucians* is 126288.

At the time of writing of Gerhardt (1965), this species had not yet been reported from South Dakota but "undoubtedly occurs here." If it is to be found in South Dakota, it is likely to occur in the Black Hills, since in the United States it prefers shaded pools in forested areas; although apparently in Canada, the presence or absence of shading had little influence on the presence of this very common species (Wood et al. 1979). Vector status of this species is unknown.

Ochlerotatus increpitus (Dyar)

Descriptions/diagnoses: Gerhardt 1965 (as Aedes increpitus), Wood et al. 1979 (as Aedes increpitus), Darsie & Wood 2004 (larva, adult), Lanzaro & Eldridge 1992 (larva, pupa, adult). Distributed from British Columbia south to California and east through the Rocky Mountains to Montana, South Dakota, Colorado, and New Mexico. The TSN for this species as *Aedes increpitus* is 126314.

Gerhardt (1965) reported that this common species appears to be restricted in South Dakota to the Black Hills, being reported sporadically from Custer, Fall River, and Pennington counties. It has been collected from a variety of aquatic habitats, including roadside ditches, snowmelt pools, and residual floodwater pools. Vector status of this species is unknown.

Ochlerotatus nigromaculis (Ludlow)

Descriptions/diagnoses: Gerhardt 1965 (as *Aedes nigromaculis*), Wood et al. 1979 (as *Aedes nigromaculis*), Darsie & Wood 2004 (larva, adult). Distributed in western North America from Alberta east to Saskatchewan, south through the western half of the United States (as far east as Iowa, Kentucky, and Texas) into Mexico. The TSN for this species as *Aedes nigromaculis* is 126334.

Gerhardt (1965) reported that this species is distributed throughout South Dakota, with maximum abundance in the semi-arid portion of the state west of the Missouri River. This species is known as the "pasture mosquito" because it is particularly adapted to habitats associated with agricultural irrigation (Gerhardt 1965), so it may be of importance especially in the foothills of the Black Hills but not at the higher elevations. Vector status of this species is unknown, although the females can be vicious biters.

Ochlerotatus spencerii (Theobald)

Descriptions/diagnoses: Gerhardt 1965 (as *Aedes spencerii*), Wood et al. 1979 (as *Aedes spencerii*), Darsie & Wood 2004 (larva, adult). The nominate subspecies occurs from British Columbia east to Manitoba, south to Utah, Wyoming, Kansas, Iowa, and Michigan. The TSN for this species as *Aedes spencerii* is 126423.

Gerhardt (1965) reported this species from Butte and Fall River counties, which places it in and near the Black Hills, as well as scattered localities across the state. A common Rocky Mountain subspecies, *Ae. spencerii idahoensis*, was listed as a separate species in Gerhardt (1965) and treated as a synonym in Wood et al. (1979) and most subsequent papers, with records from Charles Mix County, just southeast of the Black Hills. Wood et al. (1979) characterized *Ae. spencerii* as one of the worst spring-time pests in the Prairie provinces of Canada because it is among the earliest emerging species and is multivoltine, making it present throughout the spring. Vector status of this species is unknown.

Ochlerotatus sticticus (Meigen)

Descriptions/diagnoses: Gerhardt 1965 (as *Aedes sticticus*), Wood et al. 1979 (as *Aedes sticticus*), Darsie & Wood 2004 (larva, adult). Holarctic species across northern Europe and Siberia in the Palaearctic, Alaska and British Columbia east to the Maritimes, south to California, Utah, Colorado, Texas, and the Gulf Coast to Florida. The TSN for this species as *Aedes sticticus* is 126368.

Gerhardt (1965) reported this species from across the State of South Dakota, including Butte County, just north of the Black Hills. This species is commonly known as the "floodwater mosquito" because of its tendency to inhabit off-channel impoundments after flooding, so it may be less common at higher elevations in the Black Hills due to the steeper terrain discouraging that habitat. Vector status of this species is unknown.

Ochlerotatus trivittatus (Coquillett)

Descriptions/diagnoses: Gerhardt 1965 (as *Aedes trivittatus*), Wood et al. 1979 (as *Aedes trivittatus*), Darsie & Wood 2004 (larva, adult). Ontario and Nova Scotia west to Manitoba, Idaho, Colorado, and New Mexico, south to Georgia, Tennessee, Louisiana, and Texas into Mexico. The TSN for this speices as *Aedes trivittatus* is 126395.

Gerhardt (1965) reported this species from across the state, including Butte and Fall River counties in and near the Black Hills.

Larvae occur in temporary lentic habitats, particularly after summer rains. Females are persistent biters any time of day. According to the WRBU, *Ae. trivittatus* is a vector of Trivittatus virus, which can cause febrile illness in humans, and it is a possible vector of filarioid dog heartworm, *Dirofilaria immitis* (Leidy).

Psorophora Robineau-Desvoidy

Psorophora is restricted to the Western Hemisphere and contains 48 species. Several species are vectors of arboviruses. Two species of *Psorophora* occur in the Black Hills.

Psorophora ferox (Humboldt)

Descriptions/diagnoses: Gerhardt 1965, Darsie & Wood 2004 (larva, adult). Distributed from Ontario south through the eastern United States (west to the Dakotas, Nebraska, Kansas, Oklahoma, and Texas), Mexico, and Central America into South America and the West Indies.

Gerhardt (1965) reported this species from Butte County, just north of the Black Hills.

Larvae specialize in temporary lentic habitats. Gerhardt (1965) stated that the females are persistent, aggressive biters but are probably not sufficiently abundant in South Dakota to constitute a problem. According to the WRBU, this species is a vector of Ilheus virus and Venezuelan equine encephalitis and can carry eggs of human bot flies (*Dermatobia hominis* (Linnaeus, Jr.)) in Central and South America – none of these are a concern in the Black Hills. Terrell et al. (2005) rate this species as little to no risk for transmission of West Nile virus.

Psorophora signipennis (Coquillett)

Descriptions/diagnoses: Gerhardt 1965, Darsie & Wood 2004 (larva, adult). With what is probably the most restricted distribution of the mosquitoes known from the Black Hills, this species occurs from Montana and North Dakota south to Arizona, New Mexico, Texas, and Missouri into Mexico.

Gerhardt (1965) reported this species from Fall River County in the Black Hills, noting that its sparse distribution across South Dakota renders it of little importance. Vector status of this species is unknown.

Dixidae

Literature records apparently do not exist placing this family in the Black Hills. Biomonitoring efforts have revealed the presence of *Meringodixa chalonensis* and an unidentified species of *Dixa* in the Black Hills. Shearer (2006) also reported both *Dixa* and *Meringodixa* from the Black Hills. The genus *Dixella* probably also occurs in the Black Hills, although I have not seen it from there.

Dixa Meigen

As mentioned above, Shearer (2006) reported *Dixa* from the Black Hills.

Meringodixa Nowell

Meringodixa chalonensis Nowell

As mentioned above, Shearer (2006) reported *Meringodixa* from the Black Hills; being monospecific, it would have been *M. chalonensis*.

Dolichopodidae

In a catalog of the Dolichopodidae of North America, Pollet et al. (2004) reported that at least 58 species of long-legged flies occur in the State of South Dakota, and at least 85 species occur in Wyoming. The data were mostly at the state level, so the number of species that occur within the Black Hills remains unknown.

While most dolichopodids are at least associated with moist or damp habitats, some are found in freely flowing or standing water, while others require no aquatic habitat whatsoever. Robinson and Vockeroth (1981) provide keys for 51 genera of adults, 12 genera of larvae, and 16 genera of pupae. Since the identification literature for dolichopodids is so sparse, particularly for larvae and pupae, which are the most likely to be encountered in aquatic sampling efforts, it is probably best to stop identifications at the family level for Dolichopididae.

Three species, however, do have type localities in or near the Black Hills, so I do want to at least mention those records. *Dolichopus penicillatus* Van Duzee is the correct name for *Dolichopus ciliatus* (Aldrich, 1893 – preoccupied), which was described from specimens collected in Custer, South Dakota. *Liancalus hydrophilus* Aldrich was described from specimens collected in the Black Hills. Most, if not all, members of the genus *Liancalus* are hygrophilous and often found in association with vertical seepages (Corpus 1986; Pollet et al. 2004), which suggests a possible habitat in which to find specimens in the Black Hills. Finally, *Parasyntormon hinnulus* Wheeler, whose aquatic associations are unknown, was described from specimens collected in Lusk, Wyoming.

Empididae

There are very few records of empidid flies from the Black Hills, even at the family or genus level. However, as noted below, they can be quite abundant in Black Hills streams. Species level identifications have not been made for *Clinocera*, *Hemerodromia*, or *Oreogeton*, and it is likely that additional species of the other genera may occur in the Black Hills.

Chelifera Macquart

Shearer (2006) reported larvae of *Chelifera* from the Black Hills. MacDonald (1994) reported no species of *Chelifera* from the Black Hills (or South Dakota, for that matter). The nearest localities to the Black Hills for *Chelifera* species are *C. caliga* Lavallee in north-central Colorado, *C. mana* Lavallee in southcentral Wyoming, *C. cirrata* Melander in Yellowstone National Park, and *C. palloris* (Coquillett) in central Colorado.

In biomonitoring efforts that I have been involved with, there was no distinction between larvae of *Chelifera* and *Metachela*, because only recently (Brammer et al. 2009) had characters been identified to separate them. Even so, the characters are generally subjective and require side-by-side specimens for accurate separation. Consequently, the organisms were reported from most biomonitoring samples as a "slashed" taxon (i.e., *Chelifera/Metachela*).

Clinocera Meigen

This genus is not reported from the Black Hills in the literature, but I have seen it from biomonitoring samples.

Hemerodromia Meigen

Shearer (2006) reported larvae of *Hemerodromia* from the Black Hills.

Metachela Coquillett

As stated above, there was no effort at GEI to distinguish between *Chelifera* and *Metachela* in production taxonomy analyses of biomonitoring samples.

Metachela collusor (Melander)

MacDonald (1989) reported this species from Spearfish Creek. Based on the distribution map in MacDonald (1989), *M. collusor* is the only species of *Metachela* in the Black Hills. The distribution of *M. albipes* is widespread (central British Columbia and Alberta south to southern California, with scattered localities in Manitoba, Ontario, Quebec, and New Hampshire), so it is possible that *M. albipes* could also eventually be found in the Black Hills.

Neoplasta Coquillett

Neoplasta paramegorchis MacDonald & Turner

This is the only *Neoplasta* species with a distribution record from the Black Hills, South Dakota, or Wyoming in the published literature. MacDonald and Turner (1993) included three male specimens from Spearfish Creek and one female specimen from Beaver Creek in the paratype series in their description of this species. Other species with distributions near the Black Hills include *N.*

scapularis (Loew), which is widespread across the United States and southern Canada (except the Great Plains), and *N. hansoni* MacDonald & Turner, which is distributed from Washington and California eastward to the Nebraska panhandle. As of 2018 ITIS had not yet issued a TSN for *N. paramegorchis*.

Oreogeton Schiner

Shearer (2006) reported larvae of *Oreogeton* from the Black Hills.

Trichoclinocera Collin

Shearer (2006) reported larvae of *Trichoclinocera* from the Black Hills. I have not seen *Trichoclinocera* in biomonitoring samples from the Black Hills, and Sinclair (1994) did not report any species from there. Furthermore, Brad Sinclair (pers. comm.) told me that he had not seen records of *Trichoclinocera* from the Black Hills.

Wiedemannia Zetterstedt

Wiedemannia may occur in the Black Hills, but extensive records have not yet been published for that genus; the closest record for the genus is *W. undulata* Sinclair in the Big Horn Mountains of north central Wyoming (Sinclair 1997). I have not seen *Wiedemannia* in biomonitoring samples from the Black Hills.

Ephydridae

Very few published records exist for Ephydridae in the Black Hills. Most reports exist only at the family level. Ephydrid larvae generally utilize habitats that are not frequently sampled in most stream biomonitoring protocols, including sometimes very saline lentic habitats or aquatic plants, so they are also rarely reported in biomonitoring studies. A literature review of insect herbivores of aquatic and wetland plants in the United States was prepared by Harms and Grodowitz (2009) and could be useful for future researchers looking for ephydrids on aquatic plants in the Black Hills.

Allotrichoma Becker

Allotrichoma occidentale Mathis & Zatwarnicki

Description/Diagnosis: Mathis and Zatwarnicki 2012 (adult); larva unknown. Distributed from Washington south to Baja California and southern Arizona, east to the Dakotas, Nebraska, and Utah, with isolated populations in Alaska, Saskatchewan, Manitoba, and Michigan. As of 2018, this species had not been entered into ITIS, so no TSN exists for it.

Mathis and Zatwarnicki (2012) described this species from Oregon but reported two male and three female specimens collected in 1969 from Beaver Creek, Lawrence County, South Dakota, among the other material examined.

Ephydra Fallén

Ephydra hians Say

Description/Diagnosis: Wirth 1971 (adult). Distributed from British Columbia south to California, east to Minnesota, Nebraska, Colorado, and New Mexico. Also south into central Mexico.

Wirth (1971) reported this species from a locality 40 miles north of Lusk, Wyoming, just southwest of the Black Hills, and I have seen it in the Nebraska panhandle. This species is one of the famous brine flies of the Great Basin and Great Salt Lake. In his biographical travelogue *Roughing It*, Mark Twain (1872) made the observation about this species at Mono Lake, California: "You can hold them under water as long as you please – they do not mind it – they are only proud of it. When you let them go, they pop up to the surface as dry as a patent office report, and walk off as unconcernedly as if they had been educated especially with a view to affording instructive entertainment to man in that particular way."

Ephydra pectinulata Cresson

Description/Diagnosis: Wirth 1971 (adult). Distributed from the Northwest Territories south to Washington, Utah, Colorado, Nebraska, and Minnesota.
Just outside of the Black Hills proper, Wirth (1971) reported this species from a locality 40 miles north of Lusk, Wyoming.

Ephydra riparia Fallén

Description/Diagnosis: Wirth 1971 (adult). Distributed "Northern Europe; northern North America from Alaska and British Columbia east and south to Nova Scotia, Ohio, and Oklahoma (Wirth 1971).
Wirth (1971) reported this species from Beaver Creek, Lawrence County, South Dakota.

Parydra Stenhammar

The biology of *Parydra* is poorly known (Clausen and Cook 1971). The adults frequent muddy shorelines of streams and lakes, and the larvae are presumably semiaquatic. Four species are known from distribution maps in Clausen and Cook (1971) to occur in or around the Black Hills.

Parydra appendiculata Loew

Description/Diagnosis: Clausen and Cook 1971 (adult). Distributed from the Northwest Territories and Labrador south to Baja California, Arizona, Colorado, Texas, Illinois, and Pennsylvania.
Distribution maps in Clausen and Cook (1971) show occurrence of this species in and near the Black Hills.

Parydra aquila (Fallén)

Description/Diagnosis: Clausen and Cook 1971 (adult). The most common species of the genus, P. aquila is distributed from Alaska, south to California and Arizona, east to the Atlantic coast from Labrador and Quebec south to North Carolina. It is also known from Europe. There are numerous subspecies that each have rather broad, overlapping distributions.
Distribution maps in Clausen and Cook (1971) show occurrence of this species in and near the Black Hills.

Parydra parasocia Clausen and Cook

Description/Diagnosis: Clausen and Cook 1971 (adult). Clausen and Cook (1971) described this species from Sidney, Iowa, and reported the distribution of 410 specimens from Alaska east to Quebec, and south to California, Utah, Colorado, Kansas, and Indiana.
Distribution maps in Clausen and Cook (1971) show occurrence of this species in and near the Black Hills.

Parydra quadrituberculata Loew

Description/Diagnosis: Clausen and Cook 1971 (adult). This is an eastern species, distributed from Alberta, the Dakotas, Nebraska, Kansas, and Texas, east to the Atlantic coast from Quebec to North Carolina.

Distribution maps in Clausen and Cook (1971) show occurrence of this species in and near the Black Hills.

Scatella Robineau-Desvoidy

I have seen *Scatella* in biomonitoring samples collected in the Black Hills, even though I do not have literature-based records of its presence there.

Muscidae

Nearly all species of Muscidae are terrestrial; however, a few genera in the Limnophorini have aquatic or semiaquatic representatives, including *Lispe*, *Limnophora*, *Lispoides*, and *Spilogona*. Historically, most aquatic biomonitoring efforts have left the Muscidae at the family level, where Merritt et al. (2008) also leaves off with identifications. Keys to the larvae are available in Johannsen (1935). Revisions of these genera have provided very few records of these flies from the Black Hills, but at least some species from these four genera are likely to be present in the Black Hills, based on the widespread geographic presence of the genera.

Limnophora Robineau-Desvoidy

Description/diagnosis: Huckett 1965 (adults), Johannsen 1935 (larvae). The genus is Holarctic, with numerous species restricted to the Nearctic

This genus has not yet been reported from the Black Hills, but, given the widespread distributions of a couple Nearctic species – particularly *L. discreta* Stein and *L. uniseta* Stein (Huckett 1965), I believe it is likely to occur there.

I have larvae of *Limnophora* (identified according to the key in Johannsen 1935) in my personal collection from Whitewood Creek a few miles north of the Black Hills.

Lispe Latrielle

Description/diagnosis: Huckett 1965 (adults), Johannsen 1935 (larvae). This genus occurs in the Holarctic and the Neotropics, with some genera specific in the Nearctic.

This genus has not yet been reported from the Black Hills, but, given the widespread distributions of several of the Nearctic species (Huckett 1965), I believe it is likely to occur there. *Lispe nasoni* Stein was described, in part, from specimens collected in South Dakota.

Lispoides Malloch

Lispoides aequifrons Stein

Description/diagnosis: Huckett 1932 (adults); Johannsen 1935 (larva); both as *Limnophora aequifrons*).

Stein (1898) had two type localities when he described this species, one of which was Custer, South Dakota; the other was Moscow, Idaho. With the aquatic larvae found in rocky streams, Huckett (1932) suggested that *L. aequifrons* is "undoubtedly present in nearly all parts of North America."

I have *Lispoides aequifrons* in my personal collection from Annie Creek, Deadwood Creek, and Rapid Creek.

Spilogona Schnabl

Description/diagnosis: Huckett 1965 (adults), Johannsen 1935 (larvae). This genus occurs in the Holarctic and the Neotropics, with some genera specific in the Nearctic. Based on the widespread distributions of several of the Nearctic species (Huckett 1965), its presence in the Black Hills should not be surprising.

I have seen one biomonitoring sample with larvae of *Spilogona*, as identified using the key in Johannsen (1935).

Psychodidae

The moth and sand flies are a group of handsome and distinctive insects. The subfamily Phlebotominae, well known for spreading several tropical diseases, does not occur in most of temperate North America, including the Black Hills; other groups do.

I have found records of three genera of psychodids that occur in aquatic habitats of the Black Hills. There are likely other terrestrial psychodid genera that also occur there (Quate 1955).

Characters have not yet been found to distinguish larvae at the species level for any of the genera of aquatic Psychodidae. *Pericoma* and *Telmatoscopus* are also currently indistinguishable morphologically, but some production taxonomists in biomonitoring programs have assumed that *Pericoma* is fully aquatic while *Telmatoscopus* is semiaquatic to terrestrial, so identifications have sometimes included only *Pericoma*. However, it is not uncommon to find them reported as a slashed taxon, *Pericoma/Telmatoscopus*, either.

Maruina Muller

Shearer (2006) reported *Maruina* from the Black Hills.

Pericoma Walker

Shearer (2006) reported *Pericoma* from the Black Hills. I have collected larvae that I called *Pericoma/Telmatoscopus* in Spearfish Creek.

Psychoda Latreille

Jergens (1968) reported *Psychoda* at the genus level from Rapid Creek and Spearfish Creek. Larvae of *Psychoda* are not frequently encountered in the Black Hills.

Ptychopteridae

Ptychoptera Meigen

Shearer (2006) reported *Ptychoptera* from the Black Hills, and this is the only record I have found for this family from the Black Hills.

Simuliidae

Nearby areas that have recently conducted faunal surveys of the black flies have included Alberta (Currie 1986, 51 species), Nebraska (Pruess and Peterson 1987, 13 species), Colorado (Peterson and Kondratieff 1995, 29 species), and Michigan (Bright 2006, 50 species + 16 additional species expected to occur but not yet found). Adler et al. (2004) included many of these records at the county level in their monograph on the Simuliidae of North America.

Even so, actual records of black flies from the Black Hills are scarce, so the inclusion of many of these taxa in this list is my expectation based on distributions. Several species that are widespread in the Rocky Mountain cordillera or across the western Great Plains, or occur broadly across the continent, are included or mentioned.

Because the list of species is so tentative, readers are directed to the excellent keys in Adler et al. (2004) or Merritt et al. (2008) for identification of black flies collected in the Black Hills. The keys to genera are generally workable using a regular dissecting microscope, except a few characters are included in Adler et al. (2004) which require chromosomal analysis. Several of the keys to species in Adler et al. (2004) require chromosomal analysis.

Prosimulium **Roubaud**

Pruess and Peterson (1987) discussed the report by Smith and Rapp (1987) of an unidentified species of *Prosimulium* from Dawes County, Nebraska. *Prosimulium* species that are widespread in the Rocky Mountain cordillera include *P. onychodactylum*, *P. exigens*, *P. frohnei*, and *P. fulvum* (Adler et al. 2004), and these may eventually be found in the Black Hills.

Prosimulium exigens **Dyar & Shannon**

Descriptions/diagnoses: Peterson and Kondratieff 1995, Adler et al. 2004. Distribution extends from British Columbia and Alberta south to central California, Arizona, and New Mexico. The maps in Adler et al. (2004) indicate a record from Lawrence County, South Dakota.

I have collected one adult of *Prosimulium exigens* from Spearfish Creek.

Simulium **Latreille**

Shearer (2006) reported larvae of *Simulium* from the Black Hills. I have also collected larvae of *Simulium* in the Fall River near Hot Springs and in Whitewood Creek, and I saw *Simulium* larvae in biomonitoring samples from nearly every stream sampled in the northern black Hills. Some *Simulium* species that are widespread in the Rocky Mountain cordillera or the western Great Plains include *S. arcticum*, *S. meridionale*, *S. pugetense*, *S. tuberosum* complex, *S. venustum*.

Simulium bivittatum **Malloch**

Descriptions/diagnoses: Peterson and Kondratieff 1995, Adler et al. 2004. Distributed from Alberta and Saskatechewan south to Oregon and Arizona, east to South Dakota, Nebraska, and Texas.

Stains and Knowlton (1943) reported this species from South Dakota. Moulton (1998) reported *Simulium bivittatum* from the Belle Fourche River at Devil's Tower [National Monument], and Pruess and Peterson (1987) reported adults of that species from a light trap in neighboring Sioux County, Nebraska. In addition to these records, Adler et al. (2004) also indicated its presence in Lawrence County, South Dakota.

Simulium canadense **Hearle**

Descriptions/diagnoses: Peterson and Kondratieff 1995, Adler et al. 2004. The Black Hills of South Dakota represent the northeastern extent of this species, being distributed from British Columbia and South Dakota south to California, Arizona, and New Mexico and into Mexico. Adler et al. (2004) also report one extralimital record from central Florida.

Adler et al. (2004) indicate this species' presence in Custer and Lawrence counties. Coscarón et al. (2004) reported specimens of this species from Spearfish, South Dakota.

Simulium decorum Walker

Descriptions/diagnoses: Peterson and Kondratieff 1995, Adler et al. 2004. Transcontinental in distribution from Alaska, the Yukon, and Northwest Territories south to Washington, Utah, Colorad, Kansas, Missouri, Mississippi, and Florida, east to the Canadian Maritimes and the Atlantic seaboard. Although Adler et al. (2004) present no localities in (or even near) the Black Hills, Pruess and Peterson (1987) posit that *S. decorum* may occur throughout the state of Nebraska, and it is reported from scattered localities up the Rocky Mountain cordillera.

Simulium griseum Coquillett

Descriptions/diagnoses: Peterson and Kondratieff 1995, Adler et al. 2004. Distributed primarily from Alberta and Saskatchewan south to Utah and New Mexico, Adler et al. (2004) also reports populations near San Francisco, Colorado, and the Colorado River Valley between Arizona and California.

Stains and Knowlton (1943) reported this species from Cascade Springs, South Dakota (about 6 miles south of Hot Springs).

Simulium hunteri Malloch

Descriptions/diagnoses: Peterson and Kondratieff 1995, Adler et al. 2004. Distributed in North America from Alaska and the Yukon south to California, Arizona, New Mexico, and the Big Bend Region of Texas, east to southwestern Saskatchewan, the Black Hills of South Dakota, and the Colorado Front Range.

Stains and Knowlton (1943) reported this species from Spearfish, South Dakota.

Simulium meridionale Riley

Descriptions/diagnoses: Peterson and Kondratieff 1995, Adler et al. 2004. Widely distributed from northern California and Nevada north to Alberta, Saskatchewan, and Manitoba, south to New Mexico and the Gulf Coast from Mexico to the Florida panhandle, and east to the Mississippi River. Common enough to have a common name, the "turkey gnat" is known for its painful bite.

Although no records are known from in or near the Black Hills, the distribution suggests that this species may certainly occur there. Stains and Knowlton (1943) report specimens of this species (as *Simulium occidentale*) from eastern South Dakota.

Simulium piperi Dyar & Shannon

Descriptions/diagnoses: Peterson and Kondratieff 1995, Adler et al. 2004. In North America from the Queen Charlotte Islands along the Pacific Coast to southern California, east to Saskatchewan, South Dakota, Nebraska, Colorado, New Mexico, and western Texas. Pruess and Peterson (1987) reported *S. piperi* from several locations in Dawes, Sheridan, and Sioux counties in the panhandle of Nebraska.

Stains and Knowlton (1943) reported this species (as *S. sayi*) from Spearfish, South Dakota, and Adler et al. (2004) presented a record of this species from Custer County, as well.

Simulium virgatum Coquillett

Descriptions/diagnoses: Adler et al. 2004. This southern species occurs from the Black Hills south through Wyoming, Colordao, Utah, Arizona, New Mexico, Texas, and Mexico to Guatemal and Panama (Adler et al. 2004). Adler et al. (2004) suspect that *S. virgatum* actually represents a species complex given its variability and widespread distribution.

Stains and Knowlton (1943) repeated the report of Dyar and Shannon (1927) of this species' existence in South Dakota. Adler et al. (2004) show reports in Custer and Hot Springs counties in the South Dakota Black Hills.

Simulium vittatum Zetterstedt

Descriptions/diagnoses: Adler et al. 2004. Peterson and Kondratieff (1995) and Adler et al. (2004) discussed the distribution of this species, which includes nearly all of North America, Greenland, Iceland, and Mexico. It is common enough to have a common name, the "striped black fly." Pruess and Peterson (1987) report that the *S. vittatum* complex (two cytospecies/ecospecies – Rothfels and Featherston 1981; Adler and Kim 1984) occurs in nearly every stream in Nebraska.

Stains and Knowlton (1943) reported the existence of this species in both South Dakota and Wyoming, suggesting that it possibly occurs in the Black Hills.

Stegopterna Enderlein

The genus *Stegopterna* has not yet been reported from the Black Hills; however, the distribution of the *Stegopterna mutata* species complex extends transcontintally from Alaska and Newfoundland, south to California, Utah, and Alabama (Adler et al. 2004), potentially including the Black Hills.

Stratiomyidae

James (1987) points out that only the Stratiomyinae (which would include the genera *Oxycera*, *Glariopsis*, *Euparyphus*, *Caloparyphus*, *Nothomyia*, *Hedriodiscus*, *Odontomyia*, and S*tratiomys*) have aquatic larvae, and the other subfamilies are terrestrial or semiaquatic.

I could not find references to descriptions of larvae of either *Anoplodonta* or *Psellidotus*.

Anoplodonta James

Anoplodonta nigrirostris (Loew)

Description/diagnosis: Johnson 1895 (adult, as *Odontomyia nigrirostris*). Distributed from Colorado to South Dakota.

Johnson (1895) provided a record of this species from Custer, South Dakota.

Caloparyphus James

There are nine species of *Caloparyphus* in North America, none of which have specifically been reported from South Dakota or Wyoming. Richards (2014) pointed out that determination of larval *Caloparyphus* from *Euparyphus* in most keys is based on a seemingly simple character (the prothoracic spiracles are on long stalks in *Euparyphus*, but on short stalks or nearly sessile in *Caloparyphus*), but also cited Sinclair's (1989) observation that the long stalks of *Euparyphus* do not develop until the last instar. Therefore, determinations of *Caloparyphus* in the larvae are questionable unless it is certain that the specimens are last instar, which may not be possible to know. These organisms would be better called "*Caloparyphus/Euparyphus*."

Shearer (2006) reported the genus *Caloparyphus* from the Black Hills, and I have seen it in biomonitoring samples.

Euparyphus Gerstaeker

Description/diagnosis: Quist and James 1973 (adults). As discussed above under *Caloparyphus*, the prothoracic spiracles in larval *Euparyphus* are located at the ends of long stalks in the last instar, so their presence is indicative of this genus. Earlier instars cannot be separated from *Caloparyphus*.

Based on the location of the prothoracic spiracles on long stalks, I have seen the genus *Euparyphus* in biomonitoring samples from the Black Hills.

Euparyphus mutabilis Adams

Quist and James (1973) reported that this species is found in South Dakota and Wyoming but did not specically identify the Black Hills. The type series was from Lusk, Wyoming.

Hedriodiscus Enderlein

Hedriodiscus truquii (Bellardi)

Description/diagnosis: James 1936, Sorenson & Fluke 1953 (adult, as *Odontomyia truquii*). Distributed from the Great Lakes west and south to Manitoba, Oregon, California, and Texas, into Mexico.

James (1936) reports that this species is "common" in South Dakota but did not indicate if it was found in the Black Hills.

Hedriodiscus vertebratus (Say)

Description/diagnosis: McFadden 1967 (larva), James 1936, Sorenson & Fluke 1953 (adult, as *Odontomyia vertebrata*). Distributed from New England and Quebec east to Wisconsin, South Dakota, Colorado, and Missouri.

Johnson (1895) and James (1936) reported this species from South Dakota, not indicating if the records were from the Black Hills.

McFadden (1967) reported that he was told by M. T. James that the larvae live among floating vegetation in small streams, feeding on microorganisms, algae and soft parts of plants.

Odontomyia Meigen

Odontomyia arcuata Loew

Description/diagnosis: James 1936 (adult, as *Odontomyia alticola*). Distributed from South Dakota to New Mexico, California, and Oregon.

Johnson (1895) reported this species from Rapid City, and James (1936) reported this species (as *O. alticola*) from Hot Springs, South Dakota.

Odontomyia inaequalis Loew

Description/diagnosis: James 1936 (adult). Manitoba and the Great Lakes to Idaho, Utah, California, and Texas.

James (1936) reported that this species is common in late July and August in South Dakota but didn't indicate if the records included the Black Hills.

Odontomyia pilimana Loew

Description/diagnosis: James 1936, Sorenson & Fluke 1953 (adult). West central states and provinces from Idaho and Manitoba south to Colorado, Kansas, and Iowa, with a disjunct record from Virginia.

James (1936) reported this species from Custer, White, and Waubay, South Dakota.

Odontomyia pubescens Day

Description/diagnosis: McFadden 1967 (larva), James 1936, Sorenson & Fluke 1953 (adult). Distributed transcontinentally through southern Canada south to California, Arizona, Colorado, Kansas, Indiana, and New Jersey.

James (1936) reported "numerous records" from South Dakota but did not indicate if any were from the Black Hills.

Odontomyia virgo Wiedemann

Description/diagnosis: McFadden 1967 (larva), James 1936, Sorenson & Fluke 1953 (adult). Distributed from New England west to Manitoba and Oregon, south to Georgia, Ohio, Illinois, Texas, Colorado, and Utah.

Johnson (1895) reported this species from Custer, South Dakota.

Psellidotus Rondani

Psellidotus viridis (Bellardi)

Description/diagnosis: Sorenson & Fluke 1953 (adult, as *Labiostigmina viridis*). Distributed from South Dakota east to Indiana, and south through Texas into Mexico.

James and Steyskal (1952) include South Dakota in the distribution of this species, but do not indicate if the records are from the Black Hills.

Stratiomys Geoffroy

Based on distributions given in James and Steyskal (1952), the following species of *Stratiomys* could also possibly occur in the Black Hills: *S. barbata*, *S. discaloides*, *S. griseata*, *S. nevadae*, *S. normula*, and *S. obesa*. The species listed below have actual records of their presence in the Black Hills.

Stratiomys bruneri Johnson

Description/diagnosis: Johnson 1895, Sorenson & Fluke 1953 (adult). Distributed from Alberta to the Northwest Territories, south to Utah, South Dakota, Minnesota, and Michigan.

This species was originally described from Custer, South Dakota.

Stratiomys norma Wiedemann

Description/diagnosis: Sorenson & Fluke 1953 (adult). Distributed from Quebec and Ontario to Pennsylvania, Illinois, Kansas, and South Dakota.

James and Steyskal (1952) report this species from South Dakota at the state level only.

Stratiomys normula (Loew)

Description/diagnosis: Johnson 1895, Sorenson & Fluke 1953 (adult, as *Stratiomyia unilimbata*). Distributed from Montana east to Wisconsin, south to Colorado, Nebraska, and Illinois.

Johnson (1895) has a record for this species (as *Stratiomys unilimbata*) from Custer, South Dakota.

Syrphidae

This family is extremely common in the Black Hills, and adults can be found readily on flowers; however, only a few species in the family have truly aquatic or semiaquatic larvae. I have not seen records of the aquatic and semiaquatic representatives of this family in the Black Hills, but it is likely that they occur there.

Tabanidae

The horse flies and deer flies are currently represented by 19 species within the Black Hills (Easton 1983). An additional species, *Tabanus stonei*, has been collected from Belle Fourche, just north of the Black Hills, and is discussed below. Given the widespread distribution of some Tabanidae, several additional species in genera such as *Atylotus* and *Haematopota*, which are known from eastern South Dakota (Easton 1983), are likely to occur in the Black Hills.

Larvae of Tabanidae can only be identified with any degree of confidence to the genus level at this time, and only in the later instars.

Chrysops Meigen

Jergens (1968) reported Chrysops at the genus level from French Creek, Spring Creek, Battle Creek, Castle Creek, and the South Fork of Rapid Creek. Likewise, Shearer (2006) also reported *Chrysops* at the genus level from the Black Hills.

Chrysops aestuans van der Wulp

Description/diagnosis: Teskey 1969 (larva, pupa), Pechuman et al. 1983 (larva, adult), Thomas and Marshall 2009 (adult photos). Distributed from British Columbia east to Manitoba, and south to Oregon, New Mexico, and Iowa; eastward through the Great Lakes Region to Ontario, New York, and Pennsylvania, with a disjunct population in the Canadian Maritime provinces. Larvae are found around cattails, in mud, and in rotting vegetation and in marshes bordering large bodies of water (Middlekauf and Lane 1980, Pechuman et al. 1983).

Easton (1983) stated that this species is "the most commonly distributed deer fly in the state [of South Dakota] and probably the the most important fly species affecting man in recreational areas." Easton (1983) reported this species from Stockade Lake in Custer State Park and from Lake Mitchell one mile north of Hill City.

Chrysops carbonarius Walker

Description/diagnosis: Teskey 1969 (larva, pupa), Pechuman et al. 1983 (larva, adult), Thomas and Marshall 2009 (adult photos). This species is normally distributed in the eastern Nearctic from Wisconsin through Quebec to the Maritime Provinces, and south to Ohio, Tennessee, Louisiana, Florida's panhandle, and Georgia. Pechuman et al. (1983) indicated that larvae are found in saturated mud on banks of streams and ponds.

Easton (1983) reported this species from "10 miles southwest of Belle Fourche", Sylvan Lake, and Bear Butte State Park in and around the Black Hills. Given the normal distribution of this species in the east and the fact that the closest reported collections would be in central Wisconsin, these records should be considered questionable until the specimens are verified.

Chrysops fulvaster Osten Sacken

Description/diagnosis: Cameron 1926 (larva), Middlekauf and Lane 1980 (adult), Thomas and Marshall 2009 (adult photos). Distributed from in the central Nearctic from Alberta to Minnesota, south to California, New Mexico, and Oklahoma. Larvae are known from swampy ravines and banks of small, sluggish streams.

Easton (1983) reported this species from Custer, Hot Springs, "10 miles southwest of Spearfish", Tinton, and "2 miles northwest of Rapid City."

Chrysops furcatus Walker

Description/diagnosis: Teskey 1969 (larva, pupa), Middlekauf and Lane 1980 (adult), Thomas and Marshall 2009 (adult photos). Distributed in a broad transcontinental band from Alaska to Newfoundland, south to California, Colorado, North Dakota, Ontario, and Quebec. Larvae have been collected from saturated soils and wet moss in swampy meadows.

Easton (1983) reported this species from Custer.

Chrysops mitis Osten Sacken

Description/diagnosis: Teskey 1969 (larva), Pechuman et al. 1983 (adult), Thomas and Marshall 2009 (adult photos). This species is broadly distributed from Alaska to Labrador and Newfoundland, south to Idaho, Colorado, South Dakota, Minnesota, Michigan, Pennsylvania, and New Jersey. The larva of this species is found in highly organic substrates along ponds, streams, and swampy areas (Pechuman et al. 1983).

Easton (1983) reported this species from Custer and Cheyenne Crossing.

Chrysops pikei Whitney

Description/diagnosis: Pechuman et al. 1983 (larva, adult), Teskey 1969 (larva, pupa), Thomas and Marshall 2009 (adult photos). This is a southeastern Nearctic species, ranging from South Dakota to New York, south to Texas and the Gulf Coast northern Florida, and north along the Atlantic seaboard to Virginia. While the main Appalachian cordillera is excluded from its range, this species also has disjunct populations in the Black Hills and in southern Florida. Pechuman et al (1983) reported that larvae of this species are found in debris and in the banks of ponds and streams.

Easton (1983) reported this species from Edgemont.

Chrysops sequax Williston

Description/diagnosis: Pechuman et al. 1983 (larva, adult). Pechuman et al. (1983) reported that a larva of this species was collected in "soft, slimy muck on the margin of a livestock watering pond."

Easton (1983) reported this species from Bear Butte.

Hybomitra Enderlein

Shearer (2006) reported *Hybomitra* at the genus level from the Black Hills.

Hybomitra criddlei (Brooks)

Description/diagnosis: Teskey 1969 (larva, pupa), Thomas 2011 (adult photos). Mostly an upper Midwest species, with records from Ontario, Manitoba, Saskatchewan, Wisconsin, and South Dakota.

Easton (1983) reported this species from Spearfish and "5 miles northwest Rockford, Road 231"

Hybomitra epistates (Osten Sacken)

Description/diagnosis: Teskey 1969 (larva, pupa), Pechuman et al. 1983 (larva, adult), Thomas 2011 (adult photos). This is a northern species, ranging from Alaska to Hudson Bay to the Canadian Maritime Provinces, south to Oregon, Colorado, Saskatchewan, Iowa, Indiana, Ohio, West Virginia, and Delaware. Pechuman et al. (1983) reported that larvae are usually found in wet moss within swampy areas.

Easton (1983) reported this species from Spearfish.

Hybomitra frontalis (Walker)

Description/diagnosis: Teskey 1969 (larva, pupa), Pechuman et al. 1983 (larva, adult), Thomas 2011 (adult photos). *Hybomitra frontalis* is a northern species, ranging from Alaska south along the Rocky Mountains to Colorado and east to Quebec and Labrador, Nova Scotia, and Vermont. Larvae of this species are found in wet moss in swampy areas (Pechuman et al. 1983).
Easton (1983) reported this species from Custer.

Hybomitra lasiophthalma (Macquart)

Description/diagnosis: Teskey 1969 (larva, pupa), Pechuman et al. 1983 (larva, adult), Thomas 2011 (adult photos). This species is broadly distributed east of the Mississippi River to South Carolina and north to Nova Scotia and Quebec; west of the Mississippi River, its distribution extends to the Dakotas and Alberta, south into Idaho and Nevada. Larvae of this species can tolerate a wide range of moisture conditions, from dry forest floor to moist/wet sod and marshes (Pechuman et al. 1983).
Easton (1983) reported this species from Custer.

Hybomitra rhombica (Osten Sacken)

Description/diagnosis: Middlekauf and Lane 1980 (adult), Thomas 2011 (adult photos). Larvae are unknown, but Cameron (1926) described the pupa. Distributed in the west from Alaska to Saskatchewan and Wisconsin, south to California, Arizona, Colorado, and South Dakota.
Easton (1983) reported this species from Custer and Crooks Tower Lake.

Hybomitra rupestris (McDonnough)

Description/diagnosis: Thomas 2011 (adult photos). A western species, with distribution from British Columbia and South Dakota south to New Mexico along the Rocky Mountain cordillera.
Easton (1983) reported this species from Custer.

Hybomitra tetrica (Marten)

Description/diagnosis: Cameron 1926 (larva, as *Tabanus hirtulus*), Pechuman et al. 1983 (adult), Thomas 2011 (adult photos). The Black Hills are within the range of the *hirtula* subspecies, which is found from Manitoba south to Colorado and Utah, west to British Columbia and Oregon, extending south along the California/Nevada border to about central California.
Easton (1983) reported this species from Custer.

Silvius Meigen

Silvius quadrivittatus (Say)

Description/diagnosis: Burger 1977 (larva), Middlekauf and Lane 1980 (adult). Distributed from "California east to Tennessee, south to Mexico and Mississippi" (Middlekauf and Lane 1980). Larvae are found in damp silt along dryland rivers.
Easton (1983) reported this species from Hot Springs.

Silvius pollinosus Williston

Description/diagnosis: Williston 1887 (adult). Distributed in the High Plains from the Dakotas south to Colorado and Kansas.

Easton (1983) reported this species from Wasta, along the Cheyenne River.

Tabanus Linnaeus

Jergens (1968) reported tabanus at the genus level from Spring Creek, Battle Creek, and the South Fork of Rapid Creek. Likewise, Shearer (2006) also reported *Tabanus* at the genus level from the Black Hills.

Tabanus atratus Fabricius – Black Horse Fly

Description/diagnosis: Pechuman et al. 1983 (larva, adult), Thomas 2011 (adult photos). Distributed in the east from the entire Atlantic seaboard west to Ontario, the Dakotas, Nebraska, Kansas, Oklahoma, and Texas. Pechuman et al. (1983) indicated that larvae can be found across a wide range of moisture conditions from open water to moist earth. Life cycle is usually univoltine, but can extend to at least two years.
Easton (1983) reported this species from Lead in the Black Hills, as well as Newell and Belle Fourche just north of the Black Hills. I have collected this species in the panhandle of Nebraska, just south of the Black Hills.

Tabanus punctifer Osten Sacken

Description/diagnosis: Burger 1977 (larva), Middlekauf and Lane 1980 (adult). Distributed in the west from British Columbia to Kansas, south to California, Mexico, Texas, and Oklahoma.
Easton (1983) reported this species from Spearfish.

Tabanus similis Macquart

Description/diagnosis: Burger 1977 (larva), Pechuman et al. 1983 (larva, adult), Thomas 2011 (adult photos). Distributed in a broad, transcontinental band from Oregon and northern California to the Canadian Maritime Provinces and Virginia, with extensions north into Saskatchewan and south into New Mexico and Arizona. Pechuman et al. (1983) reported that this species can tolerate a wide range of moisture conditions, from dry agricultural land to semiaquatic habitats along streams, ponds, marshes, and bogs.
Easton (1983) reported this species from "10 miles southwest of Spearfish", as well as Newell and Bear Butte.

Tabanus stonei Philip

Description/diagnosis: Burger 1977 (larva), Middlekauf and Lane 1980 (adult). Distributed from British Columbia and Alberta, south to California, Arizona, and Texas. Larvae have been collected from rotting logs and may not be truly aquatic or semiaquatic.
Easton (1983) reported this species from Belle Fourche.

Tanyderidae

This family likely occurs in the Black Hills; however, I could find, as yet, no records of its presence there.

Tipulidae

There are about 1,500 species of Tipulidae in North America (approximately 14,000 species worldwide, making it the largest family in the Diptera), with habitats ranging across the North American ecotones (Alexander and Byers 1981) with both terrestrial and aquatic forms. Even though few genera of Tipulidae have been reported from the Black Hills, they vary considerably in their species traits, as described below and in the species traits table.

Numerous and widespread in distribution, the craneflies are commonly found in the Black Hills. Most larvae can only be identified to the genus level; those of *Tipula* can be identified to the subgenus level, and *Antocha* may be identified to *A. monticola*, since it is monospecific in the area. The terrestrial adults are also difficult to identify to species, primarily because they are so speciose, and species-level records are rare.

Among Alexander and Byers (1981), Oosterbroek and Theowald (1991), and Byers and Gelhaus (2008), one can find excellent illustrations of most genera. Additionally, most genera have been investigated to the point where there are numerous published larval descriptions; see Oosterbroek and Theowald (1991) for a bibliography of the descriptions of species in the limoniine genera. I believe there are likely several more genera of aquatic tipulids to be found. Byers and Gelhaus (2008) is a good key for identification to genus, and Gelhaus (1986) is a good key for identification of the subgenera of *Tipula*.

Antocha Osten Sacken

Antocha is monospecific in the western Nearctic, with *A. monticola* the only species reported; thus, it is rare among the Tipulidae in biomonitoring samples in being able to be identified to the species level. Six other species occur in the east.

Antocha monticola Alexander

Description/Diagnosis: Needham and Christiansen 1927 (larva), Alexander 1917 (adult). This species is distributed in the western Nearctic.

Shearer (2006) reported *Antocha* at the genus level from the Black Hills. I have numerous specimens of *Antocha monticola* from Annie Creek, Deadwood Creek, and Whitewood Creek in my personal collection.

Arctoconopa Alexander

Arctoconopa pahasapa (Alexander)

Description/Diagnosis: Alexander 1955 (adult). This species was described from three specimens collected in the Black Hills – a dead giveaway is the specific epithet! I have not found additional records of this species beyond those in the original description.

The type locality of this species is a small stream with beaver dams near Hill City, South Dakota. I have not seen this genus or species in biomonitoring samples or otherwise.

Dicranota Zetterstedt

Description/Diagnosis: Alexander 1920, Alexander and Byers 1981 (larva); Oosterbroek and Theowald (1991) list 21 citations for descriptions of *Dicranota* immature stages. This genus is widely distributed, with approximately 60 species in North America.

Shearer (2006) reported *Dicranota* at the genus level from the Black Hills.

Dolichopeza Curtis

I have not seen any specimens of this genus from samples collected in the Black Hills, but Byers (1961) reported three species from there. As described in the Introduction, Byers (1961) concluded that the genus inhabited the boreal forest at the front of the Wisconsin glaciations. The retreat of the glaciers left the forest intact only at the higher, cooler, wetter elevations in the Black Hills, isolated as the lower elevations became dry forests and eventually grasslands. Furthermore, the lack of speciation was due to inherent variability within the genus and its species, since they are distributed from Alaska and northwestern Canada to the Florida peninsula, from sea level to over 1,800 m elevation.

Dolichopeza americana Needham

Description/Diagnosis: Byers 1961 (adult). Widespread from Alaska and Alberta south and east to South Dakota, Iowa, Illinois, and Georgia, up the Atlantic seaboard to New Brunswick, Quebec, and Labrador.

Byers (1961) reported this species from two locations in the Black Hills: Harney Peak in Custer County, and Horsethief Lake in Pennington County.

Dolichopeza dakota Alexander

Description/Diagnosis: Alexander 1944 (adult). Described from "a small stream about midway between Sylvan Lake and the summit of Harney Peak, in a dense growth of Western White Spruce."

Dolichopeza dorsalis (Johnson)

Description/Diagnosis: Byers 1961 (adult, larva). Widespread distribution, but poorly collected, this species ranges from northern British Columbia east to New England and south to the Black Hills in South Dakota, Minnesota, Indiana, and the panhandle of Florida.

Byers (1961) reported the collection of this species in Pennington County, South Dakota.

Dolichopeza wallyae (Alexander)

Description/Diagnosis: Byers 1961 (adult, larva). Distributed from British Columbia east to New England and south to South Dakota, Kansas, Arkansas, and Florida.

Byers (1961) reported this species from Pennington County, South Dakota.

Erioptera Meigen

Description/Diagnosis: Alexander 1920, Alexander and Byers 1981 (larva); Oosterbroek and Theowald (1991) list 25 citations for descriptions of *Erioptera* immature stages. This genus is widely distributed, with approximately 90 species in North America.

I have seen specimens of *Erioptera* in biomonitoring samples from the Black Hills.

Hexatoma Latreille

Description/Diagnosis: Alexander 1920, Alexander and Byers 1981 (larva); Oosterbroek and Theowald (1991) list 31 citations for descriptions of *Hexatoma* immature stages. This genus is widely distributed, with approximately 35 species in North America.

Jergens (1968) reported *Hexatoma* at the genus level from Spring Creek and Battle Creek. Likewise, Shearer (2006) also reported *Hexatoma* at the genus level from the Black Hills.

Limonia Meigen

Description/Diagnosis: Alexander 1920, Alexander and Byers 1981 (larva); Oosterbroek and Theowald (1991) list 28 citations for descriptions of *Limonia* immature stages. This genus is distributed worldwide, with approximately 125 species in North America.

I have seen specimens of *Limonia* in biomonitoring samples from the Black Hills, even though I do not know of literature records reporting its presence there.

Ormosia Rondani

Description/Diagnosis: Alexander and Byers 1981 (larva); being in the Tipulinae, Oosterbroek and Theowald (1991) do not address the literature on the immature stages of *Ormosia*. This genus is widely distributed, with approximately 100 species in North America.

I have seen specimens of *Ormoisa* in biomonitoring samples from the Black Hills, even though I do not know of any literature records reporting its presence there.

Pedicia Latreille

Description/Diagnosis: Alexander and Byers 1981 (larva); Oosterbroek and Theowald (1991) list 29 citations for descriptions of *Pedicia* immature stages. This genus is widely distributed, with approximately 60 species in North America. The species *Pedicia pahasapa* Alexander was described from Midway Creek on Harney Peak in the Black Hills (Alexander 1958).

Pseudolimnophila Alexander

Description/Diagnosis: Alexander 1920, Alexander and Byers 1981 (larva); Oosterbroek and Theowald (1991) list 16 citations for descriptions of *Pseudolimnophila* immature stages.

I have seen specimens of *Pseudolimnophila* in biomonitoring samples from the Black Hills, even though I do not know of any literature records reporting its presence there.

Tipula Linnaeus

Description/Diagnosis: Gelhaus 1986 (larva). This genus is widely distributed, with approximately 450 species in North America.

Even though it is possible to identify late instar North American *Tipula* larvae to the subgenus level (Gelhaus 1986), most identifications during biomonitoring studies have left the identification at the genus level. For biomonitoring purposes, that is likely to be acceptable, unless multiple subgenera could be identified within a sample, indicating greater biodiversity.

Jergens (1968) reported *Tipula* at the genus level from Spearfish Creek, Battle Creek, and Castle Creek. Likewise, Shearer (2006) also reported *Tipula* at the genus level from the Black Hills. I have, in my personal collection, larval specimens of *Tipula* from the Fall River near Hot Springs and from Whitewood Creek at State Highway 79.

Aquatic Diptera Traits Table

Table 13:   Aquatic Diptera Traits Table

Species	TSN	Seen	Tolerance Value	FFG	Habit
Athericidae	130928	+	7	--	--
Atherix	130929	+	2	PR	SP
Atherix variegata	130932	+	2	PR	SP
Ceratopogonidae	127076	+	6	PR	SP
Atrichopogon	127113	+	6	PR	SP
Bezzia	127778	/	6	PR	BU
Culicoides	127340	/	6	PR	BU
Culicoides coquerelli	127371	/	6	PR	BU
Culicoides sonorensis	--	/	6	PR	BU
Culicoides variipennis	127523	/	6	PR	BU
Parabezzia	127591	/	6	PR	SP
Parabezzia williamsi	127607	/	6	PR	SP
Probezzia	127729	/	6	PR	SP
Chaoboridae	125886	-	7	PR	SP
Chaoborus	125904	-	7	PR	SP
Chaoborus americanus	125906	-	7	PR	SP
Chaoborus flavicans	125917	-	7	PR	SP
Chaoborus punctipennis	125923	-	7	PR	SP
Eucorethra	125888	-	7	PR	SW
Eucorethra underwoodi	125889	-	7	PR	SW
Mochlonyx	125893	-	7	PR	SP
Mochlonyx velutinus	125899	-	7	PR	SP
Chironomidae	127917	+	6	GC	BU
Chironominae	129228	+	6	GC	BU
Chironomus	129254	+	10	GC	BU
Cladopelma	129350	-	6	GC	BU
Cladotanytarsus	129873	-	7	GC	CB
Cryptochironomus	129368	+	8	PR	SP
Cryptotendipes	129394	-	6	GC	SP
Demicryptochironomus	129421	-	6	GC	BU
Dicrotendipes	129428	+	8	GC	BU
Einfeldia	129459	-	8	GC	BU
Endochironomus	129470	-	10	SH	CN
Glyptotendipes	129483	-	8	SH	BU
Harnischia	129516	-	8	GC	CN
Kiefferulus	129522	-	8	GC	BU
Lauterborniella	129525	-	8	GC	CN
Microchironomus	129532	-	6	GC	BU
Micropsectra	129890	+	7	GC	CB
Microtendipes	129535	+	6	FC	CN
Parachironomus	129564	-	8	PR	SP
Paracladopelma	129597	+	8	GC	SP
Paralauderborniella	129616	-	8	GC	CN
Paratanytarsus	129935	-	6	GC	SP

Species	TSN	Seen	Tolerance Value	FFG	Habit
Paratendipes	129623	+	8	GC	BU
Phaenopsectra	129637	+	7	SC	CN
Polypedilum	129657	+	6	SH	CB
Pseudochironomus	129851	+	5	GC	BU
Rheotanytarsus	129952	+	6	FC	CN
Stenochironomus	129746	-	8	GC	BU
Stictochironomus	129785	+	8	GC	BU
Tanytarsus	129978	+	6	FC	BU
Tribelos	129820	-	8	GC	BU
Xenochironomus	129837	-	8	PR	BU
Zavrelia	130038	-	8	GC	CB
Diamesinae	128341	+	6	GC	SP
Boreoheptagyia	128343	-	6	GC	SP
Diamesa	128355	+	5	GC	SP
Pagastia	128401	+	1	GC	SP
Pseudodiamesa	128416	+	6	GC	SP
Orthocladiinae	128457	+	5	GC	BU
Acricotopus	128463	-	6	GC	SP
Brillia	128477	+	5	SH	BU
Bryophaenocladius	128488	-	6	GC	BU
Camptocladius	128507	-	6	GC	BU
Cardiocladius	128511	+	5	PR	BU
Chaetocladius	128520	+	6	GC	SP
Corynoneura	128563	+	7	GC	SP
Cricotopus	128575	+	7	SH	CN
Cricotopus bicinctus	128583	+	7	OM	CN
Cricotopus nostocicola	128628	+	7	SH	CN
Cricotopus trifascia	128659	+	7	OM	CN
Diplocladius	128670	-	6	GC	SP
Epoicocladius	128682	-	5	GC	BU
Eukiefferiella	128689	+	8	GC	SP
Heleniella	128730	+	6	GC	SP
Heterotrissocladius	128737	+	4	GC	SP
Hydrobaenus	128750	+	8	SC	SP
Krenosmittia	128771	+	1	GC	SP
Limnophyes	128776	+	8	GC	SP
Metriocnemus	128821	+	6	GC	BU
"Microcricotopus"	--	-	6	GC	BU
Nanocladius	128844	+	3	GC	SP
Orthocladius	128874	+	6	GC	SP
Orthocladius obumbratus	128923	/	6	GC	SP
Parakiefferiella	128968	+	6	GC	SP
Parametriocnemus	128978	+	5	GC	SP
Paraphaenocladius	128989	+	5	GC	SP
Parorthocladius	129011	+	6	GC	SP
Psectrocladius	129018	+	8	GC	SP
Pseudosmittia	129071	-	6	GC	SP
Psilometriocnemus	129083	+	6	GC	SP

Species	TSN	Seen	Tolerance Value	FFG	Habit
Psilometriocnemus triannulatus	129084	/	6	GC	SP
Rheocricotopus	129086	+	6	GC	SP
Smittia	129110	-	6	GC	BU
Synorthocladius	129161	+	2	GC	SP
Thienemanniella	129182	+	6	GC	SP
Tvetenia	129197	+	5	GC	SP
Podonominae	127952	+	6	GC	BU
Lasiodiamesa	127962	-	6	GC	BU
Paraboreochlus	127977	+	6	GC	BU
Paraboreochlus stahli	127977	+	6	GC	BU
Prodiamesinae	128437	+	6	GC	BU
Odontomesa	128446	+	4	GC	SP
Prodiamesa	128452	+	3	GC	BU
Prodiamesa olivacea	128454	/	3	GC	BU
Tanypodinae	127994	+	7	PR	SP
Ablebesmyia	128079	+	6	PR	SP
Alotanypus	206646	-	7	PR	BU
Clinotanypus	127996	-	7	PR	BU
Conchapelopia	128130	+	6	PR	SP
Coelotanypus	128010	-	7	PR	BU
Guttipelopia	128161	-	6	GC	BU
Labrundinia	128173	-	6	PR	SP
Larsia	128183	+	6	PR	SP
Macropelopia	128034	+	6	PR	SP
Natarsia	128070	+	7	PR	SP
Nilotanypus	128202	+	6	PR	SP
Paramerina	128207	+	6	PR	SP
Pentaneura	128215	+	6	PR	SP
Procladius	128277	+	9	PR	SP
Psectrotanypus	128048	-	10	PR	SP
Radotanypus	--	+	7	PR	SP
Tanypus	128324	-	6	PR	SP
Telopelopia	128234	-	6	PR	SP
Thienemannimyia	128236	-	6	PR	SP
Zavrelimyia	128259	+	8	PR	SP
Culicidae	125930	-	8	FC	SW
Aedes	126234	-	8	GC	SW
Aedes cinereus	126264	-	8	GC	SW
Aedes intrudens	126319	-	8	GC	SW
Aaedes triseriatus	126392	-	8	GC	SW
Aedes vexans	126403	-	8	GC	SW
Anopheles	125956	-	8	FC	SW
Anopheles earlei	125963	-	8	FC	SW
Anopheles punctipennis	125977	-	8	FC	SW
Anopheles quadrimaculatus	125980	-	8	FC	SW
Anopheles walkeri	125982	/	8	FC	SW
Culex	126455	-	8	FC	SW
Culex pipiens	126488	-	8	FC	SW

Species	TSN	Seen	Tolerance Value	FFG	Habit
Culex restuans	126493	-	8	FC	SW
Culex salinarius	126495	-	8	FC	SW
Culex tarsalis	126498	-	8	FC	SW
Culex territans	126501	-	8	FC	SW
Culiseta	126429	-	8	GC	SW
Culiseta impatiens	126434	-	8	GC	SW
Culiseta incidens	126438	-	8	GC	SW
Culiseta inornata	126440	-	8	GC	SW
Culiseta morsitans	126446	-	8	GC	SW
Mansonia	126527	-	8	FC	SW
Mansonia perturbans	126531	-	8	FC	SW
Ochlerotatus	126621	-	8	GC	SW
Ochlerotatus dorsalis	--	-	8	GC	SW
Ochlerotatus excrucians	--	-	8	GC	SW
Ochlerotatus increpitus	--	-	8	GC	SW
Ochlerotatus nigromaculis	--	-	8	GC	SW
Ochlerotatus spencerii	--	-	8	GC	SW
Ochlerotatus sticticus	--	-	8	GC	SW
Ochlerotatus trivittatus	--	-	8	GC	SW
Psorophora	126538	-	8	FC	SW
Psorophora ferox	126556	-	8	FC	SW
Psorophora signipennis	126570	-	8	FC	SW
Dixidae	125809	+	1	GC	SW
Dixa	125810	+	1	GC	SW
Meringodixa	125873	+	2	GC	SW
Meringodixa chalonensis	125874	+	2	GC	SW
Dolichopodidae	136824	+	4	PR	SP
Dolichopus	137953	/	4	PR	SP
Dolichopus penicillatus	138394	/	4	PR	SP
Liancalus	138729	/	4	PR	SP
Liancalus hydrophilus	138731	/	4	PR	SP
Parasyntormon	137791	/	4	PR	SP
Parsyntormon hinnulus	137800	/	4	PR	SP
Empididae	135830	+	6	PR	SP
Chelifera	136305	+	6	GC	SP
Clinocera	135849	+	6	PR	CN
Hemerodromia	136327	+	6	PR	SP
Metachela	136347	+	6	GC	SP
Metachela collusor	136350	/	6	PR	SP
Neoplasta	136352	+	6	GC	SP
Neoplasta paramegorchis	--	/	6	GC	SP
Oreogeton	136377	+	5	PR	SP
Trichoclinocera	135903	-	6	PR	SP
Wiedemannia	135920	-	6	PR	CN
Ephydridae	146893	+	6	GC	BU
Allotrichoma	146896	-	6	GC	BU
Allotrichoma occidentale	--	-	6	GC	BU
Ephydra	147486	-	6	GC	BU
Ephydra hians	147494	-	6	GC	BU

Species	TSN	Seen	Tolerance Value	FFG	Habit
Ephydra pectinulata	147504	-	6	GC	BU
Ephydra riparia	147505	-	6	GC	BU
Parydra	147388	-	6	GC	BU
Parydra appendiculata	147394	-	6	GC	BU
Parydra aquila	147395	-	6	GC	BU
Parydra parasocia	147417	-	6	GC	BU
Parydra quadrituberculata	147427	-	6	GC	BU
Scatella	147568	+	6	GC	BU
Muscidae	150025	+	6	PR	SP
Limnophora	150730	+	6	PR	BU
Lispe	150756	-	6	PR	SP
Lispoides	150805	+	6	PR	SP
Lispoides aequifrons	150806	+	6	PR	SP
Spilogona	150808	+	6	PR	SP
Psychodidae	125351	+	10	GC	BU
Maruina	125392	+	1	SC	CN
Pericoma	125514	+	4	GC	BU
Psychoda	125468	+	10	GC	BU
Ptychopteridae	125763	-	7	GC	BU
Ptychoptera	125786	-	7	GC	BU
Simuliidae	126640	+	6	FC	CN
Prosimulium	126703	+	3	FC	CN
Prosimulium exigens	126718	/	3	FC	CN
Simulium	126774	+	6	FC	CN
Simulium bivittatum	126790	/	6	FC	CN
Simulium canadense	126795	/	6	FC	CN
Simulium decorum	126808	/	6	FC	CN
Simulium griseum	126825	/	6	FC	CN
Simulium hunteri	126827	/	6	FC	CN
Simulium meridionale	126841	/	6	FC	CN
Simulium piperi	126862	/	6	FC	CN
Simulium virgatum	126901	/	6	FC	CN
Simulium vittatum	126903	/	6	FC	CN
Stegopterna	126761	-	6	FC	CN
Stratiomyidae	130150	+	8	GC	--
Anoplodonta	130484	-	8	GC	--
Anoplodonta nigrirostris	130485	-	8	GC	--
Caloparyphus	130409	+	7	GC	SP
Euparyphus	130436	+	8	GC	SP
Hedriodiscus	13488	-	8	GC	--
Hedriodiscus truquii	130504	-	8	GC	--
Hedriodiscus vertebratus	130510	-	8	GC	--
Odontomyia	130573	+	8	GC	SP
Odontomyia arcuata	130625	/	8	GC	SP
Odontomyia inaequalis	627058	/	8	GC	SP
Odontomyia pilimana	130606	/	8	GC	SP
Odontomyia pubescens	130613	/	8	GC	SP
Odontomyia virgo	130573	/	8	GC	SP

Species	TSN	Seen	Tolerance Value	FFG	Habit
Psellidotus	625655	-	8	GC	--
Psellidotus viridis	628005	-	8	GC	--
Stratiomys	130627	-	8	FC	SP
Stratiomys bruneri	130639	-	8	FC	SP
Stratiomys norma	130678	-	8	FC	SP
Stratiomys normula	628306	-	8	FC	SP
Syrphidae	139621	-	10	GC	--
Tabanidae	130934	+	8	PR	SP
Chrysops	131078	-	8	PR	SP
Chrysops aestuans	131082	/	8	PR	SP
Chrysops carbonarius	131107	/	8	PR	SP
Chrysops fulvaster	131145	/	8	PR	SP
Chrysops furcatus	131147	/	8	PR	SP
Chrysops mitis	131169	/	8	PR	SP
Chrysops pikei	131184	/	8	PR	SP
Chrysops sequax	131193	/	8	PR	SP
Hybomitra	131321	-	8	PR	SP
Hybomitra criddlei	131354	/	8	PR	SP
Hybomitra epistates	131361	/	8	PR	SP
Hybomitra frontalis	131367	/	8	PR	SP
Hybomitra lasiophthalma	131390	/	8	PR	SP
Hybomitra rhombica	131429	/	8	PR	SP
Hybomitra rupestris	131436	/	8	PR	SP
Hybomitra tetrica	131452	/	8	PR	SP
Silvius	131062	-	8	PR	SP
Silvius quadrivittatus	131074	/	8	PR	SP
Silvius pollinosus	131073	/	8	PR	SP
Tabanus	131527	+	5	PR	SP
Tabanus atratus	131540	/	5	PR	SP
Tabanus punctifer	131688	/	5	PR	SP
Tabanus similis	131746	/	5	PR	SP
Tabanus stonei	131716	/	5	PR	SP
Tanyderidae	125799	-	7	--	SP
Tipulidae	118840	+	3	SH	BU
Antocha	119656	+	3	GC	CN
Antocha monticola	119660	+	3	GC	CN
Arctoconopa	120398	-	3	SH	BU
Arctoconopa pahasapa	120411	-	3	SH	BU
Dicranota	121027	+	3	PR	SP
Dolichopeza	118861	-	3	SH	BU
Dolichopeza americana	118861	/	3	SH	BU
Dolichopeza dakota	118861	/	3	SH	BU
Dolichopeza dorsalis	118861	/	3	SH	BU
Dolichopeza walleyi	118861	/	3	SH	BU
Erioptera	120503	+	3	GC	BU
Hexatoma	120094	+	2	PR	BU
Limonia	119704	+	6	SH	BU
Ormosia	120830	+	3	GC	BU

Species	TSN	Seen	Tolerance Value	FFG	Habit
Pedicia	121118	+	6	PR	BU
Pseudolimnophila	120365	+	3	PR	BU
Tipula	119037	+	4	SH	BU

Species	Temperature Range	Lotic	Lentic	Current	Substrate	Voltinism
Athericidae	--	--	--	--	--	--
Atherix	E: warm	4.1	--	--	6.0	Uni
Atherix variegata	E: warm	4.0	--	--	4.0	Uni
Ceratopogonidae	E: warm	--	--	--	0.5	Uni
Atrichopogon	E: warm	4.0	1.5	--	0.5	Uni
Bezzia	E: warm	4.0	1.5	--	7.0	Uni
Culicoides	E: warm	4.0	3.0	--	0.5	Multi
Culicoides coquerelli	E: warm	4.0	3.0	--	0.5	Multi
Culicoides sonorensis	E: warm	4.0	3.0	--	0.5	Multi
Culicoides variipennis	E: warm	4.0	3.0	--	0.5	Multi
Parabezzia	E: warm	--	--	--	0.5	Uni
Parabezzia williamsi	E: warm	--	--	--	0.5	Uni
Probezzia	E: warm	3.0	1.5	--	0.5	Uni
Chaoboridae	--	--	--	--	--	--
Chaoborus	E: warm	4.0	1.3	--	0.0	Multi
Chaoborus americanus	E: warm	4.0	1.3	--	0.0	Multi
Chaoborus flavicans	E: warm	4.0	1.3	--	0.0	Multi
Chaoborus punctipennis	E: warm	4.0	1.3	--	0.0	Multi
Eucorethra	--	4.0	--	--	--	--
Eucorethra underwoodi	--	4.0	--	--	--	--
Mochlonyx	--	--	2.0	--	--	--
Mochlonyx velutinus	--	--	2.0	--	--	--
Chironomidae	--	--	--	--	4.3	--
Chironominae	--	--	--	--	--	Mero
Chironomus	E: warm	4.0	1.2	--	3.2	Mero
Cladopelma	E: warm	--	5.0	--	--	Mero
Cladotanytarsus	E: warm	3.9	1.9	--	4.0	Multi
Cryptochironomus	E: warm	3.9	1.3	--	2.8	Mero
Cryptotendipes	E: warm	4.0	4.0	--	--	Mero
Demicryptochironomus	E: warm	4.0	5.0	--	4.0	Mero
Dicrotendipes	E: warm	3.5	1.2	--	4.0	Mero
Einfeldia	E: warm	--	4.0	--	--	Uni
Endochironomus	E: warm	3.0	1.9	--	4.0	Mero
Glyptotendipes	E: warm	3.5	1.2	--	2.8	Uni
Harnischia	E: warm	--	1.4	--	2.5	Mero
Kiefferulus	--	--	1.4	--	4.0	Mero
Lauterborniella	--	--	4.0	--	1.0	Mero
Microchironomus	E: warm	--	4.0	--	1.0	Mero
Micropsectra	E: warm	3.3	1.3	--	2.6	Mero
Microtendipes	E: warm	4.0	1.9	--	4.0	Mero
Parachironomus	E: warm	4.0	1.9	--	4.0	Mero
Paracladopelma	E: warm	4.0	1.9	--	4.0	Mero
Paralauderborniella	E: warm	--	--	--	--	Mero
Paratanytarsus	E: warm	3.8	1.9	--	4.0	Mero
Paratendipes	E: warm	3.8	1.9	--	4.0	Mero
Phaenopsectra	E: warm	3.0	1.5	--	4.0	Mero
Polypedilum	E: warm	3.6	1.1	--	3.5	Mero

Species	Temperature Range	Lotic	Lentic	Current	Substrate	Voltinism
Pseudochironomus	E: warm	4.0	1.3	--	2.8	Mero
Rheotanytarsus	E: warm	3.7	5.0	--	4.0	Mero
Stenochironomus	E: warm	3.7	4.0	--	--	Mero
Stictochironomus	E: warm	3.7	5.0	--	4.0	Mero
Tanytarsus	E: warm	3.8	1.3	--	4.0	Mero
Tribelos	E: warm	4.0	4.0	--	--	Mero
Xenochironomus	E: warm	4.0	4.0	--	--	Mero
Zavrelia	E: warm	4.0	5.0	--	4.0	Mero
Diamesinae	E: warm	--	--	--	--	Uni
Boreoheptagyia	S: cold	3.0	--	2.0	7.0	Uni
Diamesa	E: warm	3.3	1.2	2.5	4.8	Uni
Pagastia	E: cool	3.7	5.0	--	4.0	Uni
Pseudodiamesa	E: cool	3.0	1.2	--	--	Uni
Orthocladiinae	E: cool	--	--	--	--	Uni
Acricotopus	E: cool	4.0	4.0	--	--	Uni
Brillia	E: cool	3.4	--	--	5.5	Uni
Bryophaenocladius	E: cool	4.0	--	--	--	--
Camptocladius	E: cool	--	--	--	--	--
Cardiocladius	E: warm	3.3	--	2.3	--	Uni
Chaetocladius	--	3.5	1.7	--	4.0	Uni
Corynoneura	E: warm	3.7	1.4	--	3.0	Uni
Cricotopus	E: warm	3.6	1.1	--	3.9	Uni
Cricotopus bicinctus	E: warm	4.0	5.0	--	4.0	Uni
Cricotopus nostocicola	E: warm	3.0	--	--	3.9	Uni
Cricotopus trifascia	E: warm	3.7	5.0	--	4.0	Uni
Diplocladius	S: cold	4.0	1.7	--	4.0	Uni
Epoicocladius	E: cool	3.8	1.6	--	4.0	Uni
Eukiefferiella	E: cool	3.3	1.3	2.0	4.6	Uni
Heleniella	S: cold	4.0	--	--	--	Uni
Heterotrissocladius	E: cool	3.5	1.3	--	4.0	uni
Hydrobaenus	E: cool	3.5	1.3	--	2.5	Uni
Krenosmittia	E: warm	4.0	5.0	--	4.0	Uni
Limnophyes	E: warm	4.0	1.9	--	4.0	Uni
Metriocnemus	E: cool	3.7	1.8	--	--	Uni
"*Microcricotopus*"	--	--	--	--	4.3	--
Nanocladius	E: warm	3.7	1.9	--	4.0	Uni
Orthocladius	E: warm	3.3	1.2	2.5	5.1	Multi
Orthocladius obumbratus	E: warm	3.3	1.2	2.5	5.1	Multi
Parakiefferiella	E: warm	3.5	1.4	--	4.0	Uni
Parametriocnemus	E: cool	3.5	--	--	--	Uni
Paraphaenocladius	E: cool	3.7	1.0	--	4.0	uni
Parorthocladius	E: warm	4.0	--	--	--	Uni
Psectrocladius	E: warm	4.0	1.5	--	4.0	Uni
Pseudosmittia	E: cool	4.0	5.0	--	4.0	Uni
Psilometriocnemus	S: cold	--	--	--	--	Uni
Psilometriocnemus triannulatus	S: cold	--	--	--	--	Uni

Species	Temperature Range	Lotic	Lentic	Current	Substrate	Voltinism
Rheocricotopus	E: cool	3.3	5.0	--	4.8	uni
Smittia	E: warm	--	1.9	--	4.0	Uni
Synorthocladius	E: warm	3.7	1.7	--	4.0	Uni
Thienemanniella	E: warm	3.8	1.5	--	4.0	--
Tvetenia	E: warm	3.5	5.0	--	--	Uni
Podonominae	--	--	--	--	4.3	--
Lasiodiamesa	--	--	--	--	4.3	--
Paraboreochlus	--	--	--	--	4.3	--
Paraboreochlus stahli	--	--	--	--	4.3	--
Prodiamesinae	--	--	--	--	4.3	--
Odontomesa	E: cool	3.8	--	1.0	1.0	Uni
Prodiamesa	E: cool	3.7	5.0	--	4.0	Uni
Prodiamesa olivacea	E: cool	3.0	5.0	--	4.0	Uni
Tanypodinae	--	--	--	--	0.5	Uni
Ablebesmyia	E: warm	3.7	1.9	--	4.0	Uni
Alotanypus	--	4.0	--	--	--	Uni
Clinotanypus	--	3.7	1.4	0.5	0.0	Uni
Conchapelopia	E: warm	3.5	1.4	0.5	--	Uni
Coelotanypus	--	--	1.4	--	0.0	Uni
Guttipelopia	--	--	--	--	4.3	--
Labrundinia	--	3.7	1.0	1.0	4.0	Uni
Larsia	--	4.0	1.9	--	4.0	Uni
Macropelopia	E: warm	4.0	4.0	--	--	Uni
Natarsia	E: warm	4.0	--	--	--	Uni
Nilotanypus	E: warm	3.7	5.0	--	3.4	Uni
Paramerina	E: warm	4.0	5.0	--	4.0	Uni
Pentaneura	E: warm	3.5	1.4	--	--	Uni
Procladius	E: warm	3.8	1.1	--	3.0	Uni
Psectrotanypus	E: warm	4.0	--	--	--	Uni
Radotanypus	E: cool	4.0	--	--	--	Uni
Tanypus	E: warm	4.0	1.2	--	4.0	Uni
Telopelopia	E: warm	4.0	--	--	1.0	Uni
Thienemannimyia	E: warm	3.8	4.0	--	--	Uni
Zavrelimyia	S: cold	4.0	4.0	--	--	Uni
Culicidae	E: warm	--	--	--	--	--
Aedes	E: warm	3.7	1.0	0.5	--	Multi
Aedes cinereus	E: warm	3.7	1.0	0.5	--	Multi
Aedes intrudens	E: warm	3.7	1.0	0.5	--	Multi
Aaedes triseriatus	E: warm	3.7	1.0	0.5	--	Multi
Aedes vexans	E: warm	3.7	1.0	0.5	--	Multi
Anopheles	E: warm	3.7	1.4	--	--	--
Anopheles earlei	E: warm	3.7	1.4	--	--	--
Anopheles punctipennis	E: warm	3.7	1.4	--	--	--
Anopheles quadrimaculatus	E: warm	3.7	1.4	--	--	--
Anopheles walkeri	E: warm	3.7	1.4	--	--	--
Culex	E: warm	--	1.4	--	--	Multi
Culex pipiens	E: warm	--	1.4	--	--	Multi
Culex restuans	E: warm	--	1.4	--	--	Multi

Species	Temperature Range	Lotic	Lentic	Current	Substrate	Voltinism
Culex salinarius	E: warm	--	1.4	--	--	Multi
Culex tarsalis	E: warm	--	1.4	--	--	Multi
Culex territans	E: warm	--	1.4	--	--	Multi
Culiseta	E: warm	--	1.2	--	--	--
Culiseta impatiens	E: warm	--	1.2	--	--	--
Culiseta incidens	E: warm	--	1.2	--	--	--
Culiseta inornata	E: warm	--	1.2	--	--	--
Culiseta morsitans	E: warm	--	1.2	--	--	--
Mansonia	E: warm	--	--	--	--	--
Mansonia perturbans	E: warm	--	--	--	--	--
Ochlerotatus	E: warm	3.7	1.0	0.5	--	Multi
Ochlerotatus dorsalis	E: warm	3.7	1.0	0.5	--	Multi
Ochlerotatus excrucians	E: warm	3.7	1.0	0.5	--	Multi
Ochlerotatus increpitus	E: warm	3.7	1.0	0.5	--	Multi
Ochlerotatus nigromaculis	E: warm	3.7	1.0	0.5	--	Multi
Ochlerotatus spencerii	E: warm	3.7	1.0	0.5	--	Multi
Ochlerotatus sticticus	E: warm	3.7	1.0	0.5	--	Multi
Ochlerotatus trivittatus	E: warm	3.7	1.0	0.5	--	Multi
Psorophora	E: warm	--	--	--	--	--
Psorophora ferox	E: warm	--	--	--	--	--
Psorophora signipennis	E: warm	--	--	--	--	--
Dixidae	E: warm	--	--	--	7.0	--
Dixa	E: warm	3.7	--	--	7.0	--
Meringodixa	E: cool	4.0	--	--	7.0	--
Meringodixa chalonensis	E: cool	4.0	--	--	7.0	--
Dolichopodidae	E: warm	--	--	--	--	--
Dolichopus	E: warm	4.0	4.0	--	0.0	--
Dolichopus penicillatus	E: warm	4.0	4.0	--	0.0	--
Liancalus	E: warm	--	--	--	--	--
Liancalus hydrophilus	E: warm	--	--	--	--	--
Parasyntormon	E: warm	--	--	--	--	--
Parsyntormon hinnulus	E: warm	--	--	--	--	--
Empididae	E: warm	--	--	--	--	Uni
Chelifera	E: warm	4.0	3.0	--	--	Uni
Clinocera	E: warm	3.8	4.0	2.0	--	Uni
Hemerodromia	E: warm	4.0	--	--	--	Multi
Metachela	E: warm	--	--	--	--	Uni
Metachela collusor	E: warm	--	--	--	--	Uni
Neoplasta	E: warm	4.0	--	--	--	Uni
Neoplasta paramegorchis	E: warm	4.0	--	--	--	Uni
Oreogeton	S: cold	4.0	--	--	--	Uni
Trichoclinocera	E: warm	4.0	--	--	--	Uni
Wiedemannia	E: cool	4.3	--	--	7.0	Uni
Ephydridae	E: warm	--	--	--	--	--
Allotrichoma	E: warm	--	--	--	--	---
Allotrichoma occidentale	E: warm	--	--	--	--	---
Ephydra	E: warm	4.0	4.0	--	--	---
Ephydra hians	E: warm	4.0	4.0	--	--	---

Species	Temperature Range	Lotic	Lentic	Current	Substrate	Voltinism
Ephydra pectinulata	E: warm	4.0	4.0	--	--	---
Ephydra riparia	E: warm	4.0	4.0	--	--	---
Parydra	E: warm	4.0	4.0	--	--	--
Parydra appendiculata	E: warm	4.0	4.0	--	--	--
Parydra aquila	E: warm	4.0	4.0	--	--	--
Parydra parasocia	E: warm	4.0	4.0	--	--	--
Parydra quadrituberculata	E: warm	4.0	4.0	--	--	--
Scatella	E: warm	4.0	3.0	--	--	--
Muscidae	E: warm	--	--	--	--	--
Limnophora	E: warm	3.7	5.0	--	--	Multi
Lispe	E: warm	--	--	--	--	--
Lispoides	E: warm	--	--	--	--	--
Lispoides aequifrons	E: warm	--	--	--	--	--
Spilogona	E: warm	--	--	--	--	--
Psychodidae	E: warm	--	--	--	0.5	--
Maruina	E: cool	4.0	--	--	0.5	Uni
Pericoma	E: warm	3.7	4.0	2.0	0.5	Uni
Psychoda	E: warm	4.0	1.5	--	1.0	--
Ptychopteridae	E: warm	--	--	--	0.0	--
Ptychoptera	E: warm	4.0	3.0	--	0.0	Uni
Simuliidae	E: warm	--	--	--	7.0	Multi
Prosimulium	S: cold	3.5	--	2.2	7.0	Uni
Prosimulium exigens	S: cold	3.5	--	2.2	7.0	Uni
Simulium	E: warm	3.6	1.9	2.1	7.0	Multi
Simulium bivittatum	E: warm	3.6	1.9	2.1	7.0	Multi
Simulium canadense	E: warm	3.6	1.9	2.1	7.0	Multi
Simulium decorum	E: warm	3.0	1.9	2.0	7.0	Multi
Simulium griseum	E: warm	3.6	1.9	2.1	7.0	Multi
Simulium hunteri	E: warm	3.6	1.9	2.1	7.0	Multi
Simulium meridionale	E: warm	5.0	1.9	2.1	7.0	Multi
Simulium piperi	E: warm	3.6	1.9	2.1	7.0	Multi
Simulium virgatum	E: warm	3.6	1.9	2.1	7.0	Multi
Simulium vittatum	E: warm	3.0	1.9	2.0	7.0	Multi
Stegopterna	E: warm	4.0	--	--	7.0	Multi
Stratiomyidae	E: warm	--	--	--	--	--
Anoplodonta	E: warm	--	--	--	--	--
Anoplodonta nigrirostris	E: warm	--	--	--	--	--
Caloparyphus	E: warm	4.0	4.0	--	--	--
Euparyphus	E: warm	4.0	--	--	--	--
Hedriodiscus	E: warm	--	--	--	--	--
Hedriodiscus truquii	E: warm	--	--	--	--	--
Hedriodiscus vertebratus	E: warm	--	--	--	--	--
Odontomyia	E: warm	4.0	1.2	--	--	--
Odontomyia arcuata	E: warm	4.0	1.2	--	--	--
Odontomyia inaequalis	E: warm	4.0	1.2	--	--	--
Odontomyia pilimana	E: warm	4.0	1.2	--	--	--
Odontomyia pubescens	E: warm	4.0	1.2	--	--	--

Species	Temperature Range	Lotic	Lentic	Current	Substrate	Voltinism
Odontomyia virgo	E: warm	4.0	1.2	--	--	--
Psellidotus	E: warm	--	--	--	--	--
Psellidotus viridis	E: warm	--	--	--	--	--
Stratiomys	E: warm	4.0	4.0	--	0.0	--
Stratiomys bruneri	E: warm	4.0	4.0	--	0.0	--
Stratiomys norma	E: warm	4.0	4.0	--	0.0	--
Stratiomys normula	E: warm	4.0	4.0	--	0.0	--
Syrphidae	E: warm	--	--	--	1.0	--
Tabanidae	E: warm	--	--	--	0.0	--
Chrysops	E: warm	4.0	1.1	--	0.0	Uni
Chrysops aestuans	E: warm	4.0	1.1	--	0.0	Uni
Chrysops carbonarius	E: warm	4.0	1.1	--	0.0	Uni
Chrysops fulvaster	E: warm	4.0	1.1	--	0.0	Uni
Chrysops furcatus	E: warm	4.0	1.1	--	0.0	Uni
Chrysops mitis	E: warm	4.0	1.1	--	0.0	Uni
Chrysops pikei	E: warm	4.0	1.1	--	0.0	Uni
Chrysops sequax	E: warm	4.0	1.1	--	0.0	Uni
Hybomitra	E: warm	3.7	1.2	--	0.0	--
Hybomitra criddlei	E: warm	3.7	1.2	--	0.0	--
Hybomitra epistates	E: warm	3.7	1.2	--	0.0	--
Hybomitra frontalis	E: warm	3.7	1.2	--	0.0	--
Hybomitra lasiophthalma	E: warm	3.7	1.2	--	0.0	--
Hybomitra rhombica	E: warm	3.7	1.2	--	0.0	--
Hybomitra rupestris	E: warm	3.7	1.2	--	0.0	--
Hybomitra tetrica	E: warm	3.7	1.2	--	0.0	--
Silvius	E: warm	4.0	4.0	--	0.0	--
Silvius quadrivittatus	E: warm	4.0	4.0	--	0.0	--
Silvius pollinosus	E: warm	4.0	4.0	--	0.0	--
Tabanus	E: warm	4.0	1.1	--	0.3	Uni
Tabanus atratus	E: warm	4.0	3.0	--	0.0	Uni
Tabanus punctifer	E: warm	4.0	1.1	--	0.3	Uni
Tabanus similis	E: warm	4.0	1.1	--	0.3	Uni
Tabanus stonei	E: warm	4.0	1.1	--	0.3	Uni
Tanyderidae	E: cool	--	--	--	1.7	--
Tipulidae	--	--	--	--	5.5	Uni
Antocha	E: warm	3.9	--	2.5	7.0	Multi
Antocha monticola	E: warm	3.0	--	2.5	7.0	Multi
Arctoconopa	--	--	--	--	5.5	Uni
Arctoconopa pahasapa	--	--	--	--	5.5	Uni
Dicranota	E: warm	3.8	3.0	3.0	--	Uni
Dolichopeza	--	--	--	--	5.5	Uni
Dolichopeza americana	--	--	--	--	5.5	Uni
Dolichopeza dakota	--	--	--	--	5.5	Uni
Dolichopeza dorsalis	--	--	--	--	5.5	Uni
Dolichopeza walleyi	--	--	--	--	5.5	Uni
Erioptera	--	3.8	3.0	--	1.0	Uni
Hexatoma	E: warm	3.9	3.0	3.0	7.0	Uni

Species	Temperature Range	Lotic	Lentic	Current	Substrate	Voltinism
Limonia	E: warm	4.0	4.0	--	3.5	Uni
Ormosia	E: warm	3.7	4.0	--	0.0	Uni
Pedicia	E: cool	3.7	2.5	3.0	--	Uni
Pseudolimnophila	--	4.0	2.5	--	0.0	Uni
Tipula	E: warm	3.7	3.0	--	3.0	Mero

Species	Temperature Range	Lotic	Lentic	Current	Substrate	Voltinism

Literature Cited

Adler, P. H., and D. C. Currie. 2008. Simuliidae. Pp. 825-845 *in* Merritt, R. W., K. W. Cummins, and M. B. Berg (eds.). *An Introduction to the Aquatic Insects of North America*, fourth edition. Kendall/Hunt Publishing, Dubuque, IA.

Adler, P. H., D. C. Currie, and D. M. Wood. 2004. *Black Flies (Diptera: Simuliidae) of North America.* Cornell University Press, Ithaca, NY.

Adler, P. H., and K. C. Kim. 1984. Ecological characterization of two sibling species, IIIL-1 and IS-7, in the *Simulium vittatum* complex (Diptera: Simuliidae). *Canadian Journal of Zoology* 62: 1308-1315.

Alarie, Y. 1991. Description of larvae of 17 Nearctic species of *Hydroporus* Clairville (Coleoptera: Dytiscidae: Hydroporinae) with an analysis of their phylogenetic relationships. *Canadian Entomologist* 123: 627-704.

Alarie, Y., M. C. Michat, A. N. Nilsson, M. Archangelsky, and L. Hendrich. 2009. Larval morphology of *Rhantus* Dejean, 1833 (Coleoptera: Dytiscidae: Colymbetinae): descriptions of 22 species and phylogenetic considerations. *Zootaxa* 2317: 1-102.

Alexander, C. P. 1917. New Nearctic Crane-Flies (Diptera, Tipulidae), Part II. *Canadian Entomologist* 49:22-31, 61-64.

Alexander, C. P. 1920. The crane-flies of New York. Part II. Biology and phylogeny. *Cornell University Agricultural Experiment Station Memoir* 38: 691-1133, plates XII-XCVII.

Alexander, C. P. 1958. Undescribed species of Western Nearctic Tipulidae (Diptera). III. *The Great Basin Naturalist* 18: 31-36.

Allen, R. K. 1955. *Mayflies of Oregon.* M.S. Thesis, University of Utah, Salt Lake City, UT.

Allen, R. K., and G. F. Edmunds, Jr. 1962. A revision of the genus *Ephemerella* (Ephemeroptera: Ephemerellidae) V. The subgenus *Drunella*. *Miscellaneous Publications of the Entomological Society of America* 3: 145-180.

Allen, R. K., and C. M. Murvosh. 1987. Mayflies (Ephemeroptera: Tricorythidae) of the southwestern United States and northern Mexico. *Annals of the Entomological Society of America* 80: 35-40.

Alstad, D. N. 1980. Comparative biology of the common Utah Hydropsychidae (Trichoptera). *American Midland Naturalist* 10: 167-174.

Andersen, T., P. S. Cranston, and J. H. Epler (eds.). 2013. The larvae of Chironomidae (Diptera) of the Holarctic Region – Keys and Diagnoses. *Insect Sytematics and Evolution* (Supplement) 66: 1-571.

Angus, R. B. 2010. *Boreonectes* gen. n., a new genus for the *Stictotarsus griseostriatus* (De Geer) group of sibling species (Coleoptera: Dytiscidae), with additional karyosystematic data on the group. *Comparative Cytogenetics* 4: 123-131.

Arid West Water Quality Research Project (AWWQRP). 2006. *Aquatic Communities of Ephemeral Stream Ecosystems.* Technical Report. Funding provided by EPA Region IX, under Assistance Agreement XP-9992607 directed by Pima County Wastewater Management Department. Prepared by GEI Consultants, Inc., Littleton, CO, and URS Corporation, Albuquerque, NM.

Armitage, B. J. 1991. *Diagnostic Atlas of the North American Caddisfly Adults I. Philopotamidae,* 2nd edition. The Caddis Press, Athens, AL.

Askevold, I. S. 1991. Classification, reconstructed phylogeny and geographic history of the New World members of *Plateumaris* Thomson, 1859 (Coleoptera: Chrysomelidae). *Memoirs of the Canadian Entomological Society* 157: 1-175.

Baker, R. L., and H. F. Clifford. 1980. The nymphs of *Coenagrion interrogatum* and *C. resolutum* (Zygoptera: Coenagrionidae) from the boreal forest of Alberta, Canada. *Canadian Entomologist* 112: 433-436.

Banks, N. 1904. Neuropteroid insects from New Mexico. *Transactions of the American Entomological Society* 30: 97-110.

Banks, N. 1918. New neuropteroid insects. *Bulletin of the Museum of Comparative Zoology at Harvard College* 62: 3-22.

Barbour, M.T., J. Gerritsen, B.D. Snyder, and J.B. Stribling. 1999. *Rapid Bioassessment Protocols for Use in Streams and Wadeable Rivers: Periphyton, Benthic Macroinvertebrates and Fish*, Second Edition. EPA 841-B-99-002. U.S. Environmental Protection Agency, Washington, D.C.

Barman, E. H. 1972. *The Biology and Immature Stages of Dytiscidae (Coleoptera) of Central New York State.* Ph.D. Dissertation, Cornell University, Ithaca, NY.

Barman, E. H., E. D. Holmes, and G. A. Nichols. 1999. Biology of a central Georgia population of *Agabus stagninus* Say (Coleoptera: Dytiscidae) with a description of its mature larva and notes on the larva of *Agabus semivittatus* LeConte. *Georgia Journal of Science* 57: 255.

Barman, E. H., P. Wright, and J. E. Mashke. 2000. Biology of *Agabus disintegratus* (Crotch) (Coleoptera: Dytiscidae) in central Georgia with a description of its mature larva. *Georgia Journal of Science* 58: 203.

Baumann, R. W. 1994. Current research on Plecoptera and request for study specimens. *Perla* 14:27.

Baumann, R. W., A. R. Gaufin, and R. F. Surdick. 1977. The stoneflies (Plecoptera) of the Rocky Mountains. *Memoirs of the American Entomological Society* 31: 1-208.

Baumann, R. W., and B. C. Kondratieff. 2008. A review of the western North American genus *Triznaka* (Plecoptera: Chloroperlidae) with a new species from the Great Basin, U.S.A. *Proceedings of the Entomological Society of Washington* 100: 345-362.

Baumgardner, D. E. 2009. *Tricorythodes minutus* Traver, a new synonym of *Tricorythodes explicatus* Eaton (Ephemeroptera: Leptohyphidae). *Proceedings of the Entomological Society of Washington* 111: 57-67.

Baumgardner, D. E., S. K. Burian, and D. Bass. 2003. Life stage descriptions, taxonomic notes, and new records for the mayfly family Leptohyphidae (Ephemeroptera). *Zootaxa* 332: 1-12.

Beckemeyer, R. 1997. Nebraska and South Dakota Odonata – a compilation of collecting reports related to the July, 1998 Valentine, Nebraska Annual Meeting of the Dragonfly Society of the Americas. *Argia* 10(4): 27-28.

Bednarik, A. F., and G. F. Edmunds, Jr. 1980. Description of larval *Heptagenia* from the Rocky Mountain region (Ephemeroptera: Heptageniidae). *Pan-Pacific Entomologist* 56: 51-62.

Berenbaum, M. R. 2016. Mosquito screens. *American Entomologist* 62: 68-70.

Berg, C. O. 1950. Biology of certain aquatic caterpillars (Pyralidae: *Nymphula* spp.) which feed on *Potamogeton*. *Transactions of the American Microscopical Society* 69: 254-266.

Bergman, E. A., and W. L. Hilsenhoff. 1978. Parthenogenesis in the mayfly genus *Baetis* (Ephemeroptera: Baetidae). *Annals of the Entomological Society of America* 71: 167-168.

Bick, G. H. 1951. The early nymphal stages of *Tramea lacerata* Hagen (Odonata: Libellulidae). *Entomological News* 62: 293-30.

Bick, G. H., J. C. Bick, and L. E. Hornuff. 1977. An annotated list of the Odonata of the Dakotas. *Florida Entomologist* 60: 149-165.

Bick, G. H., and L. E. Hornuff. 1972. Odonata collected in Wyoming, South Dakota, and Nebraska. *Proceedings of the Entomological Society of Washington* 74: 1-8.

Bird. R. D. 1931. The nymph of *Enallagma basidens* Calvert (Odonata: Agrionidae). *Entomological News* 42: 276-277.

Blahnik, R. J., and R. W. Holzenthal. 2006. Revision of the genus *Culoptila* (Trichoptera: Glossosomatidae). *Zootaxa* 1233: 1-52.

Blahnik, R. J., and R. W. Holzenthal. 2014. Review and description of species in the *Oecetis avara* group, with the description of 15 new species (Trichoptera, Leptoceridae). *Zookeys* 376: 1-83.

Bland, R. G. 2008. Semiaquatic Orthoptera. Pp. 295-310 in Merritt, R. W., K. W. Cummins, and M. B. Berg (eds.). *An Introduction to the Aquatic Insects of North America*, fourth edition. Kendall/Hunt Publishing, Dubuque, IA.

Blatchley, W. S. 1926. *Heteroptera or True Bugs of Eastern North America with especial reference to the faunas of Indiana and Florida*. Nature Publishing Co., Indianapolis, IN.

Bleszynski, S. 1970. A revision of the World species of *Chilo* Zincken (Lepidoptera: Pyralidae). *Bulletin of the British Museum (Natural History)* 25: 99-195.

Blickle, R. L. 1963. New species of Hydroptilidae (Trichoptera). *Bulletin of the Brooklyn Entomological Society* 63: 17-22.

Blocksom, K. A., and L. Winters. 2006. *The Evaluation of Methods for Creating Defensible, Repeatable, Objective, and Accurate Tolerance Values for Aquatic Taxa*. EPA 600/R-06/045, U.S. Environmental Protection Agency, Cincinnati, Ohio.

Boesel, M. W. 1973. The genus *Atrichopogon* (Diptera, Ceratopogonidae) in Ohio and neighboring states. *The Ohio Journal of Science* 73: 202-215.

Borkent, A. 1981. The distribution and habitat preferences of the Chaoboridae (Culicimorpha: Diptera) of the Holarctic Region. *Canadian Journal of Zoology* 59: 122-133.

Bowles, D. E. 2006. Spongillaflies (Neuroptera: Sisyridae) of North America with a key to the larvae and adults. *Zootaxa* 1357: 1-19.

Brammer, C. A., J. R. Harkrider, and J. F. MacDonald. 2009. Differentiation of larvae and pupae of aquatic genera of Nearctic Hemerodromiinae (Diptera: Empididae). *Zootaxa* 2069: 59-68.

Brown, H. P. 1976. *Aquatic Dryopoid Beetles (Coleoptera) of the United States*. Water Pollution Control Series 18050 ELD04/72, U. S. Environmental Protection Agency, Cincinnati, OH.

Bryce, S.A., J. M. Omernik, D. A. Pater, M. Ulmer, J. Schaar, J. Freeouf, R. Johnson, P. Kuck, and S. H. Azevedo. 1996. *Ecoregions of North Dakota and South Dakota*, (color poster with map, descriptive text, summary tables, and photographs). U.S. Geological Survey, Reston, VA.

Bunte, K., and S. R. Abt. 2001. *Sampling Surface and Subsurface Particle-Size Distributions in Wadable Gravel- and Cobble-Bed Streams for Analyses in Sediment Transport, Hydraulics, and Streambed Monitoring*. General Technical Report RMRS-GTR-74, U.S. Forest Service, U.S. Department of Agriculture, Fort Collins, CO.

Burger, J. F. 1977. The biosystematics of immature Arizona Tabanidae (Diptera). *Transactions of the American Entomological Society* 103: 145-258.

Burian, S. K. 2001. A revision of the genus *Leptophlebia* Westwood in North America (Ephemeroptera: Leptophlebiidae: Leptophlebiinae). *Bulletin of the Ohio Biological Survey*, new series 13: 1-80.

Burks, B. D. 1953. The mayflies, or Ephemeroptera, of Illinois. *Bulletin of the Illinois Natural History Survey* 26: 1-216.

Byers, C. F. 1930. *A Contribution to the Knowledge of Florida Odonata*. University of Florida Press, Gainesville, FL.

Byers, G. W., and J. K. Gelhaus. 2008. Tipulidae. Pp. 773-800 *in* Merritt, R. W., K. W. Cummins, and M. B. Berg (eds.). *An Introduction to the Aquatic Insects of North America*, fourth edition. Kendall/Hunt Publishing, Dubuque, IA.

Cameron, A. E. 1926. Bionomics of the Tabanidae (Diptera) of the Canadian prairie. *Bulletin of Entomological Research* 17: 1-42.

Cannings, R. A. 1981. The larva of *Sympetrum madidum* (Hagen) (Odonata: Libellulidae). *Pan-Pacific Entomologist* 57: 341-346.

Canton, S. P., R. W. Gensemer, G. D. De Jong, C. F. Wolf, S. M. Pargee, N. E. Paden, S. D. Baker, and C. A. Claytor. 2011 Should there be an aquatic life water quality criterion for conductivity? *Impaired Waters Symposium* 2011: 554-565.

Carter, J. M., D. G. Driscoll, J. E. Williamson, and V. A. Lindquist. 2002. *Atlas of Water Resources in the Black Hills Area, South Dakota.* U.S. Geological Survey Hydrologic Investigations Atlas HA 747. U.S. Geological Survey, Reston, VA.

Chapin, J. W. 1978. *Systematics of Nearctic Micrasema (Trichoptera: Brachycentridae).* Ph.D. Dissertation, Clemson University, Clemson, SC.

Chapman, S. S., S. A. Bryce, J. M. Omernik, D. G. Despain, J. ZumBerge, and M. Conrad. 2004. *Ecoregions of Wyoming* (color poster with map, descriptive text, summary tables, and photographs). U.S. Geological Survey, Reston, VA.

Christiansen, K. A. 1990. Insecta: Collembola. Pp. 965-995 *in* Dindal, D. L. (ed.). *Soil Biology Guide.* John Wiley & Sons, New York, NY.

Christiansen, K. A., and R. J. Snider. 2008. Aquatic Collembola. Pp. 165-179 *in* Merritt, R. W., K. W. Cummins, and M. B. Berg (eds.). *An Introduction to the Aquatic Insects of North America*, fourth edition. Kendall/Hunt Publishing, Dubuque, IA.

Chutter, F. M. 1972. An empirical biotic index of quality of water in South African streams and rivers. *Water Resources* 6: 19-30.

Claassen, P. W. 1923. New species of North American Plecoptera. *Canadian Entomologist* 55: 281-292.

Claassen, P. W. 1931. Plecoptera nymphs of America north of Mexico. *Thomas Say Monographs* 3: 1-199.

Clausen, P. J., and E. F. Cook. 1971. A revision of the Nearctic species of the tribe Parydrini (Diptera: Ephydridae). *Memoirs of the American Entomological Society* 27: 1-150.

Clark, S. M., A. B. Olsen, and M. H. Goodman. 2008. The subfamily Donaciinae in Utah (Insecta: Coleoptera: Chrysomelidae). *Monographs of the Western North American Naturalist* 4: 1-37.

Clarke, R. T., M. T. Furse, J. F. Wright, and D. Moss. 1996. Derivation of a biological quality index for river sites: comparison of the observed with the expected fauna. *Journal of Applied Statistics* 23: 311-332.

Clarke, R. T., J. F. Wright, and M. T. Furse. 2003. RIVPACS models for predicting the expected macroinvertebrate fauna and assessing the ecological quality of rivers. *Ecological Modelling* 160: 219-233.

Clemens, B. 1860. Contributions to the study of Lepidopterology – No. 5. *Proceedings of the Academy of Natural Sciences of Philadelphia* 12: 218-219.

Cook, E. F. 1956. The Nearctic Chaoborinae (Diptera: Culicidae). *Bulletin of the Minnesota Agricultural Experiment Station* 218: 1-102.

Cook, E. F. 1981. Chaoboridae. Pp. 335-339 *in* McAlpine, J. F., B. V. Peterson, G. F. Shewell, H. J. Teskey, J. R. Vockeroth, and D. M. Wood (coords.). *Manual of Nearctic Diptera*, Volume 1. Monograph 27, Agriculture Canada, Ottawa, Canada.

Cook, P. P., Jr., and A. L. Antonelli. 1969. The nymph of *Amphiagrion abbreviatum*. *Annals of the Entomological Society of America* 62: 264-266.

Cooper, S. D., and N. H. Troelstrup, Jr. 2015. Influence of Pactola Reservoir on the aquatic insect assemblage of Rapid Creek. *Proceedings of the South Dakota Academy of Science* 94: 371.

Corkum, L. D. 1978. The nymphal development of *Paraleptophlebia adoptiva* (McDunnough) and *Paraleptophlebia mollis* (Eaton) (Ephemeroptera: Leptophlebiidae) and the possible influence of temperature. *Canadian Journal of Zoology* 56: 1842-1846.

Corpus, L. D. 1986. Immature stages of *Liancalus similus* (Diptera: Dolichopodidae). *Journal of the Kansas Entomological Society* 59: 635-640.

Coscarón, S., D. R. Miranda Esquivel, J. K. Moulton, C. L. Coscarón-Arias, and S. Ibañez Bernal. 2004. *Simulium (Hearlea)* Vargas, Martínez Palacios, & Díaz Nájera (Diptera: Simuliidae): Taxonomic revision and cladistic analysis. *Zootaxa* 396: 1-52.

Courtney, G. W., and R. W. Merritt. 2008. Aquatic Diptera. Part One. Larvae of Aquatic Diptera. Pp. 687-771 *in* Merritt, R. W., K. W. Cummins, and M. B. Berg (eds.). *An Introduction to the Aquatic Insects of North America*, fourth edition. Kendall/Hunt Publishing, Dubuque, IA.

Culver, D. C., H. H. Hobbs, III, M. C. Christman, and L. L. Master. 1999. Distribution map of caves and cave animals in the United States. *Journal of Cave and Karst Studies* 61: 139-140.

Daggy, R. H. 1945. New species and previously undescribed naiads of some Minnesota mayflies (Ephemeroptera). *Annals of the Entomological Society of America* 38: 373-396.

Darsie, R. F., and R. A. Ward (eds.). *Identification and Geographical Distribution of the Mosquitoes of North America, North of Mexico*. University of Florida Press, Gainesville, FL.

De Jong, G. D., and S. P. Canton. 2012. Report of the mayfly family Ameletidae (Ephemeroptera) from South Dakota. *Entomological News* 122: 22-26.

De Jong, G. D., and S. P. Canton. 2012. Presence of long-lived invertebrate taxa and hydrologic permanence. *Journal of Freshwater Ecology* 28: 277-282.

De Jong, G. D., S. P. Canton, and J. W. Chadwick. 2005. Macroinvertebrates occurring in Sunbeam Hot Springs, an absolutely hot spring in Idaho, USA. *Journal of Freshwater Ecology* 20: 611-613.

De Jong, G. D., S. P. Canton, J. S. Lynch, and M. Murphy. 2016. Aquatic invertebrate and vertebrate communities of ephemeral stream ecosystems in the arid southwestern United States. *Southwestern Naturalist* 60: 336-347.

De Jong, G. D., E. R. Smith, and D. J. Conklin, Jr. 2013. Riffle beetle communities of perennial and intermittent streams in northern Nevada, with a new state record for *Optioservus castaneipennis* (Fall) (Coleoptera: Elmidae). *The Coleopterists Bulletin* 67: 293-301.

Denning, D. G., and R. L. Blickle. 1972. A review of the genus *Ochrotrichia* (Trichoptera: Hydroptilidae). *Annals of the Entomological Society of America* 65: 141-151.

De Walt, R. E., D. W. Webb, and T. N. Kompare. 2001. The *Perlesta placida* (Hagen) complex (Plecoptera: Perlidae) in Illinois, new state records, distributions, and an identification key. *Proceedings of the Entomological Society of Washington* 103: 207-216.

Dodds, G. S. 1923. Mayflies from Colorado: descriptions of certain species and notes on others. *Transactions of the American Entomological Society* 69: 93-116.

Dosdall, L., and D. M. Lehmkuhl. 1979. Stoneflies (Plecoptera) of Saskatchewan. *Quaestiones Entomologicae* 15: 3-116.

Drake, C. J., and H. M. Harris. 1934. The Gerrinae of the Western Hemisphere (Hemiptera). *Annals of the Carnegie Museum* 23: 179-241.

Drake, C. J., and R. F. Hussey. 1955. Concerning the genus *Microvelia* Westwood, with descriptions of two new species and a check-list of the American forms (Hemiptera: Veliidae). *Florida Entomologist* 38: 95-115.

Driscoll, D. G., J. M. Carter, J. E. Williamson, and L. D. Putnam. 2002. *Hydrology of the Black Hills Area, South Dakota*. U.S. Geological Survey Water-Resources Investigations Report 02-4094. U.S. Geological Survey, Reston, VA.

Driscoll, D. G., J. E. O'Connor, and T. M. Harden. 2012. Results of paleoflood investigations for Spring, Rapid, Boxelder, and Elk creeks, Black Hills, western South Dakota. *Proceedings of the South Dakota Academy of Sciences* 91: 49-67.

Durfee, R. S., S. K. Jasper, and B. C. Kondratieff. 2005. Colorado Haliplidae (Coleoptera): Biogeography and identification. *Journal of the Kansas Entomological Society* 78: 41-70.

Durfee, R., and B. C. Kondratieff. 1993. Description of adults of *Baetis magnus* (Ephemeroptera: Baetidae). *Entomological News* 104: 227-232.

Easton, E. R. 1983. The horse flies and deer flies of South Dakota: New state records and an annotated checklist (Diptera: Tabanidae). *Entomological News* 94: 196-200.

Eaton, A. E. 1883-88. A revisional monograph of recent Ephemeridae or mayflies. *Transactions of the Linnaean Society of London*, 2nd series-zoology 3: 1-352.

Edmunds, G. F., Jr., and R. K. Allen. 1964. The Rocky Mountain species of *Epeorus (Iron)* Eaton (Ephemeroptera: Heptageniidae). *Journal of the Kansas Entomological Society* 37: 275-288.

Epler, J. H. 2010. *The Water Beetles of Florida: an identification manual for the families Chrysomelidae, Curculionidae, Dryopidae, Dytiscidae, Elmidae, Gyrinidae, Haliplidae, Helophoridae, Hydraenidae, Hydrochidae, Hydrophilidae, Noteridae, Psephenidae, Ptilodactylidae, and Scirtidae.* Report prepared for Florida Department of Environmental Protection, Tallahassee, Florida.

Etnier, D. A., C. R. Parker, J. T. Baxter, Jr., and T. M. Long. 2010. A review of the genus *Agapetus* Curtis (Trichoptera: Glossosomatidae) in eastern and central North America, with description of 12 new species. *Insecta Mundi* 149: 1-77.

Feldman, D. 2006. *Interpretation of New Macroinvertebrate Models by WQPB.* Draft Report. Montana Department of Environmental Quality, Helena, MT.

Ferrington, L. C., Jr., M. B. Berg, and W. P. Coffman. 2008. Chironomidae. Pp. 847-989 *in* Merritt, R. W., K. W. Cummins, and M. B. Berg (eds.). *An Introduction to the Aquatic Insects of North America*, fourth edition. Kendall/Hunt Publishing, Dubuque, IA.

Finn, D. S., and N. L. Poff. 2005. Variability and convergence in benthic communities along the longitudinal gradients of four physically similar Rocky Mountain streams. *Freshwater Biology* 50: 243-261.

Flint, O. S., Jr. 1960. Taxonomy and biology of Nearctic limnephilid larvae (Trichoptera), with special reference to species in eastern United States. *Entomological Americana* 40: 1-117.

Flint, O. S., Jr., E. D. Evans, and H. H. Neunzig. 2008. Megaloptera and Aquatic Neuroptera. Pp. 425-437 *in* Merritt, R. W., K. W. Cummins, and M. B. Berg (eds.). *An Introduction to the Aquatic Insects of North America*, fourth edition. Kendall/Hunt Publishing, Dubuque, IA.

Floyd, M. A. 1995. Larvae of the caddisfly genus *Oecetis* (Trichoptera: Leptoceridae) in North America. *Bulletin of the Ohio Biological Survey* 10: 1-85.

Forbes, W. T. M. 1910. The aquatic caterpillars of Lake Quinsigamond. *Psyche* 17: 219-227.

Garrison, R. W. 1981. Description of the larva of *Ischnura gemina* with a key and new characters for the separation of sympatric *Ischnura* larvae. *Annals of the Entomological Society of America* 74: 525-530.

Garrison, R. W. 1984. Revision of the genus *Enallagma* in the United States west of the Rocky Mountains, and identification of certain larvae by discriminant analysis. *University of California Publications in Entomology* 105: 1-129.

Garrison, R. W. 1994. A synopsis of the genus *Argia* of the United States with keys and descriptions of new species, *Argia sabino, A. leonorae,* and *A. pima* (Odonata: Coenagrionidae). *Transactions of the American Entomological Society* 120: 287-368.

Geraci, C. J., X. Zhou, J. C. Morse, and K. M. Kjer. 2010. Defining the genus *Hydropsyche* (Trichoptera: Hydropsychidae) based on DNA and morphological evidence. *Journal of the North American Benthological Society* 29: 918-933.

Gerhardt, R. W. 1966. South Dakota mosquito species. *Mosquito News* 26: 37-38.

Giersch, J., and R. Wisseman. 2013. Annotated list of *Rhyacophila* of North America with larval key and descriptions. Southwest Association of Freshwater Invertebrate Taxonomists Workshop 2012: the Trichoptera of North America.

Glover, J. B. 1996. Larvae of the caddisfly genera *Triaenodes* and *Ylodes* (Trichoptera: Leptoceridae) in North America. *Bulletin of the Ohio Biological Survey* 11: 1-89.

Goodwyn, P. J. P. 2006. Taxonomic revision of the subfamily Lethocerinae Lauck & Menke (Heteroptera: Belostomatidae). *Stuttgarter Beiträge zur Naturkunde, Ser. A* 695: 1-71.

Gordon, A. E. 1974. A synopsis and phylogenetic outline of the Nearctic members of *Cheumatopsyche*. *Proceedings of the Academy of Natural Sciences of Philadelphia* 126: 117-160.

Gould, G. E. 1931. The *Rhagovelia* of the Western Hemisphere, with notes on world distribution (Hemiptera, Veliidae). *University of Kansas Science Bulletin* 20: 5-61.

Grafe, C.S. (ed.). 2002. *Idaho Small Stream Ecological Assessment Framework: An Integrated Approach.* Idaho Department of Environmental Quality, Boise, ID.

Greene, E. A. 1993. *Hydraulic Properties of the Madison Aquifer System in the Western Rapid City Area, South Dakota.* U.S. Geological Survey Water-Resources Investigations Report 93-4008. U.S. Geological Survey, Reston, VA.

Greene, E. A., A. M. Shapiro, and J. M. Carter. 1999. *Hydrogeologic Characterization of the Minnelusa and Madison Aquifers near Spearfish, South Dakota.* U.S. Geological Survey Water-Resources Investigations Report 98-4156. U.S. Geological Survey, Reston, VA.

Grote, A. R. 1878. On the moths collected by Prof. Snow in New Mexico. *Transactions of the Annual Meetings of the Kansas Academy of Science* 8: 45-57.

Grote, A. R. 1881. New Pyralidae. *Papilio* 1: 15-19.

Grotheer, S. A., J. M. Chudd, S.-H. Shieh, N. J. Voelz, and J. V. Ward. 1994. *Developing a Biotic Index for Colorado Stream Quality.* Colorado Water Resources Research Institute, Fort Collins, CO.

Grubbs, S. A., R. W. Baumann, R. E. DeWalt, and T. Tweddale. 2014. A review of the Nearctic genus *Prostoia* (Ricker) (Plecoptera, Nemouridae), with the description of a new species and a surprising range extension for *P. hallasi* Kondratieff & Kirchner. *ZooKeys* 401: 11-30.

Guenther, J. L., and W. P. McCafferty. 2008. Mayflies (Ephemeroptera) of the Great Plains. IV: South Dakota. *Transactions of the American Entomological Society* 134:147-171.

Gundersen, R. W. 1977. *Taxonomic Revision of the Genus Enochrus subgenera Enochrus and Methydrus for the Nearctic Region (Hydrophilidae: Coleoptera).* Ph.D. Dissertation, University of Minnesota, St. Paul, MN.

Haddock, J. D. 1977. The biosystematics of the caddis fly genus *Nectopsyche* in North America with emphasis on the aquatic stages. *American Midland Naturalist* 98: 382-421.

Hansen, D. C., and E. F. Cook. 1976. The systematics and morphology of the Nearctic species of *Diamesa* Meigen, 1835 (Diptera: Chironomidae). *Memoirs of the American Entomological Society* 30: 1-203.

Hanson, J. F. 1961. Studies on the Plecoptera of North America: VIII. The identity of the species of *Paracapnia. Bulletin of the Brooklyn Entomological Society* 56: 25-30.

Harden, P. H., and C. E. Mickel. 1952. The stoneflies of Minnesota (Plecoptera). *University of Minnesota Technical Bulletin* 201: 1-84.

Hargett, E. G. 2011. *The Wyoming Stream Integrity Index (WSII) – Multimetric Indices for Assessment of Wadeable Streams and Large Rivers in Wyoming.* Document #11-0787, Wyoming Department of Environmental Quality, Cheyenne, WY.

Hargett, E. G., J. R. ZumBerge, C. P. Hawkins, and J. R. Olson. 2007. Development of a RIVPACS-type predictive model for bioassessment of wadeable streams in Wyoming. *Ecological Indicators* 7: 807-826.

Harris, H. N. 1937. Contributions to the South Dakota list of Hemiptera. *Iowa State College Journal of Science* 11: 169-176.

Harris, H. N. 1943. Additions to the South Dakota list of Hemiptera. *Journal of the Kansas Entomological Society* 16: 150-153.

Harris, S. C., P. K. Lago, and R. B. Carlson. 1980. Preliminary survey of the Trichoptera of North Dakota. *Proceedings of the Entomological Society of Washington* 82: 39-43.

Hart, C. A. 1895. On the entomology of the Illinois River and adjacent waters. *Bulletin of the Illinois State Laboratory of Natural History* 4: 149-273.

Hawkins, C. P. 2006. Quantifying biological integrity by taxonomic completeness: evaluation of a potential indicator for use in regional and global-scale assessments. *Ecological Applications* 16: 1277-1294.

Henry, B. C., Jr. 1993. A revision of *Neochoroterpes* (Ephemeroptera: Leptophlebiiadae) new status. *Transactions of the American Entomological Society* 119: 317-333.

Henry, T. J., and R. C. Froeschner. 1988. *Catalog of the Heteroptera, or true bugs, of Canada and the continental United States.* E. J. Brill, New York, NY.

Heppner, J. B. 1976. Synopsis of the genus *Parargyractis* (Lepidoptera: Pyralidae: Nymphulinae) in Flrioda. *Florida Entomologist* 59: 5-19.

Hilsenhoff, W. L. 1993. Dytiscidae and Noteridae of Wisconsin (Coleoptera). II. Distribution, habitat, life cycle, and identification of species of Dytiscinae. *The Great Lakes Entomologist* 26: 35-53.

Hilsenhoff, W. L., and W. U. Brigham. 1978. Crawling water beetles of Wisconsin (Coleoptera: Haliplidae). *The Great Lakes Entomologist* 11: 11-22.

Hodges, R. W., T. Dominick, D. R. Davis, D. C. Ferguson, J. G. Franclemont, E. G. Munroe, and J. A. Powell. 1983. *Check List of the Lepidoptera of America North of Mexico*. E. W. Classey, Ltd. And the Wedge Entomological Research Foundation, London, UK.

Huckett, H. C. 1932. The North American species of the genus *Limnophora* Robineau-Desvoidy, with descriptions of new species (Muscidae, Diptera). *Journal of the New York Entomological Society* 40: 25-76, 105-158, 279-339.

Huckett, H. C. 1965. The Muscidae of northern Canada, Alaska, and Greenland (Diptera). *Memoirs of the Entomological Society of Canada* 42: 1-369.

Huggins, D. G. 1978. Redescription of the nymph of *Enallagma basidens* Calvert (Odonata: Coenagrionidae). *Journal of the Kansas Entomological Society* 51: 222-227.

Huggins, D. G., and M. Moffett. 1988. *Proposed Biotic and Habitat Indices for Use in Kansas Streams*. Report No. 35, Kansas Biological Survey, Lawrence, KS.

Hungerford, H. B. 1948. The Corixidae of the Western Hemisphere (Hemiptera). *University of Kansas Science Bulletin* 32: 1-827.

Huntsman, B. O., R. W. Baumann, and B. C. Kondratieff. 1999. Stoneflies (Plecoptera) of the Black Hills of South Dakota and Wyoming, USA: Distribution and zoogeographic affinities. *Great Basin Naturalist* 59:1-17.

Jacobus, L. M., and W. P. McCafferty. 2003. Revisionary contributions to North American *Ephemerella* and *Serratella* (Ephemeroptera: Ephemerellidae). *Journal of the New York Entomological Society* 111: 174-193.

Jacobus, L. M., and N. A. Wiersema. 2014. The genera *Anafroptilum* Kluge, 2011 and *Neocloeon* Traver, 1932, reinstated status, in North America, with remarks about the global composition of *Centroptilum* Eaton, 1869 (Ephemeroptera: Baetidae). *Zootaxa* 3814: 385-391.

James, D. A. 2013. *A Survey for the Aquatic Invasive Species New Zealand Mudsnail Potamopyrgus antipodarum in the Black Hills of South Dakota*. Report prepared for the Aquatic Nuisance Species Coordinator, U.S. Fish and Wildlife Service, Lakewood, CO.

James, M. T. 1936. The genus *Odontomyia* in America north of Mexico. *Annals of the Entomological Society of America* 29: 517-550.

James, M. T., and G. C. Steyskal. 1952. A review of the Nearctic Stratiomyini (Diptera, Stratiomyidae). *Annals of the Entomological Society of America* 45: 385-412.

Jensen, S. L. 1966. *The Mayflies of Idaho (Ephemeroptera)*. University of Utah, Salt Lake City, UT.

Jessup, B., J. Stribling, and C. Hawkins. 2005. *Biological Indicators of Stream Condition in Montana Using Macroinvertebrates*. Draft Report. Tetra Tech, Inc., Rolling Springs, MD.

Johnson, C. 1974. Taxonomic keys and distributional patterns for Nearctic species of *Calopteryx* damselflies. *Florida Entomologist* 57: 231-248.

Johnson, P. J., K. D. Rousch, and X. Lin. 1997. A South Dakota record for *Chauliodes rastricornis* (Megaloptera: Corydalidae). *Entomological News* 108: 57-59.

Kennedy, C. H. 1916. Notes on the life history and ecology of the dragonflies (Odonata) of Washington and Oregon. *Proceedings of the United States National Museum* 49: 259-345.

Kightlinger, L. 2012. A decade of West Nile virus in South Dakota: human epidemiology 2002 – 2011. http://doh.sd.gov/WestNile/PDF/10yrWNVinSD.pdf.

Kilgore, J. I., and R. K. Allen. 1973. Mayflies of the Southwest: New species, descriptions, and records (Ephemeroptera). *Annals of the Entomological Society of America* 66: 321-332.

King, K. W. 1993. *A Laboratory Manual and Illustrated Guide to Orders of Common Wyoming Stream Macroinvertebrates*. Wyoming Department of Environmental Quality, Cheyenne, WY.

Kirk, V. M., and E. U. Balsbaugh. 1975. *A Preliminary Checklist of the Beetles of South Dakota*. Bulletin #142, South Dakota Agricultural Experiment Station, Brookings, SD.

Kluge, N. J. 2011. Non-African representatives of the plesiomorphion Protopatellata (Ephemeroptera: Baetidae). *Russian Entomological Journal* 20: 361-376.

Kondratieff, B. C. (coord.). 2000a. Dragonflies and damselflies (Odonata) of the United States. Northern Prairie Wildlife Research Center Online. http://www.npwrc.usgs.gov/resource/distr/insects/dfly/index.htm. Jamestown, ND.

Kondratieff, B. C. (coord.). 2000b. Mayflies of the United States. Northern Prairie Wildlife Research Center Online. http://www.npwrc.usgs.gov/resource/distr/insects/mfly/index.htm. Jamestown, ND.

Kondratieff, B. C., and R. W. Baumann (coords.). 2000. Stoneflies of the United States. Northern Prairie Wildlife Research Center Online. http://www.npwrc.usgs.gov/resource/distr/insects/sfly/index.htm. Jamestown, ND.

Lange, W. H., Jr. 1956. A generic revision of the aquatic moths of North America: (Lepiodptera: Pyralidae, Nymphulinae). *Wasmann Journal of Biology* 14: 59-144.

Lanzaro, G. C., and B. F. Eldridge. 1992. A classical and population genetic description of two new sibling species of *Aedes (Ochlerotatus) increpitus* Dyar. *Mosquito Systematics* 24: 85-101.

Larson, D. J., Y. Alarie, and R. E. Roughley. 2000. *Predaceous Diving Beetles (Coleoptera: Dytiscidae) of the Nearctic Region, With Emphasis on the Fauna of Canada and Alaska*. NRC Research Press, Ottawa, Canada.

Lauck, D. R. 1964. A monograph of the genus *Belostoma*. Part III. *B. triangulum, bergi, minor, bifoveolatum,* and *flumineum* groups. *Bulletin of the Chicago Academy of Sciences* 11: 102-154.

LeConte, J. L. 1855. Synopsis of the Hydrophilidae of the United States. *Proceedings of the Academy of Natural Sciences of Philadelphia* 7: 356-375.

Locklin, J. L., T. L. Arsuffi, and D. E. Bowles. 2006. Life history of *Sialis* (Megaloptera: Sialidae) in a lentic and lotic ecosystem in central Texas. *American Midland Naturalist* 155: 50-62.

MacArthur, R. H. 1972. *Geographical Ecology: Patterns in the Distribution of Species*. Harper & Row Publishers, New York, NY.

MacDonald, J. F. 1989. Review of Nearctic *Metachela* Coquillett, with description of a new species (Diptera: Empididae; Hemerodromiinae). *Proceedings of the Entomological Society of Washington* 91: 513-522.

MacDonald, J. F., and W. J. Turner. 1993. Review of the genus *Neoplasta* Coquillett of America north of Mexico (Diptera; Empididae; Hemerodromiinae). *Proceedings of the Entomological Society of Washington* 95: 351-376.

MacKay, R. J. and T. F. Waters. 1986. Effects of small impoundments on hydropsychid caddisfly production in Valley Creek, Minnesota. *Ecology* 67: 1680-1686.

Majka, C. G., B. J. van Vondel, and R. Webster. 2009. The Haliplidae of Atlantic Canada: new records, distribution, and faunal composition. *ZooKeys* 22: 249-266.

Mangum, R. A., and R. N. Winget. 1991. Environmental profile of *Drunella (Eatonella) doddsi* (Needham) (Ephemeroptera: Ephemerellidae). *Journal of Freshwater Ecology* 6: 11-22.

Mathis, W. N., and T. Zatwarnicki. 2012. Revision of New World species of the shorefly subgenus *Allotrichoma* Becker of the genus *Allotrichoma* with description of the subgenus *Neotrichoma* (Diptera, Ephydridae, Hecamedini). *Zookeys* 161: 1-101.

Matta, J. F. 1982. The bionomics of two species of *Hydrochara* (Coleoptera: Hydrophilidae) with descriptions of their larvae. *Proceedings of the Entomological Society of Washington* 84: 461-467.

Matta, J. F., and D. E. Peterson. 1985. The larvae of six Nearctic *Hydroporus* of the subgenus *Neoporus* (Coleoptera: Dytiscidae). *Proceedings of the Academy of Natural Sciences of Philadelphia* 137: 53-60.

Marx, E. J. F. 1957. A review of the subgenus *Donacia* in the Western Hemisphere (Coleoptera: Donaciidae). *Bulletin of the American Museum of Natural History* 112: 1-278.

Maschwitz, D. E., and E. F. Cook. 2000. Revision of the Nearctic species of the genus *Polypedilum* Kieffer (Diptera: Chironomidae) in the subgenera *P. (Polypedilum)* Kieffer and *P. (Uresipedilum)* Oyewo and Sæther. *Bulletin of the Ohio Biological Survey* 12(3): 1-135.

Matta, J. F. 1983. Description of the larva of *Uvarus granarius* (Aubé) (Coleoptera: Dytiscidae) with a key to the Nearctic Hydroporine larvae. *The Coleopterists Bulletin* 37: 203-207.

McCafferty, W. P. 1975. The burrowing mayflies of the United States (Ephemeroptera: Ephemeroidea). *Transactions of the American Entomological Society* 101: 447-504.

McCafferty, W. P. 1990. Biogeographic affinities of the Ephemeroptera of the Black Hills, South Dakota. *Entomological News* 101:193-199.

McCafferty, W. P. 1994. Distributional and classificatory supplement to the burrowing mayflies (Ephemeroptera: Ephemeroidea) of the United States. *Entomological News* 105: 1-13.

McCafferty, W. P. 2004. Contribution to the systematics of *Leucrocuta*, *Nixe*, and related genera (Ephemeroptera: Heptageniidae). *Transactions of the American Entomological Society* 130: 1-9.

McCafferty, W. P. 2006. New synonym and western range extension for *Heterocloeon anoka* (Daggy) (Ephemeroptera: Baetidae). *Proceedings of the Entomological Society of Washington* 108: 738.

McCafferty, W. P., and B. C. Kondratieff. 1999. Additions to the South Dakota Ephemeroptera. *Entomological News* 110:190-191.

McCafferty, W. P., R. D. Waltz, J. M. Webb, and L. M. Jacobus. 2005. Revision of *Heterocloeon* McDunnough (Ephemeroptera: Baetidae). *Journal of Insect Science* 5: 35-45.

McCafferty, W. P., D. R. Lenat, L. M. Jacobus, and M. D. Meyer. 2010. The mayflies (Ephemeroptera) of the southeastern United States. *Transactions of the American Entomological Society* 136: 221-233.

McDunnough, J. 1921. Two new Canadian may-flies (Ephemeroptera). *Canadian Entomologist* 53: 117-120.

McDunnough, J. 1923. New Canadian Ephemeridae with notes. *Canadian Entomologist* 55: 39-50.

McDunnough, J. 1925. New Canadian Ephemeridae with notes, III. *Canadian Entomologist* 57: 168-192.

McDunnough, J. 1926. New Canadian Ephemeridae with notes. IV. *Canadian Entomologist* 58: 296-303.

McDunnough, J. 1926. Notes on North American Ephemeroptera with descriptions of new species. *Canadian Entomologist* 58: 184-196.

McFadden, M. W. 1967. Soldier fly larvae in America north of Mexico. *Proceedings of the United States National Museum* 121: 1-72.

McGillivray, A. D. 1965. Aquatic Chrysomelidae and a table of the families of Coleopterous larvae. *New York State Museum Bulletin* 68: 288-327.

Meier, C. A., M. A. Ivie, J. B. Johnson, and D. R. Maddison. 2010. A new northern-most record for the family Hydroscaphidae (Coleoptera: Myxophaga), with description of a new Nearctic species. *The Coleopterists Bulletin* 64: 289-302.

Menke, A. S. 1963. A review of the genus *Lethocerus* in North and Central America, including the West Indies (Hemiptera: Belostomatidae). *Annals of the Entomological Society of America* 56:261-267.

Merricks, T. C., D. S. Cherry, C. E. Zipper, R. J. Currie, and T. W. Valenti. 2007. Coal-mine hollow fill and settling pond influences on headwater streams in southern West Virginia, USA. *Environmental Monitoring and Assessment* 129: 359-378.

Merritt, R. W., K. W. Cummins, and M. B. Berg (eds.). 2008. *An Introduction to the Aquatic Insects of North America*, fourth edition. Kenall/Hunt Publishing, Dubuque, IA.

Milne, L. J. 1934. *Studies in North American Trichoptera, 1.* Privately published, Cambridge, MA.

Molinar, D. R., and R. J. Lavigne. 1979. The Odonata of Wyoming (dragonflies and damselflies). *Science Monograph* 37: 1-142.

Moore, J. C., P. Saunders, G. Selby, H. Horton, M. K. Chelius, A. Chapman, and R. D. Horrocks. 2005. The distribution and life history of *Arropalites caecus* (Tullberg): Order: Collembola, in Wind Cave, South Dakota, USA. *Journal of Cave and Karst Studies* 67: 110-119.

Moore, J. C., J. Clarke, M. Heimbrook, and S. Mackessy. 1996. *Survey of Biota and Trophic Interactions within Wind Cave and Jewel Cave, South Dakota – Final Report.* Wind Cave National Park, National Park Service, Hot Springs, SD.

Morse, J. C. 1972. The genus *Nyctiophylax* in North America. *Journal of the Kansas Entomological Society* 45: 172-181.

Morse, J. C. 1975. A phylogeny and revision of the caddisfly genus *Ceraclea* (Trichoptera, Leptoceridae). *Transactions of the American Entomological Institute* 11(2):1-97.

Morse, J. C., and R. W. Holzenthal. 2008. Trichoptera Genera. Pp. 481-552 *in* Merritt, R. W., K. W. Cummins, and M. B. Berg (eds.). *An Introduction to the Aquatic Insects of North America*, fourth edition. Kendall/Hunt Publishing, Dubuque, IA.

Moulton, J. K. 1998. Reexamination of *Simulium (Psilopelmia)* Enderlein (Diptera: Simuliidae) of America north of Mexico. *Proceedings of the Entomological Society of Washington* 100: 50-71.

Moulton, S. R., II, S. c. Harris, and J. P. Slusark. 1999. The microcaddisfly genus *Ithytrichia* Eaton (Trichoptera: Hydroptilidae) in North America. *Proceedings of the Entomological Society of Washington* 101: 233-241.

Munroe, E. G. 1972. Pyraloidea, Pyralidae comprising subfamilies Scopariinae, Nymphulinae. Pp. 1-134 in Dominick, R. B. *The Moths of America North of Mexico*. Fascicle 13.1A, E. W. Classey and R. B. D. Publications, Inc., London, UK.

Murányi, D., M. Gamboa, and K. M. Orci. 2014. *Zwicknia* gen. n., a new genus for the *Capnia bifrons* species group, with descriptions of three new species based on morphology, drumming signals and molecular genetics, and a synopsis of the Wester Palaearctic and Nearctic genera of Capniidae (Plecoptera). *Zootaxa* 3812: 1-82.

Musgrave, P. N. 1935. A synopsis of the genus *Helichus* Erichson in the United States and Canada, with description of a new species (Coleoptera: Dryopodiae). *Proceedings of the Entomological Society of Washington* 37: 137-145.

Musser, R. J. 1962. Dragonfly nymphs of Utah (Odonata: Anisoptera). *University of Utah Biological Series* 12: 1-66.

Nebeker, A. V., and A. R. Gaufin. 1965. The *Capnia columbiana* complex of North America (Capniidae: Plecoptera). *Transactions of the American Entomological Society* 91: 467-487.

Needham, J. G. 1903. Aquatic insects of New York State. Part 3. Life histories of Odonata, suborder Zygoptera. *New York State Museum Bulletin* 68: 218-279.

Needham, J. G. 1904. New dragon-fly nymphs in the United States National Museum. *Proceedings of the United States National Museum* 27: 685-720.

Needham, J. G., and R. O. Christiansen. 1927. Economic insects in some streams of northern Utah. *Utah Agricultural Experiment Station Bulletin* 201: 1-36.

Needham, J. G., M. J. Westfall, Jr., and M. L. May. 2000. *Dragonflies of North America*, revised edition. Scientific Publishers, Gainesville, FL.

Nelso, C. R., and R. W. Baumann. 1989. Systematics and distribution of the winter stonefly genus *Capnia* (Plecoptera: Capniidae) in North America. *The Great Basin Naturalist* 49: 289-363.

Newman, E. 1838. Entomological Notes. *The Entomological Magazine* 5: 483-500.

Newman, R. 1999. *A Biological Assessment of Four Northern Black Hills Streams and the Reproductive Success of Longnose Dace (Rhinichthys cataractae)*. MS Thesis, South Dakota State University, Brookings, SD.

Newman, R. L., C. R. Berry, and W. Duffy. 1999. A biological assessment of four northern Black Hills streams. *Proceedings of the South Dakota Academy of Sciences* 78: 185-197.

Nimmo, A. P. 1971. The adult Rhyacophilidae and Limnephilidae of Alberta and eastern British Columbia and their post-glacial origin. *Quaestiones Entomologicae* 7: 3-234.

Nimmo, A. P. 1986. The adult Polycentropodidae of Canada and adjacent United States. *Quaestiones Entomologicae* 22: 143-252.

Nimmo, A. P. 1987. The adult Arctopsychidae and Hydropsychidae (Trichoptera) of Canada and adjacent United States. *Quaestiones Entomologicae* 23: 1-189.

Nimmo, A. P. 1991. Seven new species of *Limnephilus* from western North America with description of female of *L. pallens* (Banks) (Trichoptera, Limnephilidae, Limnephilinae, Limnephilini). *Proceedings of the Entomological Society of Washington* 93: 499-508.

Novelo-Gutiérrez, R. 2005. Five new *Erpetogomphus* Hagen in Selys larvae from Mexico, with a key to the known species (Anisoptera: Gomphidae). *Odonatologica* 34: 243-257.

Oliver, D. R., and M. E. Roussel. 1983. *The Genera of Larval Midges of Canada (Diptera: Chironomidae).* The Insects and Arachnids of Canada, Part 11. Agriculture Canada, Ottawa, Ontario, Canada.

Omernik, J. M. 1987. Ecoregions of the conterminous United States. Map (scale 1:7,500,00). *Annals of the Association of American Geographers* 77: 118-125.

Oosterbroek, P., and B. Theowald. 1991. Phylogeny of the Tipuloidea based on characters of larvae and pupae (Diptera, Nematocera) with an index to the literature except Tipulidae. *Tijdschrift voor Entomologie* 134: 211-267.

Oswood, M. W. 1978. Abundance patters of filter feeding caddisflies (Trichoptera- Hydropsychidae) and seston in a Montana (USA) Lake Outlet. *Hydrobiologia* 63: 177-183.

Oygur, S., and G. W. Wolfe. 1991. Classification, distribution, and phylogeny of North American (north of Mexico) species of *Gyrinus* Müller (Coleoptera: Gyrinidae). *Bulletin of the American Museum of Natural History* 207: 1-97.

Packauskas, R. J., and J. E. McPherson. 1986. Life history and laboratory rearing of *Ranatra fusca* (Hemiptera: Nepidae) with descriptions of immature stages. *Annals of the Entomological Society of America* 79: 566-571.

Parker, C. R., and G. B. Wiggins. 1985. The Nearctic caddisfly genus *Hesperophylax* Banks (Trichoptera: Limnephilidae). *Canadian Journal of Zoology* 63: 2443-2472.

Passoa, S. 1988. Systematic positions of *Acentria ephemerella* (Denis & Schiffermüller), Nymphulinae, and Schoenobiinae based on morphology of immature stages (Pyralidae). *Journal of the Lepidopterists' Society* 42: 247-262.

Paulson, D. R., and S. W. Dunkle. 2012. *A Checklist of North American Odonata, Including English name, Etymology, Type Locality, and Distribution*, 2012 edition. Occasional Paper No. 56, Slater Museum of Natural History, University of Puget Sound, Tacoma, WA.

Pechuman, L. L., D. W. Webb, and H. J. Teskey. 1983. The Diptera, or True Flies, of Illinois I. Tabanidae. *Illinois Natural History Survey Bulletin* 33: 1-120.

Pemberton, S. G., and R. W. Frey. 1982. Trace fossil nomenclature and the *Planolites-Palaeophycus* dilemma. *Journal of Paleontology* 56: 843-881.

Penny, N. D., P. A. Adams, and L. A. Stange. 1997. Species catalog of the Neuroptera, Megaloptera, and Raphidioptera of America north of Mexico. *Proceedings of the California Academy of Sciences* 50:39-114.

Perkins, P. D. 1980. Aquatic beetles of the family Hydraenidae in the Western Hemisphere: classification, biogeography, and inferred phylogeny (Insecta: Coleoptera). *Quaestiones Entomologicae* 16(1-2): 5-554.

Peterson, B. V., and B. C. Kondratieff. "1994" (but issued in 1995). The black flies (Diptera: Simuliidae) of Colorado: an annotated list with keys, illustrations and descriptions of three new species. *Memoirs of the American Entomological Society* 42: 1-121.

Poff, N. L., J. D. Olden, N. K. M. Vieira, D. S. Finn, M. P. Simmons, and B. C. Kondratieff. 2006. Functional trait niches of North American lotic insects: traits-based ecological applications in light of phylogenetic relationships. *Journal of the North American Benthological Society* 25: 730-755.

Pohl, G. R., B. Patterson, and J. P. Pelham. 2016. Annotated taxonomic checklist of the Lepidoptera of North America, north of Mexico. Working paper published online by the authors at ResearchGate.net.

Polhemus, J. T. 2008. Aquatic and Semiaquatic Hemiptera. Pp. 385-423 *in* Merritt, R. W., K. W. Cummins, and M. B. Berg (eds.). *An Introduction to the Aquatic Insects of North America*, fourth edition. Kendall/Hunt Publishing, Dubuque, IA.

Pollet, M. A. A., S. E. Brooks, and J. E. Cumming. 2004. Catalog of the Dolichopodidae (Diptera) of America north of Mexico. *Bulletin of the American Museum of Natural History* 283: 1-114.

Poole, R. W., and P. Gentili (eds.). 1996-1997. *Nomina Insecta Nearctica: A Checklist of the Insects of North America*. The Smithsonian Institition and Entomological Information Services, Baltimore, MD.

Pritchard, G. 1991. Insects in thermal springs. *Memoirs of the Entomological Society of Canada* 155: 89-106.

Provonsha, A. V., and W. P. McCafferty. 1977. Odonata from Hot Brook, South Dakota, with notes on their distribution patterns. *Entomological News* 88: 23-28.

Pruess, K. P., and B. V. Peterson. 1987. The black flies (Diptera: Simuliidae) of Nebraska: an annotated list. *Journal of the Kansas Entomological Society* 60: 528-534.

Quate, L.W. 1955. A revision of the Psychodidae (Diptera) in America north of Mexico. *University of California Publications in Entomology* 10: 103–273.

Quist, J. A., and M. T. James. 1973. The genus *Euparyphus* in America north of Mexico, with a key to the New World genera and subgenera of Oxycerini (Diptera: Stratiomyidae). *Melanderia* 11: 1-26.

Randolph, R. P., and W. P. McCafferty. 1996. First larval descriptions of two species of *Paraleptophlebia* (Ephemeroptera: Leptophlebiidae). *Entomological News* 107: 225-229.

Richards, B. 2014. *Caloparyphus* vs. *Euparyphus*. *Emergence: Newsletter of the Southwest Association of Freshwater Invertebrate Taxonomists* 7(4): 6-8.

Richardson, J. S. and R. J. MacKay. 1991. Lake outlets and the distribution of filter feeders: an assessment of hypothesis. *Oikos* 62: 370-380.

Roback, S. S. 1985. The immature chironomids of the eastern United States IV. Pentaneurini-genus *Ablabesmyia*. *Proceedings of the Academy of Natural Sciences of Philadelphia* 137: 153-212.

Robinson, H., and J. R. Vockeroth. 1981. Dolichopodidae. Pp. 625-639 *in* McAlpine, J. F., B. V. Peterson, G. E. Shewell, H. J. Teskey, J. R. Vockeroth, and D. M. Wood (coords.). *Manual of Nearctic Diptera, Volume 1*. Monograph 27, Agriculture Canada, Ottawa, Ontario, Canada.

Ross, H. H. 1937. Studies of Nearctic aquatic insects. I. Nearctic alder flies of the genus *Sialis* (Megaloptera, Sialidae). *Illinois Natural History Survey Bulletin* 21: 57-78.

Ross, H. H. 1938. Descriptions of Nearctic caddis flies (Trichoptera), with special reference to the Illinois species. *Illinois Natural History Survey Bulletin* 21: 101-183.

Ross, H. H. 1944. The caddis flies, or Trichoptera, of Illinois. *Illinois Natural History Survey Bulletin* 23: 1-326.

Ross, H. H. 1950. Synoptic notes on some Nearctic limnephilid caddisflies (Trichoptera, Limnephilidae). *American Midland Naturalist* 43: 410-429.

Ross, H. H., and D. R. Merkley. 1952. An annotated key to the Nearctic males of *Limnephilus* (Trichoptera, Limnephilidae). *American Midland Naturalist* 47: 435-455.

Rothfels, K. H., and D. Featherston. 1981. The population structure of *Simulium vittatum* (Zett.): the IIIL-1 and IS-7 sibling species. *Canadian Journal of Zoology* 59: 1857-1883.

Roughley, R. E. 1990. A systematic revision of species of *Dytiscus* Linnaeus (Coleoptera: Dytiscidae). Part 1. Classification based on adult stage. *Quaestiones Entomologicae* 26: 383-557.

Ruiter, D. E. 1995. The genus *Limnephilus* Leach (Trichoptera: Limnephilidae) of the New World. *Ohio Biological Survey Bulletin, new series* 11: 1-200.

Ruiter, D. E. 1996. Initial list of Trichoptera collected in the USA by 1995 Symposium Participants. *Braueria* 23: 10-12.

Ruiter, D. E., and R. J. Lavigne. 1985. *Distribution of Wyoming Trichoptera*. University of Wyoming Agricultural Experiment Station, Laramie, WY.

Saether, O. A. 1970. *Nearctic and Palaearctic Chaoborus (Diptera: Chaoboridae)*. Bulletin 174, Fisheries Research Board of Canada, Ottawa, Canada.

Sanderson, M. W. 1953-1954. A revision of the Nearctic genera of Elmidae (Coleoptera). *Journal of the Kansas Entomological Society* 26: 148-163, 27: 1-13.

Schefter, P. W., and G. B. Wiggins. 1986. *A Systematic Study of the Nearctic Larvae of the Hydropsyche morosa Group (Trichoptera: Hydropsychidae)*. Royal Ontario Museum, Toronto, Canada.

Schumann, D. A., M. C. Cavallaro, and W. W. Hoback. 2012. Size selective predation of fish by *Hydrophilus triangularis* (Coleoptera: Hydrophilidae) and *Lethocerus americanus* (Hemiptera: Belostomatidae). *Journal of the Kansas Entomological Society* 85: 155-159.

Schuster, G. A., and D. A. Etnier. 1978. *A Manual for the Identification of the Larvae of the Caddisfly Genera <u>Hydropsyche</u> Pictet and <u>Symphitopsyche</u> Ulmer in Eastern and Central North America (Trichoptera: Hydropsychidae)*. EPA-600/4-78-060, U.S. Environmental Protection Agency, Cincinnati, OH.

Shearer, J. 2006. *Macroinvertebrate Bioassessment of Black Hills Streams, South Dakota*. South Dakota Game, Fish, and Parks Report 2006-09, Rapid City, SD.

Shepard, J. J., T. G. Andreadis, and C. R. Vossbrinck. 2006. Molecular phylogeny and evolutionary relationships among mosquitoes (Diptera: Culicidae) from the northeastern United States based on small subunit ribosomal DNA (18S rDNA) sequences. *Journal of Medical Entomology* 43: 443-454.

Shepard, W. D. 1989. Elmidae. II. Description of *Orientelmis* gen.n. and new synonymy in *Cleptelmis* Sanderson. *Water Beetles of China* 2: 289-295.

Shepard, W. D., and K. W. Stewart. 1983. Comparative study of nymphal gills in North American stonefly (Plecoptera) genera and a new, proposed paradigm of Plecoptera gill evolution. *Miscellaneous Publications of the Entomological Society of America* 13.

Short, A. E. Z. 2004. Review of the *Enochrus* Thomson of the West Indies (Coleoptera: Hydrophilidae). *Koleopterologische Rundschau* 74: 351-361.

Sims, G. G. 2012. A Distribution of Dragonflies and Damselflies in the State of Wyoming. http://bugsofboogercounty.files.wordpress.com/2012/08/wyoming-odonate-distribution3.pdf.

Sinclair, B. J. 1989. The biology of *Euparyphus* Gerstaeker and *Caloparyphus* James occurring in madicolous habitats of eastern North America, with descriptions of adult and immature stages (Diptera: Stratiomyidae). *Canadian Journal of Zoology* 67: 33-41.

Sinclair, B. J. 1994. Revision of the Nearctic species of *Trichoclinocera* Collin (Diptera: Empididae: Clinocerinae). *The Canadian Entomologist* 126: 1007-1059.

Sinclair, B. J. 1997. Review of the Nearctic species of *Wiedemannia* Zetterstedt (Diptera: Empididae: Clinocerinae). *Studia Dipterologica* 4: 337-352.

Sites, R. W., and J. T. Polhemus. 1994. Nepidae (Hemiptera) of the United States and Canada. *Annals of the Entomological Society of America* 87: 27-42.

Sizer, R. L., E. H. Barmna, and G. Nichols. 1998. Biology of *Laccophilus fasciatus rufus* Melsheimer (Coleoptera: Dytiscidae) in central Georgia with descriptions of its mature larva and pupa. *Georgia Journal of Sciences* 56: 106-120.

Smetana, A. 1980. Revision of the genus *Hydrochara* Berth. (Coleoptera: Hydrophilidae). *Memoirs of the Entomological Society of Canada* 111: 1-100.

Smetana, A. 1984. Revision of the subfamily Sphaeridiinae of America North of Mexico (Coleoptera: Hydrophilidae). Supplementum 2. *Canadian Entomologist* 116: 555-566.

Smetana, A. 1988. Review of the family Hydrophilidae of Canada and Alaska (Coleoptera). *Memoirs of the Entomological Society of Canada* 142: 1-316.

Smith, C. L. 1980. *A Taxonomic Revision of the Genus <u>Microvelia</u> Westwood (Heteroptera: Veliidae) of North America Including Mexico*. Ph.D. dissertation, University of Georgia, Athens, GA.

Smith, C. L., and J. T. Polhemus. 1978. The Veliidae (Heteroptera) of America north of Mexico – keys and checklist. *Proceedings of the Entomological Society of Washington* 80: 56-68.

Smith, D. H., and D. M. Lehmkuhl. 1980. Analysis of two problematic North American caddisfly species: *Oecetis avara* (Banks) and *Oecetis disjuncta* (Banks) (Trichoptera: Leptoceridae). *Quaestiones Entomologicae* 16: 641-656.

Smith, J. P., and W. F. Rapp. 1987. Black flies (Diptera: Simuliidae) in Nebraska. *New Jersey Mosquito Control Association Proceedings of 72nd Annual Meeting:* 135-136.

Smith, S. D. 1968. The *Rhyacophila* of the Salmon River Drainage of Idaho with special reference to larvae. *Annals of the Entomological Society of America* 61: 655-674.

Solis, M. A. 2008. Aquatic and Semiaquatic Lepidoptera. Pp. 553-569 *in* Merritt, R. W., K. W. Cummins, and M. B. Berg (eds.). *An Introduction to the Aquatic Insects of North America*, fourth edition. Kendall/Hunt Publishing, Dubuque, IA.

Song, N., A.-P. Liang, and C.-P. Bu. 2012. A molecular phylogeny of Hemiptera inferred from mitochondrial genome sequences. *PLOS One 7: e48778*

South Dakota Department of Environment and Natural Resources. 2005a. Standard Operating Procedures for Field Samplers. Volume I: Tributary and In-Lake Sampling Techniques. South Dakota Department of Environment and Natural Resources, Pierre, SD.

South Dakota Department of Environment and Natural Resources. 2005b. Standard Operating Procedures for Field Samplers. Volume II: Biological and Habitat Sampling. South Dakota Department of Environment and Natural Resources, Pierre, SD.

Stagliano, D. M. 2006. *Aquatic Surveys and Assessment of the Slim Buttes Region of Harding and Butte Co., SD.* Report to the Montana and South Dakota TNC Field Offices, Montana Natural Heritage Program, Helena, MT.

Stains, G. S., and G. F. Knowlton. 1943. A taxonomic and distributional study of Simuliidae of western United States. *Annals of the Entomological Society of America* 36: 259-280.

Stark, B. P., and J. W. Kyzar. 2000. Systematics of Nearctic *Paraleuctra* with description of a new genus (Plecoptera: Leuctridae). *Tijdschrift voor Entomologie* 144: 119-135.

Stein, P. 1898. Nordamerikanische Anthomyiden. Beitrag zur Dipterenfauna der Vereinigten Staaten. *Berliner Entomologische Zeitschrift* 42: 161-288.

Stewart, K. W., and B. P. Stark. 2002. *Nymphs of North American Stonefly Genera (Plecoptera)*, 2nd edition. The Caddis Press, Columbus, OH.

Stewart, K. W., and B. P. Stark. 2008. Plecoptera. Pp. 311-384 *in* Merritt, R. W., K. W. Cummins, and M. B. Berg (eds.). *An Introduction to the Aquatic Insects of North America*, fourth edition. Kendall/Hunt Publishing, Dubuque, IA.

Steyskal, G. C., and L. V. Knutson. 1981. Empididae. Pp. 607-624 *in* McAlpine, J. F., B. V. Peterson, G. E. Shewell, H. J. Teskey, J. R. Vockeroth, and D. M. Wood (coords.). *Manual of Nearctic Diptera, Volume 1*. Monograph 27, Agriculture Canada, Ottawa, Ontario, Canada.

Stoakes, R. D., J. K. Neel, and R. L. Post. 1983. Observations on North Dakota sponges (Haplosclerina: Spongillidae) and sisyrids (Neuroptera: Sisyridae). *Great Lakes Entomologist* 16: 171-176.

Stonedahl, G. M., and J. D. Lattin. 1982. *The Gerridae or Water Striders of Oregon and Washington.* Technical Bulletin 144, Oregon State University, Corvallis, OR.

Szczytko, S. W., and B. C. Kondratieff. 2015. A review of the eastern Nearctic Isoperlinae (Plecoptera: Perlodidae) with the description of twenty-two new species. *Monographs of Illiesia* 1: 1-289.

Tennessen, K. J. 2008. Odonata. Pp. 237-293 *in* Merritt, R. W., K. W. Cummins, and M. B. Berg (eds.). *An Introduction to the Aquatic Insects of North America*, fourth edition. Kendall/Hunt Publishing, Dubuque, IA.

Teskey, H. J. 1969. Larvae and pupae of some eastern North American Tabanidae (Diptera). *Memoirs of the Entomological Society of Canada* 63: 1-147.

Thomas, A. W. 2011. Tabanidae of Canada, east of the Rocky Mountains 2: a photographic key to the genera and species of Tabaninae (Diptera: Tabanidae). *Canadian Journal of Arthropod Identification* 13: 1-503.

Thomas, A. W., and S. A. Marshall. 2009. Tabanidae of Canada, east of the Rocky Mountains 1: a photographic key to the species of Chrysopsinae and Pangoniinae (Diptera: Tabanidae). *Canadian Journal of Arthropod Identification* 8:1-172.

Thomson, R. E., and R. W. Holzenthal. 2015. A revision of the Neotropical caddisfly genus *Leucotrichia* Mosely, 1934 (Hydroptilidae, Leucotrichiinae). *ZooKeys* 499: 1-100.

Tinerella, P. P., and A. W. DeLorme. 2005. First records of the creeping water bug, *Ambrysus mormon* Montandon (Heteroptera: Naucoridae) from North Dakota. *Journal of the Kansas Entomological Society* 78: 176-178.

Traver, J. R. 1935. Systematic, Part II. Pp. 267-739 in Needham, J. G., J. R. Traver, and T.-C. Hsu (eds.). *The Biology of Mayflies, with a systematic account of North American species.* Comstock Publiching Co., Ithaca, NY.

Treat, A. E. 1954. *Acentropus niveus* in Massachusetts, remote from water. *Lepidopterists' News* 8: 23-25.

Tronstad, L. M., and B. R. Barber. 2014. *Wyoming's Stream Macroinvertebrates.* University of Wyoming Biodiversity Institute, Laramie, WY.

Turner, R. W. 1974. Mammals of the Black Hills of South Dakota and Wyoming. *University of Kansas Museum of Natural History Miscellaneous Publications* 60: 1-178.

Turrell, M. J., D. J. Dohm, M. R. Sardellis, M. L. O'Guinn, T. G. Andreadis, and J. A. Blow. 2005. An update on the potential of North American mosquitoes (Diptera: Culicidae) to transmit West Nile virus. *Journal of Medical Entomology* 42: 57-62.

Twain, M. 1872. *Roughing It.* Project Gutenberg.

Ulrich, G. W. 1986. The larvae and pupae of *Helichus suturalis* LeConte and *Helichus productus* LeConte (Coleoptera: Dryopidae). *The Coleopterists Bulletin* 40: 325-334.

U.S. Environmental Protection Agency. 2011. A Field-Based Aquatic Life Benchmark for Conductivity in Central Appalachian Streams. EPA/600/R-10/023F. Office of Research and Development, National Center for Environmental Assessment, Washington, DC.

Vieira, N. K. M., N. L. Poff, D. M. Carlisle, S. R. Moulton II, M. L. Koski, and B. C. Kondratieff. 2006. *A Database of Lotic Invertebrate Traits for North America.* Data Series #187. U. S. Geological Survey, Reston, VA.

Vineyard, R. N., G. B. Wiggins, H. E. Frania, and P. W. Shefter. 2005. The caddisfly genus *Neophylax* (Trichoptera: Uenoidae). *Royal Ontario Museum Contributions in Science* 2: 1-141.

Vondel, B. J. van. 1991. Revision of the Palaearctic species of *Haliplus* subgenus *Liaphlus* Guignot (Coleoptera: Haliplidae). *Tijdschrift voor Entomologie* 134: 75-144.

von Ellenrieder, N. 2003. A synopsis of the Neotropical species of "*Aeshna*" Fabricius: the genus *Rhionaeschna* Förster (Odonata: Aeshnidae) *Tijdschrift voor Entomologie* 146: 67-207.

Vorhies, C. T. 1908-09. *Studies on the Trichoptera of Wisconsin.* PhD Dissertation, University of Wisconsin; published in *Transactions of the Wisconsin Academy of Sciences, Arts, and Letters* 16: 647-738.

Wallace, J. R., and E. D. Walker. 2008. Culicidae. Pp. 801-823 *in* Merritt, R. W., K. W. Cummins, and M. B. Berg (eds.). *An Introduction to the Aquatic Insects of North America,* fourth edition. Kendall/Hunt Publishing, Dubuque, IA.

Walker, E. M. 1912. The North American dragonflies of the genus *Aeshna. University of Toronto Studies, Biology Series.* 11: 1-212.

Walker, E. M. 1913. New nymphs of Canadian Odonata. *Canadian Entomologist* 45: 161-170.

Walker, E. M. 1914. The known nymphs of the Canadian species of *Lestes* (Odonata). *Canadian Entomologist* 46: 189-200.

Walker, E. M. 1914. New and little-known nyumphs of Canadian Odonata. *Canadian Entomologist* 46: 349-357.

Walker, E. M. 1916. The nymphs of *Enallagma cyathigerum and E. calverti. Canadian Entomologist* 48: 192-196.

Walker, E. M. 1916. The nymphs of the North American species of *Leucorrhinia. Canadian Entomologist* 48: 414-422.

Walker E. M. 1917. The known nymphs of the North American species of *Sympetrum. Canadian Entomologist* 49: 409-418.

Walker, E. M. 1944. The nymphs of *Enallagma clausum* Morse and *E. boreale* Selys. *Canadian Entomologist* 76: 234-237.

Walker, E. M. 1953. *The Odonata of Canada and Alaska.* Vol. I. University of Toronto Press, Toronto, Ontario, Canada.

Walker, E. M. 1958. *The Odonata of Canada and Alaska.* Vol. II. University of Toronto Press, Toronto, Ontario, Canada.

Walker, F. 1859. *List of the Specimens of Lepidopterous Insects in the Collection of the British Museum, London*, Part XIX – Pyralides. 19: 799-1036.

Walker, F. 1866. Supplement 4. *List of the Specimens of Lepidopterous Insects in the Collection of the British Museum, London* 34: 1121-1533

Walters, A. W. 2011. Resistance of aquatic insects to a low-flow disturbance: exploring a trait-based approach. *Journal of the North American Benthological Society* 30: 346-356.

Walthall, G. E. 1962. *Insects of Wind Cave National Park*. Wind Cave National Park, National Park Service, Hot Springs, SD.

Waltz, R. D., D. E. Baumgardner, and J. H. Kennedy. 1996. An atypical larval color form of *Baetis intercalaris* (Ephemeroptera: Baetidae) from Pennsylvania and the Kiamichi River basin of southeastern Oklahoma. *Entomological News* 107: 83-87.

Waltz, R. D., and S. K. Burian. 2008. Ephemeroptera. Pp. 181-236 *in* Merritt, R. W., K. W. Cummins, and M. B. Berg (eds.). *An Introduction to the Aquatic Insects of North America*, fourth edition. Kendall/Hunt Publishing, Dubuque, IA.

Wang, Y-K, and J. H. Kennedy. 2004. Life history of *Mayatrichia ponta* Ross (Trichoptera: Hydroptilidae) in Honey Creek, Oklahoma. *Proceedings of the Entomological Society of Washington* 106: 523-530.

Ward, J.V. and J.A. Stanford. 1987. The ecology of regulated streams: past accomplishments and directions for future research. *Proceedings of the Third International Symposium on Regulated Streams*, Plenum Press, New York.

Watts, C. H. S. 1970. The larvae of some Dytiscidae (Coleoptera) from Delta, Manitoba. *The Canadian Entomologist* 102: 716-728.

Weaver, J. S., III. 1988. A synopsis of the North American Lepidostomatidae (Trichoptera). *Contributions of the American Entomological Institute* 24: 1-141.

Webb, D. W. 1977. The Nearctic Athericidae (Insecta: Diptera). *Journal of the Kansas Entomological Society* 50: 473-495.

Welch, P. S. 1916. Contribution to the biology of certain aquatic Lepidoptera. *Annals of the Entomological Society of America* 9: 160-181.

Westfall, M. J., Jr. 1990. Descriptions of larvae of *Argia munda* Calvert, *A. plana* Calvert, *A. tarascana* Calvert, and *A. tonto* Calvert. *Odonatologica* 19: 61-70.

Westfall, M. J., Jr., and M. L. May. 1996. *Damselflies of North America*. Scientific Publishers, Gainesville, FL.

White, D. S. 1978. A revision of the Nearctic *Optioservus* (Coleoptera: Elmidae), with descriptions of new species. *Systematic Entomology* 3: 59-74.

White, D. S., and R. E. Roughley. 2008. Aquatic Coleoptera. Pp. 571-671 *in* Merritt, R. W., K. W. Cummins, and M. B. Berg (eds.). *An Introduction to the Aquatic Insects of North America*, fourth edition. Kendall/Hunt Publishing, Dubuque, IA.

Whiting, M. F. 1981. A distributional study of *Sialis* (Megaloptera: Sialidae) in North America. *Entomological News* 102:50-56.

Wiggins, G. B. 1960. A preliminary systematic study of the North American larvae of the caddisfly family Phryganeidae (Trichoptera). *Canadian Journal of Zoology* 38: 1153-1170.

Wiggins, G. B. 1963. Larvae and pupae of two North American limnephilid caddisfly genera (Trichoptera: Limnephilidae). *Bulletin of the Brooklyn Entomological Society* 58: 103-112.

Wiggins, G. B. 1983. "*Neophylax* larvae of western North America." In handouts at Northwest Biological Assessment Workgroup 2nd Annual Taxonomic Workshop: Trichoptera. Oregon State University, Corvallis, OR

Wiggins, G. B. 1996. *Larvae of the North American Caddisfly Genera (Trichoptera)*, 2nd edition. University of Toronto Press, Toronto, Ontario, Canada.

Wiggins, G. B., and D. C. Currie. 2008. Trichoptera Families. Pp. 439-480 *in* Merritt, R. W., K. W. Cummins, and M. B. Berg (eds.). *An Introduction to the Aquatic Insects of North America*, fourth edition. Kendall/Hunt Publishing, Dubuque, IA.

Wiggins, G. B., and J. S. Richardson. 1986. Revision of the *Onocosmoecus unicolor* group (Trichoptera: Limnephilidae, Discosmoecinae). *Psyche* 93: 187-216.

Williston, S. W. 1887. Notes and descriptions of North American Tabanidae. *Transactions of the Kansas Academy of Sciences* 10: 129-142.

Wilson, C. B. 1923. Life history of the scavenger water beetle, *Hydrous (Hydrophilus) triangularis*, and its economic relation to fish breeding. *Bulletin of the U.S. Bureau of Fisheries* 39: 9-38.

Wirth, W. W. 1951. The genus *Probezzia* in North America. *Proceedings of the Entomological Society of Washington* 53: 25-34.

Wirth, W. W. 1965. A revision of the genus *Parabezzia* Malloch. *Proceedings of the Entomological Society of Washington* 67: 215-230.

Wirth, W. W. 1994. The subgenus *Atrichopogon (Lophomyidium)* with a revision of the Nearctic species (Diptera: Ceratopogonidae). *Insecta Mundi* 8: 17-36.

Wirth, W. W., and F. S. Blanton. 1969. North American *Culicoides* of the *pulicaris* group (Diptera: Ceratopogonidae). *Florida Entomologist* 52: 207-243.

Wold, J. L. 1974. *Systematics of the Genus Rhyacophila (Trichoptera: Rhyacophilidae) in Western North America with Special Reference to the Immature Stages*. M.S. Thesis, Oregon State University, Corvallis, OR.

Wood, D, M., P. T. Dang, and R. A. Ellis. 1979. The mosquitoes of Canada (Diptera: Culicidae). The Insects and Arachnids of Canada, Part 6. Agriculture Canada, Ottawa, Ontario.

Yamamoto, T., and G. B. Wiggins. 1964. A comparative study of the North American species in the caddisfly genus *Mystacides* (Trichoptera: Leptoceridae). *Canadian Journal of Zoology* 42: 1105-1126.

Yuan, L. L. 2006. *Estimation and Application of Macroinvertebrate Tolerance Values*. EPA/600/P-04/116F, U.S. Environmental Protection Agency, Washington, DC.

Zenger, J. T., and R. W. Baumann. 2004. The Holarctic winter stonefly genus *Isocapnia*, with an emphasis on the North American fauna (Plecoptera: Capniidae). *Monographs of the Western North American Naturalist* 2: 65-95.